Cognitive Data Science in Sustainable Computing

Cognitive Data Models for Sustainable Environment

Edited by

Siddhartha Bhattacharyya
Rajnagar Mahavidyalaya, Birbhum, India

Naba Kumar Mondal
Environmental Chemistry Laboratory, Department of Environm
Science, The University of Burdwan, Bardhaman, West Benga

Koushik Mondal
Computer Centre, IIT (ISM) Dhanbad, Dhanbad, Jharkhand, India

Jyoti Prakash Singh
Department of Computer Science and Engineering, National Institute of
Technology Patna, Patna, Bihar, India

Kolla Bhanu Prakash
Department of Computer Science Engineering, K L Deemed to be
University, Green Fields, Vaddeswaram, Guntur District, A.P., India

Series Editor

Arun Kumar Sangaiah
School of Computing Science and Engineering,
Vellore Institute of Technology (VIT), Vellore, India

ACADEMIC PRESS
An imprint of Elsevier

Academic Press is an imprint of Elsevier
125 London Wall, London EC2Y 5AS, United Kingdom
525 B Street, Suite 1650, San Diego, CA 92101, United States
50 Hampshire Street, 5th Floor, Cambridge, MA 02139, United States
The Boulevard, Langford Lane, Kidlington, Oxford OX5 1GB, United Kingdom

Notices
Knowledge and best practice in this field are constantly changing. As new research and experience broaden our understanding, changes in research methods, professional practices, or medical treatment may become necessary.

Practitioners and researchers must always rely on their own experience and knowledge in evaluating and using any information, methods, compounds, or experiments described herein. In using such information or methods they should be mindful of their own safety and the safety of others, including parties for whom they have a professional responsibility.

To the fullest extent of the law, neither the Publisher nor the authors, contributors, or editors, assume any liability for any injury and/or damage to persons or property as a matter of products liability, negligence or otherwise, or from any use or operation of any methods, products, instructions, or ideas contained in the material herein.

Library of Congress Cataloging-in-Publication Data
A catalog record for this book is available from the Library of Congress

British Library Cataloguing-in-Publication Data
A catalogue record for this book is available from the British Library

ISBN: 978-0-12-824038-0

For information on all Academic Press publications visit our
website at https://www.elsevier.com/books-and-journals

Publisher: Mara Conner
Acquisitions Editor: Sonnini R. Yura
Editorial Project Manager: Rafael G. Trombaco
Production Project Manager: Omer Mukthar
Cover Designer: Matthew Limbert

Typeset by TNQ Technologies

Working together
to grow libraries in
developing countries

www.elsevier.com • www.bookaid.org

Cognitive Data Models for Sustainable Environment

Siddhartha Bhattacharyya would like to dedicate this book to my lifelong teacher and mentor Respected Prof. Asit K. Datta, Retired Professor, University of Calcutta, Kolkata, India.

Naba Kumar Mondal would like to dedicate this book to his sweet and loving father and mother, whose affection, love, encouragement, and prayer of day and night made him achieve such success and honor.

Koushik Mondal would like to dedicate this book to his parents Mr. Kamal Chandra Mondal, Mrs. Seba Mondal, wife Mrs. Sharmistha Mondal, children Hriddhi Mondal and Hrishav Mondal, and his present Institute IIT (ISM) Dhanbad, Jharkhand, India.

Jyoti Prakash Singh would like to dedicate this book to his wife Rambha and daughter Snigdha for their love and support.

Kolla Bhanu Prakash would like to dedicate this book to his wife Mrs. M.V. Prasanna Lakshmi and to his parents.

Contents

5. Sustainability issues in upcoming wastewater treatment plants at Patna 101

Nityanand Singh Maurya, Sulagna Roy, and Astha Kumari

Contents **xiii**

Contributors

Haider Banka, Department of Computer Science and Engineering, IIT (ISM) DHANBAD, Dhanbad, Jharkhand, India

Deep Chakraborty, Environmental Chemistry Laboratory, Department of Environmental Science, The University of Burdwan, Bardhaman, West Bengal, India

Kartick Chandra Pal, Karimpur Pannadevi College, Karimpur, West Bengal, India

Priyanka Debnath, Environmental Chemistry Laboratory, Department of Environmental Science, The University of Burdwan, Bardhaman, West Bengal, India

M. Kiruthika, Department of Computer Science and Engineering, Jansons Institute of Technology, Coimbatore, Tamil Nadu, India

Ajay Kumar, Department of Architecture, National Institute of Technology, Patna, Bihar, India

Vaneet Kumar, Department of Applied Sciences, CT Institute of Engineering, Management and Technology, CT Group of Institutions Jalandhar, Punjab, India

Astha Kumari, Department of Civil Engineering, National Institute of Technology Patna, Patna, Bihar, India

Nityanand Singh Maurya, Department of Civil Engineering, National Institute of Technology Patna, Patna, Bihar, India

Gaurav Mohindru, Department of Computer Science and Engineering, IIT (ISM) DHANBAD, Dhanbad, Jharkhand, India

Arghadip Mondal, Environmental Chemistry Laboratory, Department of Environmental Science, The University of Burdwan, Bardhaman, West Bengal, India

Koushik Mondal, Computer Centre, IIT (ISM) DHANBAD, Dhanbad, Jharkhand, India

Naba Kumar Mondal, Environmental Chemistry Laboratory, Department of Environmental Science, The University of Burdwan, Bardhaman, West Bengal, India

T. Poongodi, School of Computing Science and Engineering, Galgotias University, Greater Noida, Uttar Pradesh, India

Vishal Rehani, Department of Applied Sciences, CT Institute of Engineering, Management and Technology, CT Group of Institutions Jalandhar, Jalandhar, Punjab, India

Sulagna Roy, L&T Construction, Chennai, Tamil Nadu, India

Saruchi, Department of Biotechnology, CT Institute of Pharmaceutical Sciences, CT Group of Institutions, Jalandhar, Punjab, India

Kamalesh Sen, Environmental Chemistry Lab, Department of Environmental Science, The University of Burdwan, Bardhaman, West Bengal, India

Surbhi Sharma, Department of Physics, Kanya MahaVidyalaya, Jalandhar, Punjab, India

R. Sujatha, School of Information Technology & Engineering, Vellore Institute of Technology, Vellore, Tamil Nadu, India

P. Suresh, School of Mechanical Engineering, Galgotias University, Greater Noida, Uttar Pradesh, India

Preface

With the exponential rise in industrial civilization, the most dangerous issue that has unfolded over the years is the rapid explosion of industrial waste and effluent. Not only the environmental wastes are toxic in nature, it has put the entire environmental ecology at stake. Risks of polluting potable water and presence of toxic gases in breathing air pose a threat to the existence and sustenance of humankind. Moreover, additional hazards are also evident through sound and light pollution as well. In the wake of tremendous environmental pollution due to the influx of industrial wastes, a sustainable environment is the need of the hour. Efficient techniques for mitigation of environmental pollution has been always on the helm of affairs given the vast amount of pollutants injected into the environment due to industrial effluent. However, the state-of-the-art methods mainly rely on time intensive and costly batch procedures. Traditional techniques employed by the environmental scientists are often cumbersome and expensive and lack sustainability. Hence, there is always a need for having recourse to time efficient, failsafe, cheaper intelligent technologies to address the problems and ensure long-term sustainability.

The existing literature available in this respect is really nonexistent, and this volume is intended to serve as a treatise and knowledge base for the community so as to inspire them to adapt environmentally friendly and sustainable solutions for the future. This book unveils intelligent and cognitive models to address issues related to effective monitoring of environmental pollution and ushering in a sustainable environmental design. As such, the book focuses on the overall well-being of the global environment for better sustenance and livelihood.

The volume comprises eleven well versed chapters discussing cognitive initiatives toward environmental sustainability.

Nanoparticles are playing an important role in controlling mosquito for the last 10 years. Now, this technology is applied in many fields for their very high affectivity compared to bulk particles because of their size between 100 nm. Mosquitos are spreading various deadly disease like yellow fever, zika, dengue, west nile, and filaria. But commercially available chemical, physical, and biological products or techniques cannot control these vectors at the satisfactory level. From this background now, nanotechnology is one of the main attractions of the world researchers. Silver nanoparticle activity was better as a mosquito larvicidal agent compared to other biosynthesized nanoparticles. The effect of silver nanoparticles was different on various stage of mosquito life cycle.

Chapter 1 focuses on the synthesis properties of silver nanoparticles from different biological sources like microbes, plants, and animals. The chapter also discusses the controlling properties of silver nanoparticles on different mosquito species and different stages of their life cycle. Finally, this chapter concludes with a comparative study of AgNP synthesis properties and controlling efficiency of AgNPs toward different mosquito species.

Monitoring parameters for changes and initiating actions is key to managing an environment effectively. Internet of Things (IoT)-based platforms have been widely discussed as a viable and efficient solution across different domains. Chapter 2 discusses the requirement of a robust IoT-based cognitive environment monitoring solution to combine factors such as data collection, routing, preprocessing, storage, data analysis techniques, and building intelligent environment-specific models, visualization to easily interpret the outcome, and an optimum platform and technologies. Evolution is seen across the spectrum of technologies, frameworks, and techniques. Data collection has improved owing to advancements in sensor technology and hardware. The adoption of Free and Open Source Software (FOSS) and Open-IoT ecosystem has provided the necessary impetus. Accelerated development and deployment is possible using pretrained models. Available implements offer us the capability to aggregate, analyze live data streams in the cloud in real-time using different analytical models, and instant visualizations of data posted by the devices at the far site.

Sustainable development demands a process of unbroken improvement in using less number of resources for satisfaction of consumer needs. Sometimes, people harm the environment, although they try to treat well. The imbalance in moral environment caused by human perception and behavior, is the major reason for unsustainable environment in today's era. To ensure sustainable development, various priority areas for action like energy and water conservation, pollution control, human health, waste management, biodiversity protection, ecosystem management and most importantly, cognitive approach of human have been illustrated. Chapter 3 emphasizes on the pillars of sustainable environment which includes environment protection as well as economic and social development. The chapter concludes considering a few measures to take for sustainable environment including cognitive development in people, so that their approach must be environmentally friendly.

Agricultural contamination poses a serious risk on human health and the environment, possibly organophosphorus, organochlorine, carbamate, etc., which are necessarily removal by sustainable aspects. Nanotechnology subjected to superior performance, lower energy consumption and potential to water cleaning, and remediated to apply adsorption of the convenient way. Herein, the adsorption mechanisms by models include as isotherms and kinetics study. The isotherms are interrogated with Langmuir, Freundlich, Dubinin-Radushkevich, Temkin, and Hill, according to linear or nonlinear models. Chapter 4 describes the kinetics with pseudo-first order, pseudo-second order, and intraparticle diffusion. The nanoadsorbents are also described with their efficiency as adsorptive capacity and interacting phenomena. In addition, the optimization studies are illustrated with different models, like Artificial Intelligence (AI) and response surface methodology (RSM)-based model.

Population explosion and advancement in living standards have initiated the generation of relatively more polluted and larger volume of wastewater in comparison to the olden days. Construction and sustainable functioning of wastewater treatment plants (WWTPs) is a challenge to ensure safe disposal of wastewater and thereby protect human health and ecosystem in a developing country. Chapter 5 describes in brief the causes of failure of old WWTPs in Patna (India). The focus has been given to analyse the basic sustainability concept of the upcoming projects in the city under National Mission for Clean Ganga program particularly in terms of environmental issues and proper functioning during and after the exhaustion of annuity period. It also highlights the sustainability aspects missing in the upcoming projects, followed by discussion on the provisions that should be included in the projects to ensure overall sustainability (with respect to environmental, economic, and social).

The novel coronavirus pandemic has currently emerged as the world's worst-ever disaster in history, whose nature of cause is still under debate. Community in the current study refers to group of people subjected to common risks or threat and with the capacity to recognize disruptive events. Disaster sociologists have common consensus that these communities should be directly involved in planning and preparing for response to disaster, mitigating effects, thus reducing its risk through community-based disaster management (CBDM) strategy. Community approach can help in assessing levels of disaster risk in advance and also enhances the social capacity of preparedness, response, and recovery. In Chapter 6, analysis of such programs and strategies are presented in which community participates in disaster resilience. Various case studies of disaster risk management policies adapted by Asian countries. The role of faith-based organizations working for disaster management and mitigation is discussed in the present study.

Colored waters nowadays is an emerging issue, especially the wastewaters discharging from dyeing industries, which ultimately affect the drinking water. To minimize the detrimental effects of contaminated water and to overcome the inadequacy of traditional methods, technology-based smart treatment processes are imperative for sustainable supply of drinking water. Nanoparticle is a very promising class of materials used for this purpose which can effectively act as a potential adsorbent materials for dye adsorption as well as a photocatalyst. In Chapter 7, a brief description of nanostructured ZnO along with its different synthesis methods and its remarkable efficiency toward removal of some widely used azo and non-azo dyes from aqueous system are discussed.

Studies show that good ventilation is one of the key factors which can play an important role to minimize the health risk from Indoor Air Pollution. In Chapter 8, toxic indoor air pollutants (CO, CO_2, and O_3) are selected as key response variables and the windows number, the kitchen volume, and cooking hour are selected as the factors to optimize the rural kitchen configuration. Optimization is executed in the design expert software while implementing RSM. From the ANOVA analysis, it is clear that all models applied are significant. Moreover, there are high desirability values in case of CO, CO_2, temperature, and relative humidity provided that the optimum conditions/configurations are applied. This work describes how rural villagers can optimize their kitchens with their low-cost materials to build a sustainable indoor condition which will provide them a sustainable healthy lifestyle.

IoT reshapes the incumbent industries into smart industries by bringing more opportunities with efficient decision making. Chapter 9 illustrates an IoT-based healthcare data model using blockchain. The fundamental characteristics of IoT make the system to face various challenges such as poor interoperability, decentralization, heterogeneous data, diversified devices, network complexity, and security related issues. Blockchain technology overcomes the challenges of IoT by maintaining a digital ledger in the distributed systems. The convergence of IoT and blockchain technology synthesizes a novel paradigm that improves production throughput, operation efficiency, and product quality. During the complete procedure, privacy is preserved using blockchain, which manages individual data, data sharing among hospitals, insurance companies, and medical centers. Blockchain acts as an interface or repository among IoT devices for the data which is generated by them. The patient's wearing IoT devices can be monitored in their trajectory, and the countermeasures could be provided by safeguarding the privacy of patients via blockchain.

The beauty industry is growing at a phenomenal rate throughout the world. Daily huge amount of personal care and cosmetic products (PCCPs) are being used and a significant amount of these rinsed off products flows down the household wastewater streams, which ultimately end up in the aquatic environment via treated wastewater effluents, and some are retained in sewage sludge. Perseverance, bioactivity, and bioaccumulation potential of many beauty products pose an adverse effect on ecosystem. Spherical or amorphic micro sized (<5 µm) plastic particulates, known as microplastic, is one of such emerging pollutant. This ingredient is used in PCCPs as sorbent phase for the liberation of active ingredients, film formation, viscosity regulation, exfoliation, skin conditioning, emulsion stabilizing, and many others. Once released into the environment, microplastics persist for centuries devoid of full decomposition and reenter normal biogeochemical cycles causing particle toxicity. The beauty industry jumps on the plastic bandwagon in cosmetic packaging and for intricacy in post use collections of disposed plastic PCCP containers, removing strongly contaminated residues of greasy and creamy cosmetic products, recycle of plastic packaging is rarely applied. Therefore, on the basis of utmost necessity of awareness, Chapter 10 provides a precise idea about the cosmetic related ecological hazards emphasizing on the overview of microplastic ingredients and plastic packaging of PCCPs that instigate a mounting environmental concern.

House construction in rural India is adopting traditional building material and design from early Vedic period to current period. The brick and/or mud in wall and thatch as roof covering reflect the sustainable construction practices. The brick and mud and thatch are locally available and use of waste material of agriculture product. The locally available materials are energy free and structurally suitable and stable. Use of recyclable natural building materials in rural housing involves lesser energy consumption and low CO_2 emission. The thick brick and/or mud wall keep the interior cool in summer and warm in winter. The long-sloped thatch roof overhanging protects the wall from heating and rain water by keeping away the bright sunrays and driving rain. Rural housing depends on low venture and high upkeep. Financial exchanges are minor. This support work makes standard work for all. This is one of the significant components of monetary sustainability of network. House-making in India is a socially touchy and profoundly ritualized measure. It is a get-together that includes much particular standing, and which merges the ties among neighboring rural zones and villages. The help of relations, companions, and neighbors is compensated by food, drink, and shared assistance. This is one of the main elements of the social relationship and improvement. This chapter will clarify the sustainability idea of development of rural housing. This study also examines rural character perception from the housing perception. Chapter 11 shows that local building material and traditional design are very important for perception of rural housing. To maintain rural character and sustainability in rural housing, it is recommended that design guidelines should be matched with tradition and use local building material and incorporate community participation in rural house construction. This method can apply for data modeling for future design guidelines in rural housing for sustainability.

This volume will benefit academia and researchers interested in joining interdisciplinary projects on sustainable intelligent environmental data management. It is strongly expected that academic environments will introduce this book to their graduate students for enabling them to gather knowledge about the cognitive and intelligent paradigms in handling environmental pollution. The book is also intended for the environmentalists by introducing them to both technical and nontechnical details. They will find the book instrumental for understanding the aspects (opportunities and challenges) of processing and handling environmental perspectives. The book will also help them to design, develop, and integrate futuristic technologies for efficient monitoring of environmental pollution and promoting sustainable environmental practices. In addition, graduate-level students as well as scholars in international universities wishing to learn more about this important topic might also find this book useful.

The proposed book also targets AI scientist developers worldwide who are interested in developing cognitive solutions for the glaring problems of environmental pollution. The editors feel that this attempt will be worth rewarding if the present volume comes to the benefit of the end-users.

February, 2021
India

Siddhartha Bhattacharyya
Naba Kumar Mondal
Koushik Mondal
Jyoti Prakash Singh
Kolla Bhanu Prakash

Chapter 1

Multidimensional controlling properties of biofabricated silver-nanoparticles on different mosquito species

Arghadip Mondal, Priyanka Debnath, Naba kumar Mondal
Environmental Chemistry Laboratory, Department of Environmental Science, The University of Burdwan, Bardhaman, West Bengal, India

1. Introduction

Nanoparticles are defined as very small particles which size range within 1−100 nm. Naturally, this form of particles was unstable due to a very small size, but now a new technology was invented, where produce nanoparticles from bulk particles, this modern technology known as nanotechnology [1]. In 1974, this technology was first invented but last 10−15 years many research was done in this field because this particle size was very small and the surface area is very high compared to bulk particles. So, the activity of these particles was very high. For this reason, modern researchers focus on this technology [2]. Different subjects like chemistry, physics, and biology were involve in this field. Nanoparticles were synthesized in three different ways like chemically, physically, and biologically depending on their requirement. Byproducts of the chemically synthesized nanoparticles was very hazardous in the environment because a huge amount of chemicals were require at the time of chemical synthesis. The aspect of environmental health issues regarding the chemical synthesis of nanoparticles was not ecofriendly. Besides this, another process such as physical synthesis was also used for this small particle synthesis. The physical process requires a large amount of electrical power and magnetic field use. But in the biological process neither require huge physical energy nor produce large amount of hazardous byproduct [3]. In this process only require any biological substrate, which plays both roles as a reducing and capping agent. For this reason, biologically synthesized nanoparticles are widely used in various fields like cosmetics product, drug delivery, textile industries, pharmaceuticals products, and many other purposes [4]. Beside this uses, nanoparticles were also applied to mosquito control. The mosquito is a big problem throughout the world including India. Many mosquitocidal agents are available but none of them can control to a satisfactory level. From this background, many biosynthesized nanoparticles were applied on the mosquito as an ecofriendly insecticidal agent [5].

Table 1.1 represents the comparative data of mosquitocidal activity of different synthesized nanoparticles. Fouad et al. 2017 found that *Cassia fistula*—mediated silver nanoparticles killed 50% of *Aedes albopictus* larvae at 8.3 mg/L [6]. Another side Vinoth et al. 2018 revealed an excellent result, where *Sargassum polycystum* mediated silver nanoparticles killed 50% *Culex quinquefasciatus* larvae at 0.57% mg/L [7]. But larvicidal activity of *Aegle marmelos*—mediated copper nanoparticles were very poor, where the LC_{50} value was 74.6 mg/L on *Cx. quinquefasciatus* [8]. Comparatively *Tagetes* sp.—mediated copper nanoparticles show good results against *Cx. quinquefasciatus* [7]. Dhabi and Arasu in 2018 synthesized zinc nanoparticles from *Scadoxus multiflorus*. That nanoparticles larvicidal activity more or less same as *Tagets* sp.—mediated copper nanoparticles [8]. But zinc nanoparticles which synthesized from *Myristica fragrans* shows good result against *Aedes aegypti* [9]. In case of gold nanoparticles, LC_{50} values are higher than silver nanoparticles and lower than copper nanoparticles. *Turbinaria ornate* mediated gold nanoparticles killed 50% *Anopheles stephensi* at 12.79 mg/L [10]. *Chrysosporium tropicum*—mediated gold nanoparticles also killed 50% *A. aegypti* at 12 mg/L [11]. From this data, it is concluded that silver nanoparticles is more suitable as a mosquitocidal than others synthesized different nanoparticles. Another side silver nanoparticles were synthesized from different group of biological sources such as different plant parts, different fungi, or other biological products. This biological products are very important, because they play main role behind synthesis and nanoparticles properties also deepened on this product. But the group of biological product that are most efficient toward silver nanoparticles synthesis has not yet been discovered. From this background, we will briefly describe the efficacy of different group of biological products toward silver nanoparticles synthesis and also discuss AgNP effects on different stage of mosquito life cycle in this chapter.

2. Silver nanoparticles synthesis

Silver nanoparticles were synthesized easily compare with other nanoparticles. In this chapter, we focus on the biological synthesis of silver nanoparticles. Biologically, synthesis means any biological substrate were used as capping and reducing agents. Here, biological substrate was categorized in different groups like plants, microorganisms, and animal-mediated synthesis. The detailed discussion of these categories are as follows.

TABLE 1.1 Mosquito-larvicidal activity of different nanoparticles.

Sl. No.	Nanoparticles	Sources	Target species	Mortality rate	References
1.	Silver nanoparticles	Cassia fistula	Aedes albopictus	8.3 mg/l (LC_{50})	[6]
2.	Silver nanoparticles	Sargassum polycystum	Culex quinquefasciatus	0.57 mg/L (LC_{50})	[7]
3.	Copper nanoparticles	Aegle marmelos	Culex quinquefasciatus	74.6 mg/L (LC_{50})	[8]
4.	Copper nanoparticles	Tagetes sp.	Culex quinquifasciatus	34.4 mg/L (LC_{50})	[7]
5.	Zinc nanoparticles	Scadoxus multiflorus	Aedes aegypti	34.04 mg/L (LC_{50})	[8]
6.	Zinc nanoparticles	Myristica fragrans	Aedes aegypti	3.4 mg/L (LC_{50})	[9]
7.	Gold nanoparticles	Turbinaria ornate	Anopheles stephensi	12.79 mg/L (LC_{50})	[10]
8.	Gold nanoparticles	Chrysosporium tropicum	Aedes aegypti	12 mg/L (LC_{50})	[11]

2.1 Plant-mediated synthesis

Most biologically synthesized AgNPs have used plant extract compare to other sources. Table 1.2. represented huge uses of plant extract due to low cost and easy availability. All parts of the plant were used for AgNPs synthesis depending on their phytochemical properties.

Nakkala et al. 2014 used *Alternanthera dentate* leaf extract for AgNPs synthesis. They use leaves due to the availability and their synthesis capability rate. *Alternanthera dentate*−mediated AgNPs are mostly spherical in size and range between 50 and 100 nm [12]. Mariselvam et al. 2014 also used another traditional medicinal plant like *Tribulus terrestris*. They use the fruit extract for this AgNPs synthesis. *Tribulus terrestris* fruit extract mediated AgNPs size comparatively smaller, between 16 and 28 nm, and these particles also spherical in shape [13]. Another traditional plant *Vitex negundo*−mediated AgNPs are mostly spherical and face centered cubic (fcc) structure and their size 5−30 nm. Zargar et al. use leaf extract for this synthesis [14]. Rajkumar and abdul 2011 synthesis AgNPs from leaf extract of medicinal plant *Eclipta prostrate*. Their synthesized nanoparticles shape are very versatile, such as

TABLE 1.2 Plant-mediated silver nanoparticles.

Sl. No.	Source	Plant's part	Size (nm)	Shape	References
1.	*Alternanthera dentate*	Leaves	50 −100	Spherical	[12]
2.	*Tribulus terrestris*	Fruit	16−28	Spherical	[13]
3.	*Vitex negundo*	Leaves	5−30	Spherical and fcc	[14]
4.	*Eclipta prostrate*	Leaves	35−60	Triangles, pentagons, hexagons	[15]
5.	*Nelumbo nucifera*	Leaves	25−80	Spherical, triangular	[16]
6.	*Colocasia esculenta*	Stem	13−50	Spherical	[17]
7.	*Azadirachta indica*	Leaves	60−70	Spherical	[18]
8.	*Cocos nucifera*	Leaves	14.2	Spherical, triangular	[19]
9.	*Sida retusa.*	Leaves	20−40	Spherical	[20]
10.	*Datura metel*	Leaves	16−40	Quasi-linear superstructures	[21]

triangles, pentagons, and hexagons due to different phytochemical presence and size range between 36 and 60 nm [15]. Anthos kumar et al. 2011 extracted the aquatic plant leaf extract for AgNPs synthesis. Their synthesized nanoparticles size range of 25−80 nm and mostly are spherical and triangular [16]. Mondal et al. 2019 synthesis AgNPs from very easily available, cheap plant *Colocasia esculenta*. They used plant stem for AgNPs synthesis. *Colocasia esculenta*−mediated nanoparticles are mostly spherical in size and between 13 and 50 nm [17]. Another very common medicinal plant *Azadirecta indica* was used Mankand et al. 2018. They use the leaf of this plant for the synthesis of spherical AgNPs. The size of the synthesized nanoparticles was between 60 and 70 nm [18]. Uddin et al. 2020 use *Cocos nucifera* leaf mediated AgNPs are mostly spherical and triangular and their average size of 14.2 nm [19]. Sooraj et al. 2020 use a flowering plant known as *Sida retusa*. Synthesized AgNPs are mostly spherical and size between 20 and 40 nm [20]. A toxic plant *Datura metel* also uses for AgNPs synthesis. Their leaf was used for synthesis. Synthesized AgNPs are mostly quasi-linear superstructure and size range between 16 and 40 nm [21].

2.2 Microorganism-mediated synthesis

Silver nanoparticles were synthesized through different microorganisms including bacteria, fungus, archaea, and many other microscopic organisms. Different microorganism-mediated silver nanoparticles are different in shape and size which represented at Table 1.3.

TABLE 1.3 Microorganism-mediated silver nanoparticles.

Sl. No.	Source	Size (nm)	Shape	References
1.	*Proteus mirabilis*	10−20	Spherical	[22]
2.	*Verticillium* sp.	20−30	Round	[23]
3.	*Shewanella oneidensis* MR-1	2−6	Spherical	[24]
4.	*Trichoderma viride*	5−40	Spherical	[25]
5.	*Enterobacter aerogenes*	25−35	Spherical	[26]
6.	*Cladosporium* sp.	24	Spherical	[27]
7.	*Pseudomonas aeruginosa*	6−7	Spherical, disk-shaped	[28]
8.	*Agaricus bisporus*	8−20	Spherical	[29]
9.	*Exiguobacterium* sp.	5−50	Spherical	[30]
10.	*Bacillus* spp.	77−92	Spherical	[31]

Somadi et al. 2009 synthesized silver nanoparticles from *Proteus mirabilis* and most of the synthesized nanoparticles are spherical in shape and size between 10 and 20 nm [22]. Mukherjee et al. 2001 were using a fungi species, *Verticillium* sp. for the synthesis of silver nanoparticles. This fungus mediated silver nanoparticles size is 20−30 nm and most of the round-shaped [23]. But *Shewanella oneidensis* mediated silver nanoparticles are very small size, such as 2−6 nm. These bacteria-mediated silver nanoparticles are spherical in shape [24]. In 2010, Fayaz et al. produce silver nanoparticles from *Trichoderma viride* fungi extract. The size of this synthesized nanoparticle was variables between 5 and 40 nm and most of them are spherical in shape [25]. Another bacteria-mediated synthesis was done, where *Enterobacter aerogenes* uses as a capping and reducing agent. These synthesized particles are spherical and the size range is between 25 and 35 nm [26]. Popli et al. 2018 use *Cladosporium* sp. for the synthesis of AgNPs. *Cladosporium* sp.−mediated AgNPs are spherical in shape and average size 24 nm [27]. Srivastava and Constanti 2012 use other bacteria for AgNPs synthesis. This *Pseudomonus aeruginosa* mediated AgNPs are mostly spherical and disk-shaped and their size 6−7 nm [28]. Sonbaty 2013 use comparatively large fungus species, *Agaricus bisporus* for AgNPs synthesis. *Agaricus bisporus*−mediated AgNPs are mostly spherical and size between 8 and 20 nm [29]. Tamboli and Lee 2013, synthesized AgNPs from *Exiguobacterium* sp. Their prepared nanoparticles size 5−50 nm and mostly are spherical [30]. Elbeshehy et al. 2015 synthesized silver nanoparticles shape were spherical [31].

2.3 Animal products - mediated synthesis

Animal products mediated silver nanoparticles synthesis is one of the newest concept. In this process, many researchers are uses different animal products as a biological source, which act as a reducing and capping agent. Table 1.4 shows different animal products−mediated AgNPs. Jha and Prasad (2015) uses fish gut as a reducing and capping agent on this experiment fish gut extract nicely synthesized AgNPs with spherical size within 8−40 nm size range [32].

Similarly Akintayo et al. (2020) use goat fur for the synthesis of AgNPs. Their synthesized AgNPs were mostly spherical in shape and size between 11 and 31 nm [33]. Chicken egg white was another animal product that successfully synthesized silver nanoparticles. Those AgNPs shape were different from previous experiment. Here maximum AgNPs size were biconcave with comparatively small size (2−20 nm) [34]. Liang et al. (2014) used egg shell as a synthesized agent and their synthesized nanoparticles size was very small (3 nm) compared to other biosynthesized AgNPs. Most of the synthesized nanoparticles are spherical [35]. Again, Sinha and Ahmaouzzaman used egg shell of *Anas platyrhynchos*. But in those experiment synthesized nano-particles size were large compare to the previous example where they used the

TABLE 1.4 Animal-mediated silver nanoparticles.

Sl. No.	Source	Size (nm)	Shape	References
1.	Fish gut	8–40	Spherical	[32]
2.	Goat Fur	11–31	Spherical	[33]
3.	Chicken eggs white	2–20	Biconcave shape	[34]
4.	Eggshell	3	Spherical	[35]
5.	Egg shell of *Anas platyrhynchos*	6–26	Spherical and oval	[36]
6.	Cockroach wings	<50	Spherical	[37]
7.	Bioactive bile salt	21	Spherical	[38]
8.	Cow urine	20	Spherical	[39]
9.	Milk protein	7	Spherical	[40]
10.	Donkey dungs	36	Spherical	[41]

same reducing and capping agent, and the shape of AgNPs were spherical and oval [36]. Recently Khatami et al. (2019) use a totally different product for synthesis AgNPs. The use cockroach wings for AgNPs synthesis. Their synthesized nanoparticles size was less than 50 nm with spherical shape [37]. Bioactive bile salt is another good reducing and capping agent for AgNPs synthesis. Tangavelu et al. uses this bile salt and their synthesized AgNPs average size was 21 nm with spherical shape [38]. At 2019 Vinay et al. used an excretory animal product and synthesized AgNPs. Their cow urine-mediated silver nanoparticles are mostly spherical with 20 nm average size [39]. One another animal nutrient product milk, which uses for AgNPs synthesis. Milk-mediated AgNPs size was comparatively small (7 nm) with spherical shape [40] Khattat et al. produce AgNPs by using donkey dunk. Their synthesized nanoparticles were spherical with 36 nm average size [41].

Sections 2.1–2.3 are provide a clear idea about AgNPs synthesis from three different biological sources. Previous discussion revealed that plant, microorganism as well as animal products are capable toward AgNPs synthesis. But the plant extract are more easily available than two other sources. Beside this in the large scale, plant extracts are more acceptable due to their easily availability.

3. Silver nanoparticles application on mosquito

Table 1.1 shows that silver nanoparticles are most effective against mosquitoes compared to other synthesized nanoparticles. AgNPs are applied to different stage of mosquito, like egg, larvae, pupae, and adult. The effect of these nanoparticles is different in different stages of mosquito life. Details application of AgNPs on mosquito is as follows.

3.1 AgNPs application on mosquito larvae

Most of the research work will be done in this field. Because, mosquito larvae are cannot fly and they feed on their habitat, for this reason, most of the researcher target the larval stages of the mosquito. Table 1.5 represented that Saini et al. 2019, synthesized AgNPs from a traditional plant *Cullen corylifolium*. This plant-mediated AgNPs size between 20 and 60 nm and they applied these nanoparticles on malarial vector *Anopheles stephensi*. The LC_{50} and LC_{90} values are 6.03 and 10.83 mg/L against the third instar larvae *A. stephensi* [42].

Another side a flowering plant *Leonotis nepetifolia*—mediated AgNPs were applied on dengue vector *Aedes aegypti*. The average size of the synthesized nanoparticles was 8.5 nm. The larvicidal activity of these AgNPs is lower than previous one, their LC_{50} value was 47.44 mg/L against the third instar larvae [43]. Muhammad et al. 2020 synthesized nanoparticle activity are more on less equal to *Leonotis nepetifolia*—mediated AgNPs. They synthesized nanoparticles from an oil-producing plant *Ricinus communis*. Synthesized nanoparticles they applied on second and third instar larvae. The LC_{50} and LC_{90} value were 46.22 and 85.30 mg/L and the average size of this AgNPs are 26 nm [44]. Anthos et al. 2020, use *Annona muricata* fruit tree for the synthesis of AgNPs. Their synthesized AgNPs size 20—34 nm, which applied on the third instar larvae of *A. aegypti*. The larvicidal activity of these nanoparticles was moderate, where the LC_{50} value was 45.521 mg/L [45]. But a tropical tree known as *Garcinia mangastana* mediated AgNPs shows very good result shows as a larvicidal agent against fourth instar larvae of *A. aegyepti*. The LC_{50} value of this larvicidal agent was 5.93 mg/L, which is far better than previously discussed nanoparticles [46]. Another flowering plant *Holarrhena antidysenterica*—mediated AgNPs also shows good results. This synthesized nanoparticles average size of 28 nm, which applied on fourth instar larvae of *C. quinquefasciatus*. The LC_{50} and LC_{90} value are 5.53 and 12.01 mg/L, respectively [47]. At 2018 Alyahya et al. applied *Holostemma ada-kodien* as a synthesizing agent of AgNPs. They targeted the third instar larvae of malarial vector *A. stephensi*. Where the LC_{50} value was 12.18 mg/L. The average of these synthesized nanoparticles was 20—88 nm [48]. Besides those plants, an algal species known as *Sargassum polycystum* mediated AgNPs also applied on the fourth instar larvae of *A. stephensi*. Synthesized

TABLE 1.5 Mosquito larvicidal activity of silver nanoparticles.

Sl. No.	Biological source of AgNPs	Nanoparticles size (nm)	Mosquito sp.	Larval stage	Mortality rate	References
1.	Cullen corylifolium	20–60	Anopheles stephensi	III	6.03 mg/L LC50 10.83 mg/L LC90	[42]
2.	Leonotis nepetifolia	8.5	Aedes aegyepti	III	47.44 mg/L LC50	[43]
3.	Ricinus communis	26	Aedes aegypti	II and III	46.22 mg/L LC50 85.30 mg/L	[44]
4.	Annona muricata	20–34	Aedes aegyepti	III	45.521 mg/L	[45]
5.	Garcinia mangostana	19	Aedes aegyepti	IV	5.93 mg/L	[46]
6.	Holarrhena antidysenterica	28	Culex quinquefasciatus	IV	5.53 mg/L LC50 12.01 mg/L LC90	[47]
7.	Holostemma ada-kodien	20–80	Anopheles stephensi	III	12.18 mg/L LC50	[48]
8.	Sargassum polycystum	20–88	Anopheles stephensi	IV	3.07 mg/L LC50	[49]
9.	Habenaria plantaginea	50	Anopheles stephensi Aedes aegypti	III III	12.23 mg/L LC50 13.38 mg/L LC50	[49]
10.	Aglaia elaeagnoidea	3–34	Culex quinquefasciatus Aedes aegyepti	III	24.91 22.8	[50]

nanoparticles average size range of 20–88 nm. These algae-mediated AgNPs show tremendous results, where the LC_{50} value was recorded at 3.07 mg/L [51]. Aarthi et al. 2017 use an orchid species *Habenaria plantaginea* for the synthesis of AgNPs. There synthesized nanoparticles were applied to two different mosquito species larvae like *A. stephensi* and *A. aegypti*. Both the third instar mosquito larvae more on the less same sensitivity against synthesized AgNPs. Where the LC_{50} value are 12.23 and 13.38 mg/L, respectively [49]. Benelli et al. 2017 synthesis AgNPs from *Aglaia elaeagnoidae* extract. Their synthesized nanoparticle's size was very small, ranging from 3 to 34 nm. These nanoparticles also applied to two different mosquito larval species like *C. quinquefasciatus* and *A. aegypti*. The LC_{50} value of both third instar larvae was recorded at 24.91 and 22.8 mg/L [50].

3.2 AgNPs application on mosquito pupae

Many researchers synthesized AgNPs also applied to mosquito pupae, which represented in Table 1.6. The pupal stage of mosquito cannot fly and pupae also cannot feed. The details discussion as follows.

Sundaravadivelan and Padmanabhan 2013, use a fungal species *Trichoderma harzianum* for the synthesis of AgNPs. Their synthesized AgNPs size range 10–20 nm and they applied it on *Aedes aegypti* pupae. The 50% lethal concentration was 0.26 mg/L that remarkably good result [52]. Murugan et al. 2015, were synthesized AgNPs from a flowering plant *Toddalia asiatica*. Synthesized nanoparticles average size 25–30 nm and these nanoparticles applied on filarial vector *Culex quinquefasciatus*. The LC_{50} value of this synthesized AgNPs was 31.38 mg/L [53]. Another side *Cassia fistula*–mediated AgNPs shows better pupicidal results than *Taddalia asiatica* mediated nanoparticles. In this experiment, they applied on *Culex pipiens pallens*, and the LC_{50} values are 19.9 mg/L. The average size of these nanoparticles was 7–25 nm [6]. Murugan et al. 2015 also prepared pupicidal AgNPs from *Ulva lactuca*. Synthesized nanoparticles size range between 20 and 35 nm and they applied on malarial vector *A. stephensi* pupae, and the LC_{50} value was 6.86 mg/L [54]. *Aedes albopictus* pupae also control by *Berberis tinctoria*–mediated silver nanoparticles, where the LC50 value was 14.87 mg/L. But here the nanoparticles size 65–70 nm, which comparatively larger than other but pupicidal activity is very good [55]. Murugan et al. 2015, applied a flowering shrub known as a *Datura metel* for AgNPs synthesis. They applied their synthesized nanoparticles on *A. stephensi* and the LC_{50} value of this application was 6.755 mg/L. The average size range of these nanoparticles was 40–60 nm [56]. Again Murugan et al. 2015 use an algae species, *Centroceras clavulatum*, for AgNPs synthesis. The average size of those nanoparticles was 35–65 nm and they applied on *Aedes aegypti* pupae. The LC_{50} value is 33.877 mg/L [57]. *Aristolochia indica*–mediated AgNPs also applied to *A. stephensi* pupae. The average size of those nanoparticles was 30–55 nm. The LC_{50} value of these AgNPs on pupae was 15.65 mg/L [58].

TABLE 1.6 Mosquito pupicidal activity of silver nanoparticles.

Sl. No.	Biological source of AgNPs	Nanoparticles size (nm)	Mosquito sp.	Mortality rate	References
1.	Trichoderma harzianum	10–20	Aedes aegypti	0.26 mg/L LC$_{50}$	[52]
2.	Toddalia asiatica	25–30	Culex quinquefasciatus	31.38 mg/L LC$_{50}$	[53]
3.	Cassia fistula	7–25	Culex pipiens pallens	19.9 mg/L LC$_{50}$	[6]
4.	Ulva lactuca	20–35	Anopheles stephensi	6.86 mg/L LC$_{50}$	[54]
5.	Berberis tinctoria	65–70	Aedes albopictus	14.87 mg/L LC$_{50}$	[55]
6.	Datura metel	40–60	Anopheles stephensi	6.755 mg/L LC$_{50}$	[56]
7.	Centroceras clavulatum	35–65	Aedes aegypti	33.877 mg/L LC$_{50}$	[57]
8.	Aristolochia indica	30–55	Anopheles stephensi	15.65 mg/L LC$_{50}$	[58]
9.	Euphorbia hirta	30–60	Anopheles stephensi	34.52 mg/L LC$_{50}$	[59]
10.	Chrysanthemum indicum	25–69	Anopheles stephensi	35.05 mg/L LC$_{50}$	[60]

Priyadarshini et al. 2012, synthesized AgNPs from a weed species *Euphorbia hirta*, and the average size of the nanoparticles was 30.6 nm. They applied on *A. stephensi* pupae and LC_{50} value was 34.52 mg/L [59]. Arokiyaraj et al. 2015, usage of a flowering plant is known as *Chrysanthemum indicum* for AgNPs synthesis. They applied on *A. stephensi* and the LC_{50} value 35.05 mg/L was moderate compare to others. The average size of their synthesized nanoparticles was 25–69 nm [60].

3.3 AgNPs application on mosquito egg and adult

From previous literature, we found that most of the researchers target mosquito larvae and pupae, but few researchers target mosquito eggs and adults. Arjuna et al. 2012, synthesized AgNPs from *Annona squamosal* extract. They applied AgNPs on adult and egg of *A. stephensi*. They find out a high concentration of AgNPs remarkably reduce adult longevity and they also found that many emerging adults at this time had defected wings or legs and abnormal morphologies and laid few eggs compared to normal conditions. Besides this many eggs fail to hatch [61]. Veerakumar et al. 2014 applied *Feronia elephantum* plant for AgNPs synthesis and they applied these AgNPs on three different adult mosquito species. The LC_{50} value was 18.041, 20.399, and 21.798 mg/L against *A. stephensi*, *A. aegypti*, and *C. quinquefasciatus*, respectively [62].

Comparative study between Sections 3.1–3.3 suggest that biologically synthesized AgNPs shows a good result against larval stage of mosquito compare to egg, pupae, and adult. Tables 1.5 and 1.6 shows 23.55 and 25.06 mg/L average $LC_{50,}$ values of larvae and pupae, respectively. This result also supports that larvae are more sensible than pupae. Larvae are more affected when nanoparticles were treated at a water body. Because pupae of mosquitoes can't feed anything but larvae feed on their habitat. Besides, adult are movable and egg are covered with a coat. So, this is maybe the main reason behind high larvicidal activity of biologically synthesized AgNPs.

4. Research gaps

At present many researchers use nanoparticles as a mosquitocidal agent, from those nanoparticles, AgNPs are more effective, but why silver is more effective than other NPs is still not properly understood. Besides this, synthesized AgNPs are effective at laboratory conditions but the efficacy of NPs at the natural conditions, and what the effect of this nanoparticle on other species on nature has yet to be investigated.

At the same time, this chapter summarized that AgNPs are mostly affected by larvae than other stages of the mosquito. But based on our knowledge, very limited research done on the mechanism of AgNPs on the larval stage, this area has also yet to be investigated.

5. Conclusion

This chapter mainly focuses on the AgNPs synthesis from different groups of biological product. Most of the synthesized NPs are truly affective on mosquito larvae than other stages because larvae feed on their surrounding aquatic habited and they cannot fly. This chapter also suggests that plant-mediated AgNPs are more effective due to their small size. But there are a few areas like mode of action, mechanism of AgNPs synthesis, and AgNPs activity on mosquito at natural habited are still unknown. Finally, it concluded that AgNPs can be synthesized from different biological sources, but plant-mediated synthesis is more accepted depending on their availability and low cost. Another side these synthesized AgNPs are more affected on larvae compare to egg, pupae, and adult mosquitoes, but large-scale research are needed for detail scenario.

Appendix A. Supplementary data

Supplementary data to this article can be found online at https://doi.org/10. 1016/B978-0-12-824038-0.00004-3.

References

[1] El-Nour KM, Eftaiha A, Al-Warthan A, Ammar RA. Synthesis and applications of silver nanoparticles. Arab J Chem 2010;3(3):135—40. https://doi.org/10.1016/j.arabjc. 2010.04.008.

[2] Salam HA, Kamaraj RPM, Jagadeeswaran P, Gunalan S, Sivaraj R. Plants: green route for nanoparticle synthesis. Int Res J Biol Sci 2012;1(5):85—90.

[3] Sinha T, Ahmaruzzaman M, Bhattacharjee A. A simple approach for the synthesis of silver nanoparticles and their application as a catalyst for the photodegradation of methyl violet 6B dye under solar irradiation. J Environ Chem Eng 2014;2(4):2269—79. https://doi.org/10.1016/j.jece.2014.10.001.

[4] Popescu M, Velea A, Lorinczi A. Biogenic production of nanoparticles. Digest J Nanomater Biostruct 2010;5:1035—40.

[5] Banumathi B, Vaseeharan B, Ishwarya R, Govindarajan M, Alharbi NS, Kadaikunnan S, et al. Toxicity of herbal extracts used in ethno-veterinary medicine and green-encapsulated ZnO nanoparticles against *Aedes aegypti* and microbial pathogens. Parasitol Res 2017;116(6):1637—51. https://doi.org/10.1007/s00436-017-5438-6.

[6] Fouad H, Hongjie L, Hosni D, Wei J, Abbas G, Ga'Al H, Jianchu M. Controlling *Aedes albopictus* and *Culex pipiens* pallens using silver nanoparticles synthesized from aqueous extract of Cassia fistula fruit pulp and its mode of action. Artif Cells Nanomed Biotechnol 2017;46(3):558—67. https://doi.org/10.1080/21691401.2017.1329739.

[7] Mondal NK, Hajra A. Synthesis of copper nanoparticles (CuNPs) from petal extracts of marigold (*Tagetes* sp.) and sunflower (*Helianthus* sp.) and their effective use as a control tool against mosquito vectors. J Mosq Res 2016. https://doi.org/10.5376/jmr.2016.06.0019.

[8] Angajala G, Pavan P, Subashini R. One-step biofabrication of copper nanoparticles from *Aegle marmelos* correa aqueous leaf extract and evaluation of its anti-inflammatory and mosquito-larvicidal efficacy. RSC Adv 2014;4(93):51459−70. https://doi.org/10.1039/c4ra10003d.

[9] Vinoth S, Shankar SG, Gurusaravanan P, Janani B, Devi JK. Anti-larvicidal activity of silver nanoparticles synthesized from *Sargassum polycystum* against mosquito vectors. J Cluster Sci 2018;30(1):171−80. https://doi.org/10.1007/s10876-018-1473-4.

[10] Deepak P, Sowmiya R, Balasubramani G, Aiswarya D, Arul D, Josebin MP, Perumal P. Mosquito-larvicidal efficacy of gold nanoparticles synthesized from the seaweed, *Turbinaria ornata* (Turner) J. Agardh 1848. Part Sci Technol 2017;36(8):974−80. https://doi.org/10.1080/02726351.2017.1331286.

[11] Soni N, Prakash S. Efficacy of fungus mediated silver and gold nanoparticles against *Aedes aegypti* larvae. Parasitol Res 2011;110(1):175−84. https://doi.org/10.1007/s00436-011-2467-4.

[12] Nakkala JR, Mata R, Gupta AK, Sadras SR. Biological activities of green silver nanoparticles synthesized with *Acorus calamus* rhizome extract. Eur J Med Chem 2014;85:784−94. https://doi.org/10.1016/j.ejmech.2014.08.024.

[13] Mariselvam R, Ranjitsingh A, Nanthini AU, Kalirajan K, Padmalatha C, Selvakumar PM. Green synthesis of silver nanoparticles from the extract of the inflorescence of *Cocos nucifera* (Family: arecaceae) for enhanced antibacterial activity. Spectrochim Acta Mol Biomol Spectrosc 2014;129:537−41. https://doi.org/10.1016/j.saa.2014.03.066.

[14] Zargar M, Hamid AA, Bakar FA, Shamsudin MN, Shameli K, Jahanshiri F, Farahani F. Green synthesis and antibacterial effect of silver nanoparticles using *Vitex negundo* L. Molecules 2011;16(8):6667−76. https://doi.org/10.3390/molecules16086667.

[15] Rajakumar G, Rahuman AA. Larvicidal activity of synthesized silver nanoparticles using *Eclipta prostrata* leaf extract against filariasis and malaria vectors. Acta Trop 2011;118(3):196−203. https://doi.org/10.1016/j.actatropica.2011.03.003.

[16] Santhoshkumar T, Rahuman AA, Rajakumar G, Marimuthu S, Bagavan A, Jayaseelan C, et al. Synthesis of silver nanoparticles using *Nelumbo nucifera* leaf extract and its larvicidal activity against malaria and filariasis vectors. Parasitol Res 2010;108(3):693−702. https://doi.org/10.1007/s00436-010-2115-4.

[17] Mondal N, Mondal A, Hajra A, Shaikh W, Chakraborty S. Synthesis of silver nanoparticle with *Colocasia esculenta* (L.) stem and its larvicidal activity against *Culex quinquefasciatus* and *Chironomus* sp. Asian Pac J Trop Biomed 2019;9(12):510. https://doi.org/10.4103/2221-1691.271724.

[18] Mankad M, Patil G, Patel D, Patel P, Patel A. Comparative studies of sunlight mediated green synthesis of silver nanoparticles from *Azadirachta indica* leaf extract and its antibacterial effect on *Xanthomonas oryzae* pv. oryzae. Arab J Chem 2020;13(1):2865−72. https://doi.org/10.1016/j.arabjc.2018.07.016.

[19] Uddin AK, Siddique MA, Rahman F, Ullah AK, Khan R. *Cocos nucifera* leaf extract mediated green synthesis of silver nanoparticles for enhanced antibacterial activity. J Inorg Organomet Polym Mater 2020;30(9):3305−16. https://doi.org/10.1007/s10904-020-01506-9.

[20] Sooraj MP, Nair AS, Vineetha D. Sunlight-mediated green synthesis of silver nanoparticles using Sida retusa leaf extract and assessment of its antimicrobial and catalytic activities. Chem Pap 2020;75(1):351−63. https://doi.org/10.1007/s11696-020-01304-0.

[21] Kesharwani J, Yoon KY, Hwang J, Rai M. Phytofabrication of silver nanoparticles by leaf extract of *Datura metel*: hypothetical mechanism involved in synthesis. J Bionanosci 2009;3(1):39—44. https://doi.org/10.1166/jbns.2009.1008.

[22] Samadi N, Golkaran D, Eslamifar A, Jamalifar H, Fazeli MR, Mohseni FA. Intra/extra-cellular biosynthesis of silver nanoparticles by an autochthonous strain of *Proteus mirabilis* isolated from photographic waste. J Biomed Nanotechnol 2009;5(3):247—53. https://doi.org/10.1166/jbn.2009.1029.

[23] Mukherjee P, Ahmad A, Mandal D, Senapati S, Sainkar SR, Khan MI, et al. Fungus-mediated synthesis of silver nanoparticles and their immobilization in the mycelial matrix: a novel biological approach to nanoparticle synthesis. Nano Lett 2001;1(10):515—9. https://doi.org/10.1021/nl0155274.

[24] Debabov VG, Voeikova TA, Shebanova AS, Shaitan KV, Emel'Yanova LK, Novikova LM, Kirpichnikov MP. Bacterial synthesis of silver sulfide nanoparticles. Nanotechnol Russia 2013;8(3—4):269—76. https://doi.org/10.1134/s1995078013020043.

[25] Fayaz AM, Balaji K, Girilal M, Yadav R, Kalaichelvan PT, Venketesan R. Biogenic synthesis of silver nanoparticles and their synergistic effect with antibiotics: a study against gram-positive and gram-negative bacteria. Nanomed Nanotechnol Biol Med 2010;6(1):103—9. https://doi.org/10.1016/j.nano.2009.04.006.

[26] Biosynthesis of Silver Nanoparticles. Characterization and their antimicrobial activity. Nanomed Drug Deliv 2012:107—14. https://doi.org/10.1201/b13119-11.

[27] Popli D, Anil V, Subramanyam AB, Namratha NM, Ranjitha RV, Rao SN, et al. Endophyte fungi, *Cladosporium* species-mediated synthesis of silver nanoparticles possessing in vitro antioxidant, anti-diabetic and anti-Alzheimer activity. Artif Cells Nanomed Biotechnol 2018;46(Suppl. 1):676—83. https://doi.org/10.1080/21691401.2018.1434188.

[28] Srivastava SK, Constanti M. Room temperature biogenic synthesis of multiple nanoparticles (Ag, Pd, Fe, Rh, Ni, Ru, Pt, Co, and Li) by *Pseudomonas aeruginosa* SM1. J Nanoparticle Res 2012;14(4). https://doi.org/10.1007/s11051-012-0831-7.

[29] El-Sonbaty SM. Fungus-mediated synthesis of silver nanoparticles and evaluation of anti-tumor activity. Cancer Nanotechnol 2013;4(4—5):73—9. https://doi.org/10.1007/s12645-013-0038-3.

[30] Tamboli DP, Lee DS. Mechanistic antimicrobial approach of extracellularly synthesized silver nanoparticles against gram positive and gram negative bacteria. J Hazard Mater 2013;260:878—84. https://doi.org/10.1016/j.jhazmat.2013.06.003.

[31] Elbeshehy EK, Elazzazy AM, Aggelis G. Silver nanoparticles synthesis mediated by new isolates of Bacillus spp., nanoparticle characterization and their activity against Bean Yellow Mosaic Virus and human pathogens. Front Microbiol 2015;6. https://doi.org/10.3389/fmicb.2015.00453.

[32] Jha AK, Prasad K. Synthesis of silver nanoparticles employing fish processing discard: an eco-amenable approach. J Chin Adv Mater Soc 2014;2(3):179—85. https://doi.org/10.1080/22243682.2014.930796.

[33] Akintayo GO, Lateef A, Azeez MA, Asafa TB, Oladipo IC, Badmus JA, et al. Synthesis, bioactivities and cytogenotoxicity of animal Fur-mediated silver nanoparticles. IOP Conf Ser Mater Sci Eng 2020;805:012041. https://doi.org/10.1088/1757-899x/805/1/012041.

[34] Thiyagarajan K, Bharti VK, Tyagi S, Tyagi PK, Ahuja A, Kumar K, et al. Synthesis of non-toxic, biocompatible, and colloidal stable silver nanoparticle using egg-white protein as capping and reducing agents for sustainable antibacterial application. RSC Adv 2018;8(41):23213—29. https://doi.org/10.1039/c8ra03649g.

[35] Liang M, Su R, Huang R, Qi W, Yu Y, Wang L, He Z. Facile in situ synthesis of silver nanoparticles on procyanidin-grafted eggshell membrane and their catalytic properties. ACS Appl Mater Interfaces 2014;6(7):4638−49. https://doi.org/10.1021/am500665p.

[36] Sinha T, Ahmaruzzaman M. High-value utilization of egg shell to synthesize Silver and Gold−Silver core shell nanoparticles and their application for the degradation of hazardous dyes from aqueous phase-A green approach. J Colloid Interface Sci 2015;453:115−31. https://doi.org/10.1016/j.jcis.2015.04.053.

[37] Khatami M, Iravani S, Varma RS, Mosazade F, Darroudi M, Borhani F. Cockroach wings-promoted safe and greener synthesis of silver nanoparticles and their insecticidal activity. Bioproc Biosyst Eng 2019;42(12):2007−14. https://doi.org/10.1007/s00449-019-02193-8.

[38] Muthuramalingam TR, Shanmugam C, Gunasekaran D, Duraisamy N, Nagappan R, Krishnan K. Bioactive bile salt-capped silver nanoparticles activity against destructive plant pathogenic fungi through in vitro system. RSC Adv 2015;5(87):71174−82. https://doi.org/10.1039/c5ra13306h.

[39] Vinay S, Udayabhanu, Nagaraju G, Chandrappa C, Chandrasekhar N. Novel Gomutra (cow urine) mediated synthesis of silver oxide nanoparticles and their enhanced photocatalytic, photoluminescence and antibacterial studies. J Sci Adv Mater Devices 2019;4(3):392−9. https://doi.org/10.1016/j.jsamd.2019.08.004.

[40] Pandey S, Klerk CD, Kim J, Kang M, Fosso-Kankeu E. Eco friendly approach for synthesis, characterization and biological activities of milk protein stabilized silver nanoparticles. Polymers 2020;12(6):1418. https://doi.org/10.3390/polym12061418.

[41] Mohammad MK, Hamid N, Mehrnaz R, Mahshid E, Yasaman F. Anticancer activity of silver nanoparticles synthesized using medicinal animal waste extract on Mcf-7 human breast cancer. Cell Line 2020. SSRN: https://ssrn.com/abstract=3696919.

[42] Saini H, Yadav R, Kumar D, Kumar G, Agrawal V. *Cullen corylifolium* (L.) Medik. Seed extract, an excellent system for fabrication of silver nanoparticles and their multipotency validation against different mosquito vectors and human cervical cancer cell line. J Cluster Sci 2019;31(1):161−75. https://doi.org/10.1007/s10876-019-01630-8.

[43] Manimegalai T, Raguvaran K, Kalpana M, Maheswaran R. Green synthesis of silver nanoparticle using Leonotis nepetifolia and their toxicity against vector mosquitoes of *Aedes aegypti* and *Culex quinquefasciatus* and agricultural pests of *Spodoptera litura* and *Helicoverpa armigera*. Environ Sci Pollut Control Ser 2020;27(34):43103−16. https://doi.org/10.1007/s11356-020-10127-1.

[44] Waris M, Nasir S, Abbas S, Azeem M, Ahmad B, Khan NA, et al. Evaluation of larvicidal efficacy of *Ricinus communis* (Castor) and synthesized green silver nanoparticles against *Aedes aegypti* L. Saudi J Biol Sci 2020;27(9):2403−9. https://doi.org/10.1016/j.sjbs.2020.04.025.

[45] Santhosh SB, Natarajan D, Deepak P, Gayathri B, Kaviarasan L, Naresh P, et al. Metabolic enzyme inhibitory and larvicidal activity of biosynthesized and heat stabilized silver nanoparticles using *Annona muricata* leaf extract. BioNanoScience 2020;10(1):267−78. https://doi.org/10.1007/s12668-019-00709-w.

[46] Karthiga P, Rajeshkumar S, Annadurai G. Mechanism of larvicidal activity of antimicrobial silver nanoparticles synthesized using *Garcinia mangostana* bark extract. J Cluster Sci 2018. https://doi.org/10.1007/s10876-018-1441-z.

[47] Kumar D, Kumar G, Agrawal V. Green synthesis of silver nanoparticles using *Holarrhena antidysenterica* (L.) Wall.bark extract and their larvicidal activity against dengue and filariasis vectors. Parasitol Res 2017;117(2):377−89. https://doi.org/10.1007/s00436-017-5711-8.

[48]   Alyahya SA, Govindarajan M, Alharbi NS, Kadaikunnan S, Khaled JM, Mothana RA, et al. Swift fabrication of Ag nanostructures using a colloidal solution of Holostemma ada-kodien (Apocynaceae) — antibiofilm potential, insecticidal activity against mosquitoes and non-target impact on water bugs. J Photochem Photobiol B Biol 2018;181:70—9. https://doi.org/10.1016/j.jphotobiol.2018.02.019.

[49]   Aarthi C, Govindarajan M, Rajaraman P, Alharbi NS, Kadaikunnan S, Khaled JM, et al. Eco-friendly and cost-effective Ag nanocrystals fabricated using the leaf extract of *Habenaria plantaginea*: toxicity on six mosquito vectors and four non-target species. Environ Sci Pollut Control Ser 2017;25(11):10317—27. https://doi.org/10.1007/s11356-017-9203-2.

[50]   Benelli G, Govindarajan M, Senthilmurugan S, Vijayan P, Kadaikunnan S, Alharbi NS, Khaled JM. Fabrication of highly effective mosquito nanolarvicides using an Asian plant of ethno-pharmacological interest, Priyangu (*Aglaia elaeagnoidea*): toxicity on non-target mosquito natural enemies. Environ Sci Pollut Control Ser 2017;25(11):10283—93. https://doi.org/10.1007/s11356-017-8898-4.

[51]   Vinoth S, Gowri Shankar S, Gurusaravanan P, Janani B, Karthika Devi J. Anti-larvicidal activity of silver nanoparticles synthesized from *sargassum polycystum* against mosquito vectors. J. Clust. Sci. 2019;30:171—80. https://doi.org/10.1007/s10876-018-1473-4.

[52]   Sundaravadivelan C, Padmanabhan MN. Effect of mycosynthesized silver nanoparticles from filtrate of *Trichoderma harzianum* against larvae and pupa of dengue vector *Aedes aegypti* L. Environ Sci Pollut Control Ser 2013;21(6):4624—33. https://doi.org/10.1007/s11356-013-2358-6.

[53]   Murugan K, Venus JS, Panneerselvam C, Bedini S, Conti B, Nicoletti M, et al. Biosynthesis, mosquitocidal and antibacterial properties of *Toddalia asiatica*-synthesized silver nanoparticles: do they impact predation of guppy *Poecilia reticulata* against the filariasis mosquito *Culex quinquefasciatus*? Environ Sci Pollut Control Ser 2015;22(21):17053—64. https://doi.org/10.1007/s11356-015-4920-x.

[54]   Murugan K, Samidoss CM, Panneerselvam C, Higuchi A, Roni M, Suresh U, et al. Seaweed-synthesized silver nanoparticles: an eco-friendly tool in the fight against *Plasmodium falciparum* and its vector *Anopheles stephensi*? Parasitol Res 2015;114(11):4087—97. https://doi.org/10.1007/s00436-015-4638-1.

[55]   Kumar PM, Murugan K, Madhiyazhagan P, Kovendan K, Amerasan D, Chandramohan B, et al. Biosynthesis, characterization, and acute toxicity of *Berberis tinctoria*-fabricated silver nanoparticles against the Asian tiger mosquito, *Aedes albopictus*, and the mosquito predators *Toxorhynchites splendens* and *Mesocyclops thermocyclopoides*. Parasitol Res 2015;115(2):751—9. https://doi.org/10.1007/s00436-015-4799-y.

[56]   Murugan K, Dinesh D, Kumar PJ, Panneerselvam C, Subramaniam J, Madhiyazhagan P, et al. Datura metel-synthesized silver nanoparticles magnify predation of dragonfly nymphs against the malaria vector *Anopheles stephensi*. Parasitol Res 2015;114(12):4645—54. https://doi.org/10.1007/s00436-015-4710-x.

[57]   Murugan K, Aruna P, Panneerselvam C, Madhiyazhagan P, Paulpandi M, Subramaniam J, et al. Fighting arboviral diseases: low toxicity on mammalian cells, dengue growth inhibition (in vitro), and mosquitocidal activity of *Centroceras clavulatum*-synthesized silver nanoparticles. Parasitol Res 2015;115(2):651—62. https://doi.org/10.1007/s00436-015-4783-6.

[58]   Murugan K, Labeeba MA, Panneerselvam C, Dinesh D, Suresh U, Subramaniam J, et al. *Aristolochia indica* green-synthesized silver nanoparticles: a sustainable control tool against the malaria vector *Anopheles stephensi*? Res Vet Sci 2015;102:127—35. https://doi.org/10.1016/j.rvsc.2015.08.001.

[59] Priyadarshini KA, Murugan K, Panneerselvam C, Ponarulselvam S, Hwang J, Nicoletti M. Biolarvicidal and pupicidal potential of silver nanoparticles synthesized using *Euphorbia hirta* against *Anopheles stephensi* Liston (Diptera: Culicidae). Parasitol Res 2012;111(3):997−1006. https://doi.org/10.1007/s00436-012-2924-8.

[60] Arokiyaraj S, Kumar VD, Elakya V, Kamala T, Park SK, Ragam M, et al. Biosynthesized silver nanoparticles using floral extract of *Chrysanthemum indicum* L.—potential for malaria vector control. Environ Sci Pollut Control Ser 2015;22(13):9759−65. https://doi.org/10.1007/s11356-015-4148-9.

[61] Arjunan NK, Murugan K, Rejeeth C, Madhiyazhagan P, Barnard DR. Green synthesis of silver nanoparticles for the control of mosquito vectors of malaria, filariasis, and dengue. Vector Borne Zoonotic Dis 2012;12(3):262−8. https://doi.org/10.1089/vbz.2011.0661.

[62] Veerakumar K, Govindarajan M. Adulticidal properties of synthesized silver nanoparticles using leaf extracts of *Feronia elephantum* (Rutaceae) against filariasis, malaria, and dengue vector mosquitoes. Parasitol Res 2014;113(11):4085−96. https://doi.org/10.1007/s00436-014-4077-4.

Chapter 2

Machine learning–enabled cognitive approaches for handling IoT-based environmental data

Gaurav Mohindru[1], Koushik Mondal[2], Haider Banka[1]

[1]Department of Computer Science and Engineering, IIT (ISM) DHANBAD, Dhanbad, Jharkhand, India; [2]Computer Centre, IIT (ISM) DHANBAD, Dhanbad, Jharkhand, India

1. Introduction

The advancements in computing and analytics have now given impetus to increased automation and adoption of intelligent IoT-based solutions across domains and industry verticals. There is a new trend seen toward digitization to address some core and emerging challenges and to stay relevant to the changing technology. In achieving an end-to-end digitization, the digital business models are highly dependent on IoT-enabled business platforms to provide disruptive data-driven business solutions, as mentioned in Ref. [1]. As per this survey, all the leading business houses believes that the quantum leap forward toward successful digitization is only possible when IoT-driven platforms are used significantly to better productivity at different decision-making levels. The value of the business will increase in manifolds by implementing data-driven analytical solutions that are generated through predictive modeling. Industry 4.0 framework adopts a data and analytics-driven approach with the increased use of cloud computing and IoT platform for cognitive decision making as shown in Fig. 2.1.

Machine learning (ML) cognitive solutions for environmental monitoring in different domains such as weather forecasting, agriculture, manufacturing, smart cities [2], smart buildings [3], industries [4] and so forth, which can offer novel use cases for efficient management of the ecosystem. The environmental statistics will help to build applications, based on statistical methods to the problem-areas, related to the environment. The data modeling techniques have now become more powerful with the advancement of geographic information

Cognitive Data Models for Sustainable Environment. https://doi.org/10.1016/B978-0-12-824038-0.00008-0

19

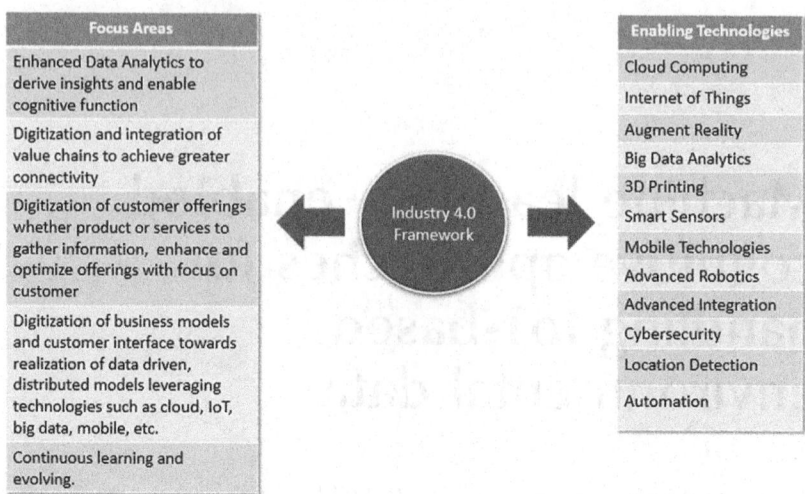

Focus Areas	Enabling Technologies
Enhanced Data Analytics to derive insights and enable cognitive function	Cloud Computing
	Internet of Things
Digitization and integration of value chains to achieve greater connectivity	Augment Reality
	Big Data Analytics
	3D Printing
Digitization of customer offerings whether product or services to gather information, enhance and optimize offerings with focus on customer	Smart Sensors
	Mobile Technologies
	Advanced Robotics
	Advanced Integration
Digitization of business models and customer interface towards realization of data driven, distributed models leveraging technologies such as cloud, IoT, big data, mobile, etc.	Cybersecurity
	Location Detection
	Automation
Continuous learning and evolving.	

FIGURE 2.1 Industry 4.0 framework with its contributing digital technologies [1].

system (GIS) and ML techniques. Unnecessary acts like deforestation and desertification have triggered global warming leading to absurd climatic patterns. Environmental risk assessments by quantifying the statistical relationships and characterization of the features offer a robust exploratory data analysis (EDA) platform to monitor environmental parameters like air or water quality, groundwater quality near a hazardous waste site, risk assessment to a potentially contaminated area, and hydrological data to predict the occurrences of floods using different visualization techniques [5]. The mathematical models available for encountering such problems need to be revisited in the light of artificial intelligence—based ML/deep learning paradigms [6]. The risk assessment models combined with vulnerability models for estimating the damage is also crucial for different environmental situations. Obtaining a thorough understanding of the human impact and developing forecasting and predictions will require a range of analytical approaches and data from different reliable sources (satellite, sensor, social media, cell phone, etc.) to enable us to design the learning rules of the numerical models from data [7].

In agriculture, the general parameters of the environment like dew, fog, hail, humidity, pressure, snow, rain, temperature, cyclone, wind direction, soil parameters like humidity, soil pH, nitrogen level, and so forth are deciding factors and can be analyzed to provide for providing prediction, better crop management, and utilization of scarce natural resources. The IoT-based ML techniques [8] help the nongovernment, academic institutes to design different models and subsequent predictions through which a large section of farmers can farm crops in time, take necessary measures during the growing phase, and producing good yields in return. The productivity of agricultural lands can increase by incorporating different cognitive intelligence through IoT-enabled services. The future lies in using data-driven knowledge and technology advancements. The adoption of the technical evolution and use of the IoT-based approach will help fill-up any digital and knowledge divide among the stakeholders across a domain or industry and in the society at large before it takes effect. The process of capacity building, knowledge dissemination, and skills development are key to protect everyone's stake associated with this industrial revolution with IoT, which is popularly known as Industrial IoT, or IIoT [9]. The objective is to deploy autonomous IoT-based sensor solutions with less human interaction to find an optimal solution by using different ML and image processing (IP) techniques. The adopted framework of an organization (logical and physical) and configuration should be fault-tolerant and distributed algorithm compatible where communication via wireless technologies offer viable low-cost solutions. The end-users' need to design an IoT-based soil pH, soil moisture, and soil nitrogen level data collection system and analyze the collected data through GIS and image processing techniques to predict crop yields and other valuable information using ML algorithms [10].

The recent advancements of IoT technologies along with different sensors for capturing data, cellular/noncellular connections for building an interconnected network, low maintenance and low-cost weather station, digital imaging, and processing have helped to create weather intelligence models for personalized weather forecasting and weather-based decision making. Weather intelligence uses a Long Range (LoRa) like Low Power Wide Area Network (LPWAN) dedicated network backbone with intelligent nodes that are connected to different gateways to send valuable information to an integrated online platform [11]. The availability of commercial-off-the-shelf (COTS) sensors [12] and integrating platforms like Raspberry Pi and others help to collect necessary weather information and integrating with the central server easily. The collected data is then fine-tuned with different preprocessing techniques and analyzed using different ML and Big Data algorithms to gain valuable insights for maintaining crop growth, crop disease identification, and other localized measures to protect the livelihood of any preidentified areas. The ICT enabled irrigation, fertilization, pest prevention, and control or even livestock management based on IoT collected data are gaining popularity due to its simplicity in implementation and easy integration with large data-processing cloud-based platforms like Amazon Web Services (AWS) [13], Azure [14], IBM Watson [15], Google Colab [16], and so forth for decision processing and relaying the results to the end-users in less or no time. The LPWAN-based technology initially offered solutions for noncritical and delay-tolerant applications but eventually, it acquired around 45% market share for IoT data transmission. LoRa is well designed for the protocol stack and sits in a Data Link Layer whereas LoRaWAN resides in the network layer in the OSI model [17]. LoRa and LoRaWAN are the integral parts of LPWAN technology for providing communication support in IoT-based solutions. The different parameters of environmental data also enable most of the "smart" activities in smart, sustainable city designing. It includes pollution and environmental degradation [18], transport congestion [19], clean water [20], energy availability [21], public lighting [22], sewage systems (wastewater treatment) [23], informal or precarious housing (high cost of real estate) [24], waste and risks, inequalities of development and social constraints [25], and tough living conditions, among others. Leveraging the advances in devices, technologies, and tools, Free and Open Source Software (FOSS) [26,27] and open-IoT environment [28] can offer viable and efficient solutions for handling different environment-related problems that human society, around the world, are facing today.

In this chapter, we intend to build and analyze environmental IoT data analysis models. It is a simple model that can be enhanced to realize production systems through required optimization and adjusting suitable parameters. It can offer cues to help create a long-term and sustainable solution for environmental data analysis and monitoring.

2. Cognitive IoT data-processing framework

The cognitive approach in IoT is intended toward adding human-like intelligence to the IoT framework. Cognitive IoT is a step toward taking the traditional IoT toward the era of artificial intelligence and augmented reality [29]. Cognitive IoT leverages technologies to simulate the human senses of viewing, hearing, listening, and interpreting. In achieving that objective it uses technologies such as ML and cognitive AI to mimic the human brain, machine vision, and pattern recognition to mimic human vision, speech recognition, natural language processing, and ontology to mimic human hearing of different languages and derive a meaning suitable to the context and tone in which the speech is given [30]. The broader goals are to make the application more aware of itself and also its surroundings, to be able to use context and semantics toward making cognitive decisions, and also to exhibit greater robustness and intelligence. The scope of cognitive intelligence spans across all entities in the data-processing pipeline as well as the network and communication. Cognition drives X-information data analytics paradigms using available knowledge base and real-time data manipulation with the help of computation driven available techniques [31]. To maintain continuous update itself and be relevant just like a human, the application strives toward constantly and dynamically updating the knowledge base. Cognitive IoT application exhibits or is expected to incorporate the below characteristics [30].

- it should learn by itself to the extent possible,
- it is constantly learning and make new revelations and rules so it follows a probabilistic approach,

- it should factor context and situation awareness in analysis and decision making,
- the involved entities should be well connected with the surrounding and share knowledge for overall awareness of the ecosystem,
- it should manage by itself, from the new learning the context awareness and be less dependent on external intervention,
- the learning proves it continuous so the cycle of sensing to learning and auctioning runs in a continuous cycle,
- it should be scalable to expand its scope,
- just like humans, it should interact with all different kinds of objects and even humans,
- it should be flexible to adapt to changing scenarios in real-time and still be relevant, and
- it should maintain the knowledge and awareness of past decisions.

Shown in Fig. 2.2 is a high-level representation of a Cognitive IoT approach. It outlines the entities involved and the data flow and their involvement in cognitive steps of sensing, understanding, learning, and acting. The vertical bars highlights the cognitive steps against the stages where they may take place. Depending on the use case, the technologies involved and resources available, and constraints at different stages from edge to cloud, the cognitive steps may be carried out at one or more levels. For example, consider that the sensors are smart enough and have resources to do some intelligent

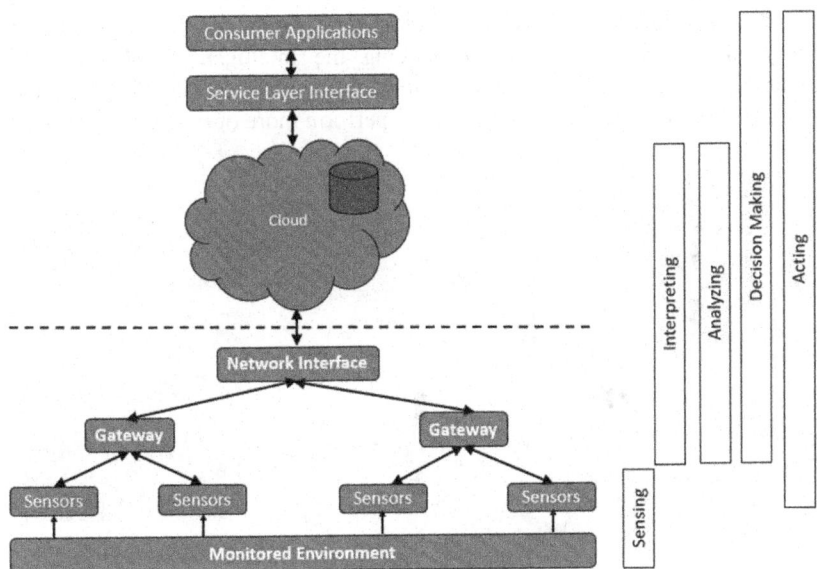

FIGURE 2.2 Cognitive approach in IoT.

sensing and control. Then to either conserve power or to alter their behavior and configuration, they can carry out a subset of operations in addition to the primary function of sensing. The same applies to the edge and fog layer. If the edge has the hardware and resource to do some preprocessing and run some analytics and algorithms, it can perform operations locally and in a distributed fashion, which is one of the goals of Cognitive IoT. Increasingly edge devices are getting more equipped with hardware and lightweight software to perform much of data analytics locally. Such behavior is essential when we think of low latency applications, for example, the self-driving car, where all the decision making should be performed locally because communication delays might have adverse effects.

Fig. 2.3 describes the high-level steps as the data flow from the sensor to the cloud. The steps may vary between different applications as per the reasons stated above. Cognitive IoT tends to make processing more distributed than centralized. ML/deep learning on cloud largely focuses on higher level abstraction and intelligence, which allowed hybrid machine learning approaches [32]. Many vendors are coming up with new and innovative platforms and offerings to enable adoption of IoT in different hybrid fields. A judicious assessment needs to be done against the objectives of a particular use case and application as no one model fits all the scenarios. Some scenarios may not need to concern too much on checking real-time availability of sensors and immediately reconfiguring the sensing if one or more sensors is not acting properly. Some cases necessitate local analysis or edge analytics, some may opt for more analysis on the cloud as they might not have the optimized edge.

The roadmap toward realizing a federated, scalable, and flexible cognitive computing ecosystem involves enhancing the intelligence and inference capability of the edge. The trend is toward designing better hardware and software for the edge and sensor nodes to perform more operations locally and

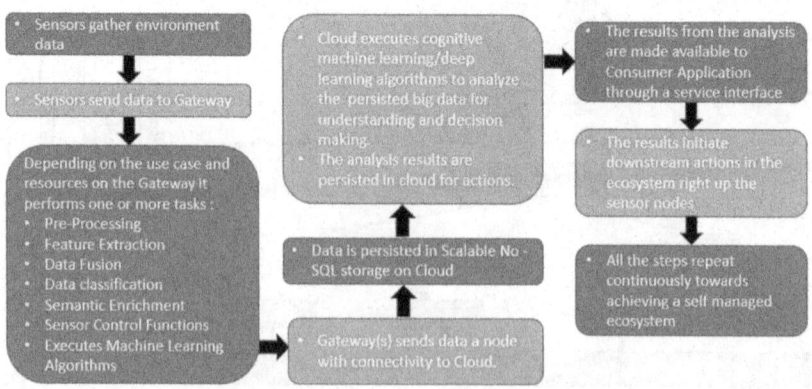

FIGURE 2.3 Data flow in a cognitive IoT application.

effectively. Edge nodes are packed with more storage, and the idea of local data centers is becoming increasingly relevant. The future Cognitive IoT ecosystem will follow a more localized and collaborative approach toward data analytics where each edge node is more enhanced to perform analysis and learning by itself and also participate and collaborate with the other surrounding nodes in the edge to share and leverage the intelligence [33]. The nodes will be more context-aware of the use case requirements in the sense that they may simultaneously participate in multiple workflows. The goal of each use case will be different and nodes will adapt themselves to achieve it. They will leverage the semantics and context in their processing. The idea is to combine the intelligence and capacity of local nodes instead of having the cloud as one centralized with massive storage and processing. The cloud will still be relevant but the nature of analytics on loud would transition to a more holistic and broader scope. Also the time to set up an IoT ecosystem would be a key factor which will reduce with collaboration and reuse. The edge will directly invoke cloud analytics API as needed. Fig. 2.4 presents a view of a future Cognitive IoT ecosystem.

Product vendors, community, and technology providers are aligning themselves to create new offerings. Frameworks and software that can run on the edge with their lightweight versions and also integrate and distribute processing are more relevant. Ready-to-use edge platforms performing a subset of the tasks in the data-processing pipeline and can be easily integrated into workflow promote the idea of reuse, collaboration, cost savings, and faster time to market.

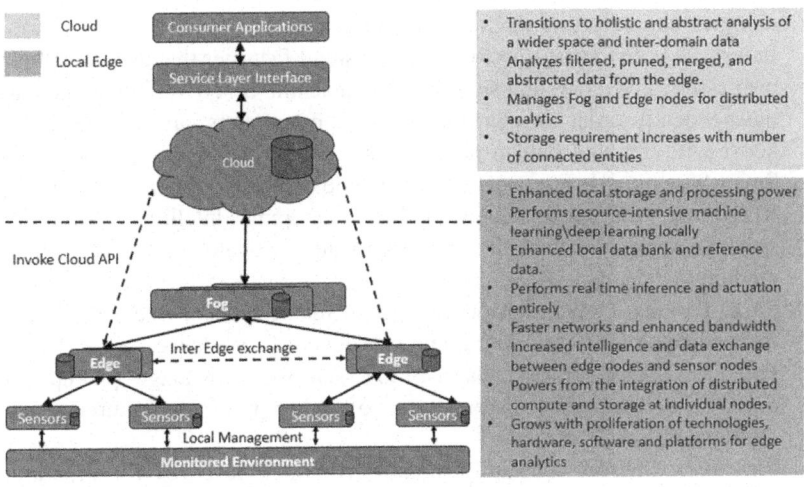

FIGURE 2.4 Future cognitive IoT application.

3. Solution and technology overview

Environment monitoring can range from monitoring a contained environment such as a factory floor or a house to a wider and holistic environment monitoring such as surface temperature, rainfall, weather, soil, hydrology, topographical changes, and atmospheric changes involves capturing the environment change indicators. A solution thus needs to be designed with the consideration and constraints of the use case involved. A generic solution would broadly contain the stages for data collection through sensors, analyzing it at the edge and/or sending it to the central location through the gateway(s), handle and provide the data to different platforms for analysis at the edge and on cloud, namely, open-IoT cloud servers like AWS, IBM Cloud, the things networks (TTN) [34], creating work/data flow, namely, node-red [35], storing the data in the NoSQL like environments, namely, InfluxDB [36], MongoDB [37], etc., analyzing the data using an optimum framework, namely, Anaconda-based R-Python framework, use of machine learning and deep learning models for data analysis and finally visualizing using visualization tools, namely, Grafana [38].

This data collection and processing pipeline can be used across IoT domains [39] like

- Smart Grid and Energy: Smart energy distribution grid enabled by IoT smart meters, perimeter access controlled by IoT sensors. Sensors can monitor emissions and radiation levels from different power plants.
- IoT Augmented Traffic Control: Traffic control systems capable of display warnings, rerouting traffic from accident sites, and ultimately offer better traffic management and decongestion.
- Driverless Cars: Developing driverless vehicles powered by many sophisticated IoT sensors measuring different statistics and wireless technologies like device-to-device communication.
- Industry 4.0: IoT sensor networks create smart factories that can track their own assets, inventory, and supplies. The machine is self-sufficient and can independently order its own materials from suppliers and manage it in-house with minimal human intervention.
- Smart Shopping: IoT sensors track valuable assets, provide access control, RFID and NEF technologies facilitate shopping. Intelligent Inventory systems communicate with suppliers to automatically restock based on predictive demand.
- Smart Parking: Optical sensors monitor available spots or free parking lots for semiautomated car parking and subsequent surveillance.
- Public Services: Trash cans use IoT sensors to manage pickup requirements, intelligent lighting turns off when it is not required using motion sensors.
- Structures: Sensors constantly monitor vibrations to check for structural abnormalities and overall integrity.

- Emergency services: Infrared thermal sensors detect sudden surges in temperature like the one caused by fire and then alert authorities.
- Construction: IoT sensors facilitate heavy machinery handling, asset control, foundation analysis, and material usage tracking.
- Intelligent Agriculture: IoT sensors check soil humidity, pests, and disease in the crops. Livestock and farm animals tracked, monitored, and cared for in an automated fashion.
- Weather prediction: IoT sensors collect data for humidity, pressure, snow, rain, temperature, cyclone, wind direction, etc. through different COTS sensors.

Cognitive IoT systems follow the approach of continuous sensing the environment, interpreting and analyzing the sensed data to derive insights, and then applying the knowledge gained in the application. This drives the steps of the handling large data-processing pipeline [40] for different applications with use case specific customizations. Considering an environment monitoring use case, the key stages in the environment data-processing pipeline and the technology and tools used are discussed in the next sections.

3.1 Data collection

At the edge of an IoT ecosystem are the device and sensors that collect data and metrics for analysis.

Some of the commonly used COTS sensors include the one to measure attributes like temperature, pressure, presence of some chemical, proximity, water quality, smoke, level, motion, acceleration, angular velocity, humidity, light, image, infrared, gas, etc. Some of them are widely used in IoT applications. The sensor and the usage across different applications domains are summarized in this section.

(a) Temperature sensor

   Common IoT applications
   - Used in manufacturing and industries, to measure the temperature of the ambiance and the machine or device.
   - Used in agriculture, to measure the temperature of the soil, a determining factor for crop growth and better yield.

   Commonly available variants
   - Types: Thermocouples, thermistors, resistor temperature detectors, infrared sensors, IC semiconductor-based, etc.
   - Names: DHT11, LM35, DS18B20, etc.

(b) Proximity sensor

   Common IoT applications
   - Used in retail and automotive domains to detect the presence, absence, and properties of a nearby object.

Commonly available variants
- Types: Inductive, capacitive, photoelectric, ultrasonic, etc.
- Names: RM18, E18-D80NK, E2B-M12KS02-WP-C2, Littelfuse-59140-020, etc.

(c) Pressure sensor

Common IoT applications
- Used to monitor systems and devices that work on some kind of gaseous or liquid pressure.

Commonly available variants
- Names: BMP180, MPX10DP, NXPMPX2300DT1, RobodoSEN38, etc.

(d) Water quality sensors

Common IoT applications
- Used to detect the various aspects to measure water quality in water distribution ecosystem.

Commonly available variants
- Types: Chlorine residual sensor, temperature sensor, turbidity sensor, conductivity sensor, dissolved organic carbon sensor, conductivity sensor, pH sensor, dissolved oxygen, and oxidation-reduction potential sensor, etc.
- Names: DS18B20, 103SR13A-1, TS-300B, AM2305, etc.

(e) Chemical sensors

Common IoT applications
- Used in pharmaceutical laboratories, space stations, environmental monitoring, and control.

Commonly available variants
- Types: Resistive sensor, Electrochemical sensor, Concentration sensor, fluorescent chloride sensor, hydrogen sulfide sensor, pH glass electrode, etc.
- Names: MQ135, HB100, STM32, etc.

(f) Gas sensors

Common IoT applications
- Used in industries such as manufacturing, oil and gas, coal mines, chemical laboratories, agriculture, and air quality monitoring to detect the presence of gases, some of them hazardous in nature.

Commonly available variants
- Types: CO_2/CO sensor, air pollution sensor, hydrogen/oxygen/nitrogen sensor, hygrometer, etc.
- Names: SGP30, SGAS701, 8G811, MQ135, etc.

(g) Optical and ionization smoke sensors

Common IoT applications
- Used in hazardous environments to sense airborne particulates and gases and their level.

(h) Infrared sensors

Common IoT applications
- Used in health care to monitor blood flow and blood pressure, for security surveillance, etc. Infers the surroundings by emitting or sensing infrared radiation.

Commonly available variants
- Types: Charged Couple Device (CCD), Complementary Meta-Oxide Semiconductor (CMOS), etc.

Additionally, some of the other commonly used sensors in IoT-based environments are motion sensors, gyroscopic sensor, humidity sensors to measure the concentration of water vapor and relative humidity, optical dust sensor (Shinyei PPD42NS) to measure the concentration of polluting particles of size 2.5 microns, using infrared scattering.

3.2 Data storage

Data gathered from different systems and sensors downstream needs to be persisted for further analysis. The characteristics of IoT data are it is real-time, heterogeneous, voluminous, and time-series and of a varying structure as per the source. It can be persisted in a NoSQL data store. We can leverage InfluxDB for the purpose. InfluxDB can handle voluminous time-series data storage and query and thus is suitable for use in IoT-based solutions. Here the timestamp is used to identify a single point in a given data series. In some ways, it is comparable to an SQL database table with the primary key preset as a time attribute but there are many differences as well.

InfluxDB is designed to perform real-time analysis on the stored data. The data model and storage engine are tailored for high performance and heavy load. The storage engine shares similarity with a Log-Structured Merge (LSM) Tree. The write ahead log (WAL) and a collection of read-only data files in InfluxDB can be compared to sorted strings tables (SSTables) in an LSM Tree. Time structured merge (TSM) tree files contain sorted compressed series data. The cache which feeds from WAL ensures a faster response each time a query is fired to the database.

3.3 Data flow orchestration

Another challenge faced in an IoT ecosystem is to efficiently and easily connect the involved entities to such devices, the end systems, the databases, the platforms, and exposed APIs for data flow. There is a need for a framework that supports such interconnection and can be easily deployed at the edge and the cloud as per the need of the use case and the technologies involved. It should be intuitive for easy configuration and management and should enable stakeholders to visualize the involved entities and data flow. It should support the protocols used in an IoT ecosystem.

Node-red is one such data flow framework that can be used for the IoT domain to efficiently manage real-time applications. It helps to model and

TABLE 2.1 Comparison of different communication protocols in IoT domains.

Protocol Feature	MQTT [41]	HTTP/ RESTful [42]	XMPP [43]	CoAP [44]
Transport	TCP/IP	TCP/IP	TCP/IP	UDP
Cellular suitability (1000s nodes)	Excellent	Excellent	Excellent	Excellent
Messaging	Publish-subscribe or request-response	Request-response	Publish-subscribe or request-response	Request-response
Orientation	Message	Web service/document	Message	Web service/document
Power used	Low	High	High	Low
Low power and lossy networks	Fair	Fair	Fair	Excellent

orchestrate the multiple involved entities and heavy inbound/outbound data flow. It can be deployed to run on the device as well as on the cloud. It has a set of prebuilt nodes that support almost all of the protocols used for IoT communication thereby help to quickly develop an IoT workflow. It is also possible to create custom nodes. The publish-subscribe pattern, popular in IoT-based applications is supported by the node-red platform through Message Queuing Telemetry Transport (MQTT) protocol, a lightweight TCP/IP based transport protocol. Some of the other popular communication protocols involved in the IoT domain shows in Table 2.1.

Additionally it supports a set of functions to perform operations in the workflow. One can incorporate ML algorithms in the flow. It also helps to track the flowing data and debug workflows.

3.4 Data analysis

Data Analysis helps us understand the data and the trend and derive insights. The data analysis can be done using ML and deep learning algorithms executed on the cloud or on the edge depending on the resource constraints and the specifics of the use case. Algorithms can be categorized as supervised, unsupervised, semisupervised, or reinforced. Whenever labeled data is not available, the learning correlates and find patterns and insights from the data. It servers to enhance the cognitive intelligent of the IoT ecosystem. Depending on the node position in the ecosystem hierarchy where the data analysis is carried out the volume, variety and abstraction of the data varies and hence the influence of result of the analysis outcome in providing insights. The nature of data available at different points in the data flow pipeline dictate the analysis that can be carried out and the algorithms that can leveraged.

3.5 Data visualization

Visualizing the output is an important phase for any machine learning-based IoT domain for better insights into the data and the events. Visual details are easy to interpret and act upon and therefore helps greatly in controlling and managing environments. Several visualization techniques are available in open-source domains to offer the end-users' a viable solution to present experimental outputs. Grafana is one such open-source analytics and visualization application. It enabled querying, visualizing, exploring, alerting on stored data, and combining data from multiple sources. It can integrate with InfluxDB as the data source. Grafana creates a basic graph panel with the random walk data. It provides features to transform time-series-based datasets into intuitive graphs and visualizations with analytical information. It has a rich dashboard feature that allows visualizing the output of multiple queries with analytical details at the same time. The debugging workflow feature of Grafana allows it to receive an alert, drill down to get details with all metrics, and examine the log with distributed traces. Grafana specializes in querying and visualization. Another key monitoring and alerting open-source tool is Prometheus with a multidimensional data model. Grafana being more of data analysis and visualization tool integrates with Prometheus to create monitoring and alerting systems with proper visualizations. Grafana can use Prometheus as a data source.

3.6 Action

The output of data analysis will provide inputs to initiate further action. For example, if the visualization, monitoring, and alerting tool show some trend or some scenario that needs appropriate actions, either the system would act proactively or it can issue alerts.

In this subsequent section, we will discuss an experimental environment monitoring framework to capture, store, preprocess, analyze, and visualize different parameters of the weather data that are collected from different weather stations or different sensors available around the world under open-IoT initiatives by different leading organizations. This data helps to create weather intelligence models for personalized weather forecasting and weather parameters based decision-making. We will also discuss how to store and visualize different parameters of water viz. height, pH, quality, and temperature sourced from distant water management facilities. We have used the publicly available sample data from the National Oceanic and Atmospheric

Administration's [45] Center for Operational Oceanographic Products and Services (Dataset, n.d.). This data comprises 15,258 water level observations. The water level is measured in ft. The observations were taken in six-minute intervals in a date range of August 18, 2015, through September 18, 2015, at two stations: Santa Monica, CA (ID 9410840) and Coyote Creek, CA (ID 9414575). Additionally, the measurements of the average water temperature, water pH, water quality, and water temperature are stored in the database to help query functionality in schema exploration. The level description field of NOAA data specifically is captured for testing purposes. We have created a visualization model in Grafana for the NOAA data set and subsequently built a time-series temperature data model of a region in the Anaconda Python framework using pandas visualization package with speed-up feature [46]. The authors have also created a node-red workflow environment that will display the temperature and humidity of a location considering the latitude and longitude of a place during different times of the day by using the MQTT publish-subscribe protocol feature.

4. Experimental results and discussion

The node-red, InfluxDB, Grafana, Anaconda Python along with different sensors under TTN open-source project helped us build an end-to-end IoT platform to collect the data from the environment, receive and pass on through different gateways, forward the data in the TTN to node-red for flow management, persisting in InfluxDB. Node-red is lightweight and can be set up on the edge, on cloud or on local workstations. It can integrate to different nodes and support custom JavaScript for any operations on the data.

The framework allows us to visualizing the data using Grafana, after algorithmic analysis through Python, if required. The high-level flow diagram of the experimental environmental IoT monitoring is shown in depicted in Fig. 2.5. The experimental model building approach can be put in either in AWS or IBM Watson or Google Collab for large-scale IoT data modeling deployment and analysis. Such an end-to-end model can connect, monitor, and control thousands of billions of internet-connected devices, exchange information, and offer autonomous action based on continuous input processing.

These applications and solutions offer proven solutions across diverse industries, starting from the environment, manufacturing, transportation, oil and gases, healthcare [47–49], and energy sectors.

The design flow of the end-to-end solution capturing data and delivering output is depicted in Fig. 2.6. Once we have the data through MQTT publish server, we can demonstrate the subsequent output in Fig. 2.7.

Further, we worked with temperature time-series data models of monthly temperature in several African countries from 1901 to 2005. In the pre-processing phase, the country codes are added with country names based on ISO code mapping [50]. This is a popular technique in GIS to merge them to

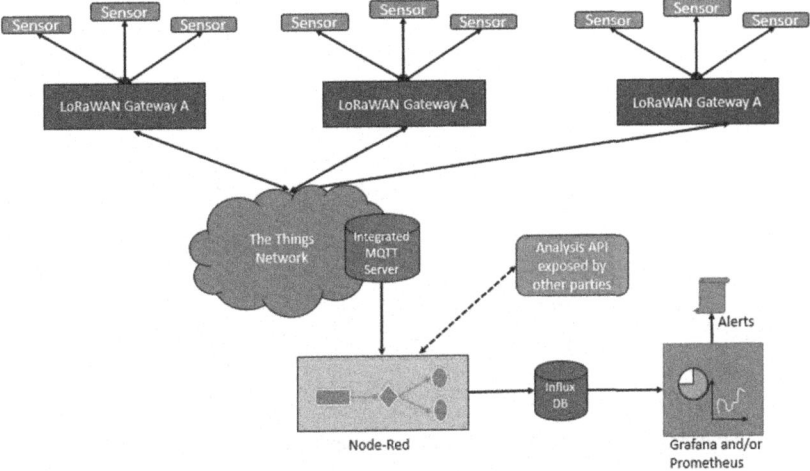

FIGURE 2.5 High level flow diagram of experimental environment IoT ecosystem.

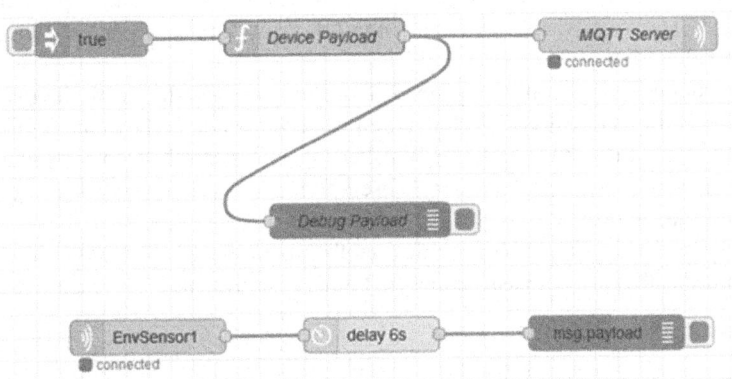

FIGURE 2.6 MQTT based environmental sensor deployment in node-red framework.

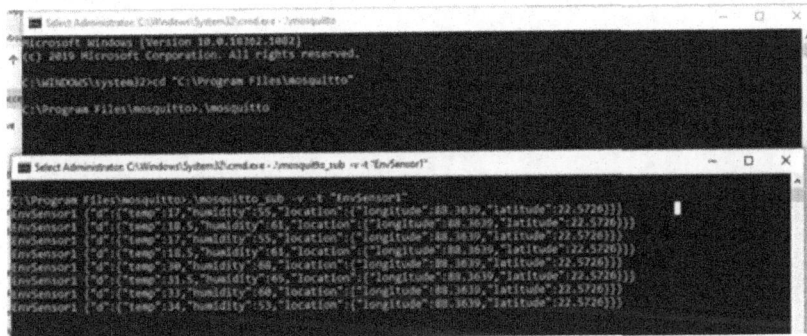

FIGURE 2.7 The subscribe message received from node-red workflow framework.

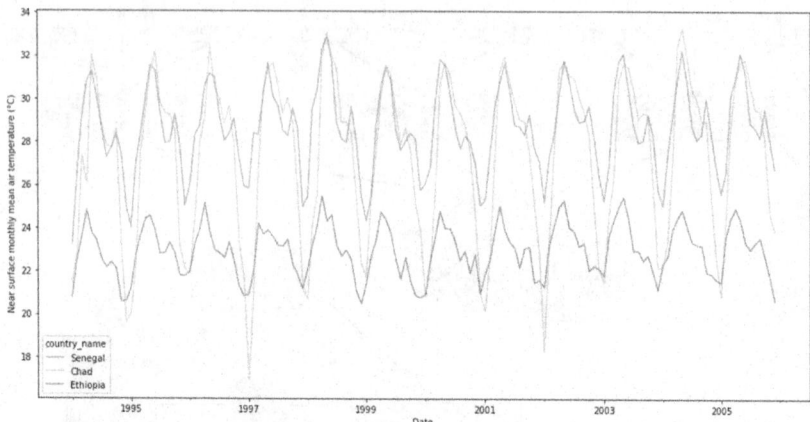

FIGURE 2.8 Monthly mean temperature comparison of temperature dataset.

get uniformity during several occasions of drawing inferences. The plot constitutes far east to far west data. The monthly mean comparison data is plotted in Fig. 2.8. The year-wise boxplot, in Fig. 2.9, provides a time-interval distribution value plot.

This plot encloses the interquartile 50% observations in a box with the two ends representing the 25th and 75th percentiles. The median is shown by a line inside the box at the 50th percentile. Whiskers at the top and bottom represent the extremes of observed data. Any observations outside the whiskers are represented using dots.

The heatmap, in Fig. 2.10 provides another perspective to our data and reveal patterns that are not easily identified otherwise. Heatmap represents the

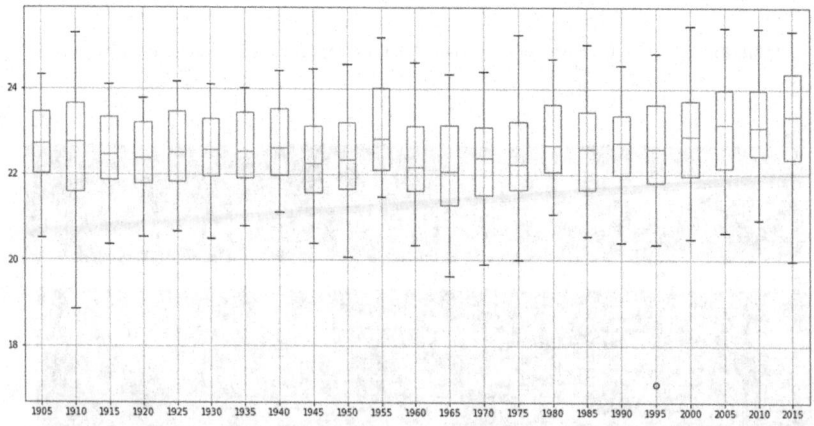

FIGURE 2.9 Year-wise temperature comparison box-plot of temperature dataset.

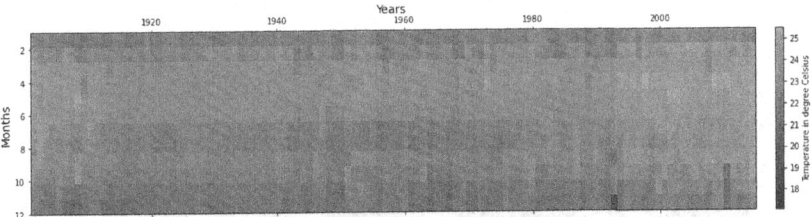

FIGURE 2.10   Monthly mean temperature heatmap of temperature dataset.

temperature in each cell (color encoded) at a given year and month. The TimeGrouper library of pandas is really helpful to handle time-series data in a Python environment.

The last part of the experiment is associated with a water level dataset that includes 15,258 observations of water levels (in ft.), collected every six minutes at two stations in one month, in the year 2015. The storing of this type of dataset in a NoSQL database like InfluxDB needs some special intervention to maintain its time-series characteristics. We have to add the "-precision rfc3339" option [51] before inserting data into the database. The advantage of InfluxDB is that it can easily integrate with the visualizing tool Grafana. Grafana is a powerful data visualization tool that helps create, explore, and share data in a graphical representation, thus making it easy to interpret. It is used to visualize time-series data as shown in Fig. 2.11.

The water level for the given time span is depicted in Fig. 2.12 and water quality-related time-series data is depicted in Fig. 2.13.

5. Last note

The IoT applications and their growth in recent times, due to the development of different types of sensors and end-to-end designing frameworks, enabled us to design different environmental data models for prediction and subsequent decision making almost instantly. Now stakeholders can design futuristic solutions to enable faster growth. With the advancement of technologies that can be leveraged in IoT-based solutions, there is a trend toward an increased adoption of IoT technologies across different domains. It is anticipated that the world will have 43 billion IoT-enabled devices by 2023. Together with high performance, IoT device manufacturers are prioritizing low-cost and energy-efficiency features in providing value to the end-users. Additionally, the advancements in processing hardware and software to process and analyze data, offer more flexibility and freedom in choosing different frameworks to build IoT applications. The efforts of different stakeholders, the manufacturers, the platform providers, the IoT communities, the researchers, and the users of the IoT applications are leading toward the realization of more cost-effective, inclusive, flexible, and futuristic use cases with a better return on investment

FIGURE 2.11 Time-series data representation using InfluxDB and Grafana for NOAA water dataset.

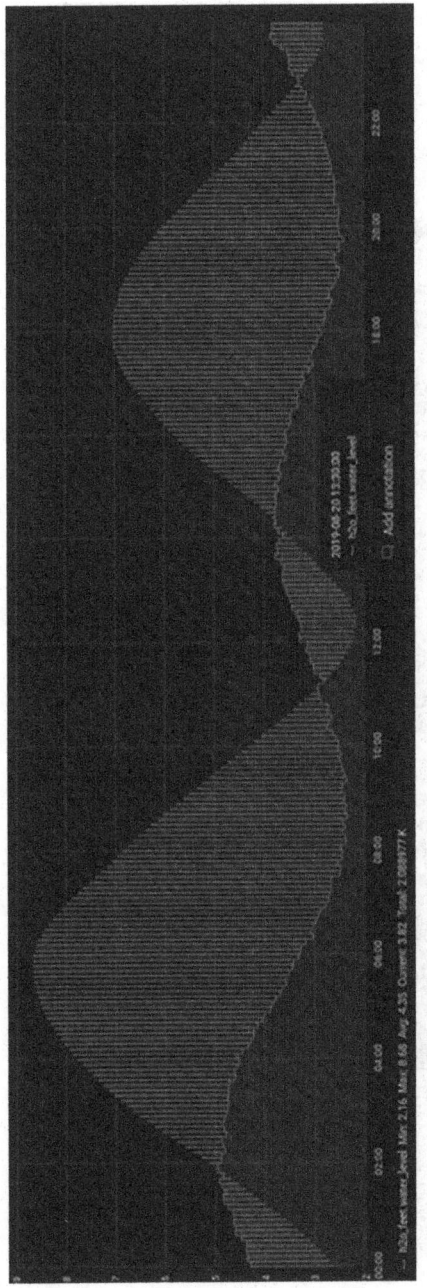

FIGURE 2.12   Time-series representation for water level (in ft.) in water management facilities.

FIGURE 2.13 Time-series representation for water quality in water management facilities.

and reduction in time to market. The trend is toward greater use of open platforms and technologies, which will address some of the key issues that arise out of proprietary platforms and technologies, together with the expanding scale and heterogeneity of the entities and data involved.

6. Future directions

Cognitive IoT aims to add human-like intelligence into conventional IoT to achieve the goals of the next generation of artificial intelligence and robotics. The idea is to make the IoT system not just act following a preconfigured set of rules and but keep evolving and add context-awareness in decision making, similar to humans. It adds more resilience and robustness in the IoT ecosystems, influences new use cases simulating human behavior. But it is not an easy walk and has its own challenges that can evoke interest for future exploration.

Some of the challenges and future directions are discussed:

- Cognitive IoT adds more intelligence and a continuous learning cycle. This would need more processing and hence the requirement of more power and resources. It is a challenge to meet the power requirements so that devices do not get drained fast. The algorithms that run on the edge as well the entire IoT ecosystem needs to be fine-tuned to optimally manage the balance of power utilization while meeting the goals of decentralized processing. This process of monitoring the ecosystem and reconfiguring needs to be continuous as the IoT ecosystem is dynamic with the addition of new objects and the exit of the old ones from the network.
- Cognitive IoT integrates more information into the ecosystem for making context-aware and dynamic decisions. It needs to gather knowledge about the state of every object in the ecosystem, the health of the ecosystem, the real-time data, be cognitive of factors impacting decision as well as learn from previous decisions under similar situations. This raises challenges in terms of infrastructure, resources, and cost for hosting data, networking, and communication. Studies should focus on the impact and challenges to maintain such models and come up with ideas to make it more efficient and business worthy.
- Cognitive IoT is intrusive in nature and decentralized. It may raise concerns about privacy, security, and ethical use. Use cases and platforms must be designed with the consideration so the users do not breach the norms. Privacy breach, compromise of device security with cyber-attacks, data theft, and use for malicious means are potential risks. Appropriate anonymization should be in place and strict regulations and controls should be observed during the design of such systems. Researchers and stakeholders should come up with ideas so that use of Cognitive IoT is ethical and secure. They should ensure that the right intelligence is imparted so that it serves to augment humans and help them and not otherwise to create conflicts. The necessary checks and balances should be ensured.

- As the world is moving toward miniaturization, the trend is reduction in size and packing more memory and resources in a relatively smaller space. The same applies to processing and sensors. Also, networks have become faster. The use of 5G technology will help in transmitting more information faster. This opens new and exciting areas of research and adoption of IoT where the sensors can derive information from areas not feasible earlier.
- Cognitive IoT opens up avenues of growth across different domains. It will change the established pattern of doing things. To embrace this change, the challenge is for individual stakeholders and organizations is to gain awareness and equip themselves. Organizations need to chart a road map for the adoption of this disruptive technology change.

Appendix A. Supplementary data

Supplementary data to this article can be found online at https://10.1016/B978-0-12-824038-0.00008-0.

References

[1] Geissbauer R, Vedso J, Schrauf S. Industry 4.0: building the digital enterprise. 2016. Retrieved from PwC Website, https://www.pwc.com/gx/en/industries/industries-4.0/landing-page/industry-4.0-building-your-digital-enterprise-april-2016.pdf.

[2] Mohammadi M, Al-Fuqaha A. Enabling cognitive smart cities using big data and machine learning: approaches and challenges. IEEE Commun Mag 2018;56(2):94−101.

[3] Ploennigs J, Amadou B, Barry M. Materializing the promises of cognitive IoT: how cognitive buildings are shaping the way. IEEE Internet Things J 2017;5(4):2367−74.

[4] Foukalas F. Cognitive IoT platform for fog computing industrial applications. Comput Electr Eng 2020;87:106770.

[5] Mondal K. Different visualization issues with big data. In: Published in "Smart Innovations, Systems and Technologies (SIST)" by springer in international conference on information and communication technology for intelligent systems (ICTIS 2015), ISSN: 2190-3018; November 2015. p. 555−62.

[6] Mondal K, Dutta P. Big data parallelism: challenges in different computational paradigms. In: Proceedings of the 2015 third international conference on computer, communication, control and information technology (C3IT). IEEE; February 2015. p. 1−5.

[7] Jayasinghe U, Lee HW, Lee GM. A computational model to evaluate honesty in social internet of things. In: Proceedings of the symposium on applied computing; April 2017. p. 1830−5.

[8] Wolf M. Machine learning+ distributed IoT= edge intelligence. In: 2019 IEEE 39th international conference on distributed computing systems (ICDCS). IEEE; July 2019. p. 1715−9.

[9] Chen B, Wan J, Lan Y, Imran M, Li D, Guizani N. Improving cognitive ability of edge intelligent IIoT through machine learning. IEEE Netw 2019;33(5):61−7.

[10] Thompson LM. Weather and technology in the production of corn and soybeans. CARD Reports, Book 17; 1963. http://lib.dr.iastate.edu/card_reports/17.

[11] Wixted AJ, Kinnaird P, Larijani H, Tait A, Ahmadinia A, Strachan N. Evaluation of LoRa and LoRaWAN for wireless sensor networks. In: 2016 IEEE sensors. IEEE; October 2016. p. 1—3.

[12] Hera MR, Rahman A, Afrin A, Uddin MYS, Venkatasubramanian N. AQBox: an air quality measuring box from COTS gas sensors. In: 2017 international conference on networking, systems and security (NSysS). IEEE; January 2017. p. 191—4.

[13] Ahmad T, Morelli U, Ranise S, Zannone N. A lazy approach to access control as a service (ACaaS) for IoT: an AWS case study. In: Proceedings of the 23rd ACM on symposium on access control models and technologies; June 2018. p. 235—46.

[14] Klein S. IoT Solutions in Microsoft's Azure IoT Suite. Berkeley, CA: Apress; 2017.

[15] Petnik J, Vanus J. Design of smart home implementation within IoT with natural language interface. IFAC-PapersOnLine 2018;51(6):174—9.

[16] Rashid RA, Chin L, Sarijari MA, Sudirman R, Ide T. Machine learning for smart energy monitoring of home appliances using IoT. In: 2019 eleventh International Conference on Ubiquitous and Future Networks (ICUFN). IEEE; July 2019. p. 66—71.

[17] Marais JM, Malekian R, Abu-Mahfouz AM. LoRa and LoRaWAN testbeds: a review. In: 2017 IEEE AFRICON. IEEE; September 2017. p. 1496—501.

[18] Watts J. Human society under urgent threat from loss of Earth's natural life. Guardian 2019;6.

[19] Rizvi SSR, Abbass S, Rehman AU, Zia TJ, Younas M, AsadUllah M. Socio-IoT enabled smart drive system for smart cities. Int J Comput Sci Netw Secur. 2018;18(8):1—12.

[20] Gill SS, Chana I, Buyya R. IoT based agriculture as a cloud and big data service: the beginning of digital India. J Organ End User Comput 2017;29(4):1—23.

[21] Su G, Moh M. Improving energy efficiency and scalability for IoT communications in 5G networks. In: Proceedings of the 12th international conference on ubiquitous information management and communication; January 2018. p. 1—8.

[22] Rossi C, Gaetani M, Defina A. AURORA: an energy efficient public lighting IoT system for smart cities. Perform Eval Rev 2016;44(2):76—81.

[23] Rajendran VAP, Nadarajah YAL, Ramasamy RKAL. A cloud-based implementation of IoT system in recycle industry. In: Proceedings of the 3rd international conference on big data and internet of things; August 2019. p. 88—92.

[24] Dave B, Buda A, Nurminen A, Främling K. A framework for integrating BIM and IoT through open standards. Autom ConStruct 2018;95:35—45.

[25] Cleaver F. The inequality of social capital and the reproduction of chronic poverty. World Dev 2005;33(6):893—906.

[26] Evers S. Foss platform for cloud based IoT solutions. Bosch software innovations GmbH. FOSDEM 2018. 2018. p. 1—31. https://archive.fosdem.org/2018/schedule/event/eclipse_iot/.

[27] Pfeil M, Bartoschek T, Wirwahn JA. OpenSenseMap-a citizen science platform for publishing and exploring sensor data as open data. Free and Open Source Software for Geospatial (FOSS4G) Conference Proceedings. 15; 2015. p. 39 (1).

[28] Sun L, Li Y, Memon RA. An open IoT framework based on microservices architecture. China Commun 2017;14(2):154—62.

[29] Wu Q, Ding G, Xu Y, Feng S, Du Z, Wang J, et al. Cognitive internet of things: a new paradigm beyond connection. IEEE Internet Things J 2014;1(2):129—43.

[30] Pramanik PKD, Pal S, Choudhury P. Beyond automation: the cognitive IoT. artificial intelligence brings sense to the Internet of Things. In: Cognitive computing for big data systems over IoT. Cham: Springer; 2018. p. 1—37.

[31] Mondal K. Big data parallelism: issues in different X-information paradigms. In: Published in Elsevier procedia computer science, vol. 50, special issue on Big data, cloud and computing challenges, ISSN 1877-0509; October 2015. p. 395–400. https://doi.org/10.1016/j.procs.2015.04.028.

[32] Mondal K. Application design and analysis of different hybrid intelligent techniques. Published in international journal of hybrid intelligent techniques (IOS Press), vol. 13 Issue 3, ISSN 1448-5869. April 2016. p. 173–81. https://doi.org/10.3233/HIS-160234.

[33] Tiwari N, Mondal K. NCS based ultra-low power optimized machine learning techniques for image classification. In: Published in the proceedings of IEEE R10 symposium TEN-SYMP-2019, ISBN 978-1-7281-0296-2. p. 750–3. https://doi.org/10.1109/TENSYMP 46218.2019.8971238.

[34] Howerton JM, Schenck BL. The deployment of a LoRaWAN-based IoT air quality sensor network for public good. In: 2020 systems and information engineering design symposium (SIEDS). IEEE; April 2020. p. 1–6.

[35] Diaz AL. IBM and partners launch JS Foundation – cloud computing news. IBM; October 17, 2016.

[36] Miller R. InfluxData scores $35 million series C to expand time series database business. TechCrunch releases in Github – influxdata/influxdb. 2016.

[37] Kingsbury K. MongoDB 3.4.0-rc3. 2017. https://jepsen.io/analyses/mongodb-3-4-0-rc3. Jepsen.

[38] Jones A. Open source monitoring stack: Prometheus and Grafana. Bizety; 2019. https://grafana.com/.

[39] Mohindru G, Mondal K, Banka H. Internet of Things and data analytics: a current review. Wiley Interdiscip Rev Data Min Knowl Discov 2020;10(3):e1341.

[40] Mondal K. Design Issues of Big Data Parallelism, published in Springer AISC series. In: Third international conference on international system design and intelligent applications (India 2016), ISSN 2194-5357; February 2016.

[41] Stanford CA, Hong LT. SO/IEC 20922:2016 information technology – MQ Telemetry Transport (MQTT) v3.1.1. iso.org. International Organization for Standardization; 2016.

[42] Meng J, Mei S, Yan Z. Restful web services: a solution for distributed data integration. In: 2009 international conference on computational intelligence and software engineering. IEEE; December 2009. p. 1–4.

[43] Ozturk O. Introduction to XMPP protocol and developing online collaboration applications using open source software and libraries. In: 2010 international symposium on collaborative technologies and systems. IEEE; May 2010. p. 21–5.

[44] Shelby Z, Hartke K, Bormann C. The Constrained Application Protocol (CoAP). 2014.

[45] NoAA Dataset. National Oceanic and Atmospheric Administration's (NOAA) Center for Operational Oceanographic Products and Services. https://s3.amazonaws.com/noaa.water-database/NOAA_data.txt and https://tidesandcurrents.noaa.gov/stations.html?type=Water+Levels.

[46] World Bank Temperature Datasets. https://climateknowledgeportal.worldbank.org/.

[47] Alhussein M, Muhammad G, Hossain MS, Amin SU. Cognitive IoT-cloud integration for smart healthcare: case study for epileptic seizure detection and monitoring. Mobile Netw Appl 2018;23(6):1624–35.

[48] Mezghani E, Exposito E, Drira K. A model-driven methodology for the design of autonomic and cognitive IoT-based systems: application to healthcare. IEEE Trans Emerg Top Comput Intell 2017;1(3):224–34.

[49] Scrugli MA, Loi D, Raffo L, Meloni P. A runtime-adaptive cognitive IoT node for healthcare monitoring. Proceedings of the 16th ACM International Conference on Computing Frontiers. 2019. p. 350—7.

[50] Davis D. Taming the engineering of information services websites with standards. Computer 2015;48(8):84—9.

[51] Simpkin C, Taylor I, Harborne D, Bent G, Preece A, Ganti RK. Dynamic distributed orchestration of Node-Red IoT workflows using a vector symbolic architecture. In: 2018 IEEE/ACM workflows in support of large-scale science (WORKS). IEEE; November 2018. p. 52—63.

Chapter 3

Evolution of sustainable environment: a cognitive outlook

Vaneet Kumar[1], Saruchi[2], Vishal Rehani[1]

[1]*Department of Applied Sciences, CT Institute of Engineering, Management and Technology, CT Group of Institutions Jalandhar, Jalandhar, Punjab, India;* [2]*Department of Biotechnology, CT Institute of Pharmaceutical Sciences, CT Group of Institutions Jalandhar, Jalandhar, Punjab, India*

1. Introduction

It becomes a difficult task of defining a smart material with a cognitive outlook. At present time, there are many applications of so-called smart materials, the latest revolution in the field of materials science as cognitive outlook for sustainable development. Contrasting any stationary or a dead material, a smart material is alive because it has the capability to change in an environment. They are also referred as multifunctional substances, having a particular physical as well as chemical features, which can be changed in response to some external stimuli. Smart material systems are useful for bringing improvement in the functioning of devices. This chapter reviews and provides the literature in the field of smart materials as cognitive outlook utilized in different fields for sustainable development.

Developments of structure capable of conforming their shape according to the change in the environment are considered to be the promising materials in the field of research. Smart materials are referred to as intelligent materials capable of exhibiting change in their properties on receiving a stimuli like temperature, pH, moisture content, magnetic field, radiations, and so forth. They respond to the change in the environment at the optimized conditions, revealing their owned features in accordance with the environment. Operational application of such materials simplifies designs, thereby reducing part counts and hence increasing the lifecycle of an object. Smart materials ultimately substitute outmoded tools employed in designing building, vehicle, and other consumable products. Low weight and size accompanied with

Cognitive Data Models for Sustainable Environment. https://doi.org/10.1016/B978-0-12-824038-0.00002-X
45

improvised designing makes smart materials a striking option. Such materials receive, transmit, and process stimuli and give responses, generating some beneficial influence.

Smart behavior of any material explains self-adaptation, sensing ability, memory, and numerable functions of the substance. It is the most progressive field of research. The present system have complexity as well as are costly. Therefore developing a smart and intelligent material is a transforming thought as they have large number of benefits. It has enabled well-integrated as well as actuation system without affecting weight.

2. Classification of smart materials

Smart materials have the properties which can be changed in the organized manner by some external stimulus. There are various kinds of smart materials as discussed in the chapter [1−4].

2.1 Piezoelectric substances

There are some piezoelectric substances which produces voltage on application of stress undergoing certain mechanical changes. Structures prepared using piezoelectric material exhibits properties like bending, compression and expansion. They can be used to design keyboards, microphones, speakers, transducers, and so forth.

2.2 Thermosensitive substances

There are some thermosensitive substances that can adapt different shapes depending on variability in temperature. Such materials can be used as a denting material for cars apart from their biomedical applications. They find their use in thermostats and air vehicles also.

2.3 pH-sensitive substances

Some pH-sensitive materials have also been used in paints. They act as indicators of corrosion when added to the paints.

2.4 Chromogenic substances

Chromogenic materials have also been employed in paints, which have a tendency to exhibit color change upon heating or exposing it to light. Photograph of laboratory made hydrogel is indicated in (Fig. 3.1).

FIGURE 3.1 Picture of hydrogel.

2.5 Hydrogels

Hydrogels are three-dimensional polymer networks that are highly cross-linked, highly porous, and have been formulated for slab, film, coating, and microparticles. Hydrogels can absorb water 1000 times their own weight and show swelling ability. They are also referred as smart as well as an intelligent material because of its capability to receive the change in stimuli and respond to it by exhibiting change in its properties. They are biodegradable and biocompatible in their nature and have been utilized in a number of fields such as in food packaging material, dye sorbents, sensing devices, drug delivery, diapers, in field of agriculture, and many more.

2.6 Magnetoresponsive substances

These are the smart materials, which have the tendency to change the shape and size in presence of magnetic field, but when magnetic field is removed it regains its original form.

2.7 Optical fiber

These materials are premeditated for guiding light. They are lighter and highly flexible with lower attenuation. The emerging optical communications results in the development of optical sensors. Such smart substances have tendency of matching its color with the background.

2.8 Active and passive substances

They can also be categorized as active and passive.

Active materials: They have tendency to undergo alteration in their geometries as well as in its characteristics on heating or apply electric fields. Piezoelectric materials behaves as an active smart material.

Passive materials: Materials that do not behave actively are referred as a passive material. They do not have the ability of energy transduction. Such substances are utilized in the form of sensors. Table 3.1 is indicating factor affecting smart material.

3. Properties of smart materials

- They are highly biocompatible in its nature
- They are highly compressed and simple in its structure
- They have excellent mechanical strength
- Self-detection and diagnosis
- Sensing as well as actuation device

4. Application of smart materials

There is wide scope of smart materials in the field of engineering due to its ability to respond to change in external stimulus. It has such a mechanical as well as physical properties that they attract the interest of scientist and research workers toward it. Different fields of applications of intelligent materials are discussed ahead.

TABLE 3.1 Factors affecting smart material.

S.No	Type of material	Factors affecting
1.	Piezoelectric material	Electric voltage
2.	Photochromic material	Radiations
3.	Electrochemical material	Electrical voltage
4.	Thermochromic material	Temperature
5.	Shape memory alloys	Temperature
6.	Magnetoresponsive material	Presence of magnetic field
7.	Chromogenic material	Exhibits change in its color because of electrical-thermal-radiative stimulus

4.1 Application in the field of nanotechnology and acoustics

Zinc oxide nanoparticles were synthesized with the help of a precipitation method. It was further processed into nanofluids. Nanofluid so synthesized was characterized by X-ray diffraction (XRD), electron disperpersive studies (EDS) as well as scanning electron microscopy (SEM). Acoustic as well as optical analysis of the prepared nanofluid was conducted which showed the random behavior of the particles in the fluid. It showed good dispersion and the particles in the nanofluid were well associated. The behavior of the fluid was determined to be non-Newtonian [5]. A beam-forming technique, which is proficient to localize emission sources, was determined to be time consuming. To overcome the time consumption, a technique was developed to form a high speed beam formation having the capability of localizing an aural radiation source [6].

A high energy storable material was developed by modification of ferro-electric material, exhibiting enhanced electric strength. The breakdown strength of the developed smart material was determined to be improvised by doping ferroelectric material with $Bi_{1.5}ZnNb_{1.5}O_7$. The characterization of the doped material was done through SEM and XRD. Energy storage as well as its efficiency of the smart material was also assessed [7].

Synthesis of nanocomposite by the electrospinning process has also reached greater heights. Graphene has more exposed surface area and it also exhibit good electrical and thermal conduction. It also exhibits higher young modulus. The conductivity and the strength can be increased by merging it with graphene. Electrospinning process assists in forming very thin polymer fibers in a nanometric range. A nanocomposite fiber was formed from poly-vinylpyrrolidone and graphene. On increasing the content of graphene thick fibers were formed. It was also determined that on increasing graphene amount it doesn't lowers the high sound absorption capacity. The impact of addition of graphene on thickness of fiber and conformation of polymeric chain was also studied. Nanocomposite so formed was characterized by Raman micro-spectroscopy, atomic force microscopy, and SEM [8].

4.2 Application in the field of piezoelectric and electrochromic device

The pulse mode of radiation-receiving system was studied by taking piezo-ceramic plates separated by glycerine as a medium. The signals obtained by the excitation of emitters was calculated. Moreover, the length of signal was also determined [9]. Improvement in the piezoelectric materials were brought by preparing a polymer using polyvinylidene fluoride and fluoroethylene. Semicrystalline smart material was resulted having excellent piezoelectric properties [10]. Due to tremendous piezoelectrical property, lead zirconate titanate was used as a ceramic. A crystalline sample based on it was synthesized in presence of lead oxide, titanium oxide zirconium oxide and manganese carbonate using Columbite-precursor technique. Dielectric as well as ferroelectric features were also analyzed, exhibiting diffused peaks. The synthesized sample showed high piezoelectric coefficient as well as higher electromechanical coupling factor. It was considered to be highly appropriate for pioezoelectric transformer as well as high power device [11]. A macrofiber were generated and utilized in the form of an intelligent substance for piezoelectric transducer, which proved to be more advantageous than the outmoded ceramic substances. The transducer so produced was employed to monitor health structure [12]. An electrochromic substances have capability to exhibit reversible color changes on varying voltage, is catching a great attention because of its application in form of smart windows, data storage as well as display. A hybrid film based on vanadium pentasulfate and polystyrene sulfonate were prepared which proved to be a favorable material for enhancing electrochromic device as well as hybrid electrochromic material process ability [13]. Fig. 3.2 is indicating use of smart material in building.

4.3 Application in field of civil engineering

A unique cement material was prepared which further developed advanced concrete and reinforced concrete material with the prolonged durability. With the time span microcracks starts appearing which weakens the concrete material by the ingression as well as transportation of liquid. Cracks should be organized or auto-healed by the precipitation of calcium carbonate in the cracked area in presence of moisture. But it occurs at the site of small cracks and takes much longer time. Fig. 3.3 is indicating treatment of cracks with bacterium.

FIGURE 3.2 Use of smart material in buildings.

FIGURE 3.3   Treatment of cracks with bacterium.

Lysinibacillus sphaericus bacteria in the presence of calcium carbonate ($CaCO_3$) was effectually used as a crack healer as well as crack sealer. Digital microscopy and ultrasonic pulse velocity techniques were utilized for the analysis of the process. Some sorptive tests were also employed to check the reduction in the absorption [14].

Mending of the cracks in the cement material is most importance in order to enhance stability. Therefore a green approach using microorganisms was applied to increase the life of cementing material. Microbes induced calcium carbonate precipitation was utilized for treatment of cracks in the material and mortar was considered to be an effective healing agent. The cracks were observed to seal within few days. The healing of the cracks were confirmed through the thickness of cracks as well as from decrease in volume [15]. Lot of efforts are being made by research workers for the development of self-healing concrete material. Green approach is the most popular technique utilized these days for self-healing of concrete. The use of bioconcrete was determined to be more advantageous than $CaCO_3$ precipitation method. This method has produced remarkable results. Bacteria were cultured, processed, and obtained in the form of granules. Later, it was dried as well as stored and then biomortars were prepared using mortar. Self-healing was proved by crack size reduction. The cracks were observed to close within time span of 28 days and liquid permeation was also determined to be decreased [16].

It is very important to protect the buildings and the structures from natural calamities and disasters such as strong blowing wind and earthquake. The maintenance of the civil structures also becomes the priorities. Therefore lot of work has been done on development of magnetic-rheological elastomeric devices [17]. The permeable nature of concrete decides its durability. The concrete having very low permeability exhibits higher protection and greater durability. Sulfates are reported to be highly deteriorating toward concrete material. Therefore mineral additive such as rice waste was added to it which improved the resistance of cement material toward the attack of sulfates [18]. Artificial measuring instrument meant for determination of cracks in the structure is inopportune as well as expensive. But using smart film technique, helps to monitor cracks commendably for longer duration. In comparison to traditional method, smart film covers the whole area over a large scale of the structure which reveals initial development of cracks relating its position and size [19].

Magnetic-rheological device is used to control damping force. Iron suspended into an oil, worked as a medium for the damper device. Dampers were introduced in civil engineering in 1990. This device has great significance due to its simple structure, dynamic nature, lower operational cost, higher force capability. But the drawback was its installation. To overcome the drawback a smart system was developed by merging damper device with an electromagnetic inductive system. Permanent magnet as well as a solenoid coils was joined with damper device. Therefore a smart system was generated to avoid the negative impact of the passive systems [20].

4.4 Application in field of electronics

Solution-gel spin coating technique was used to develop zinc oxide seed layer over glass substrates. Dihydrate form of sodium acetate was used as main material whereas, ethanol and monoethanolamine. Zinc oxide layer so obtained was dried. Ta and Al were doped with zinc oxide so as to develop nanorods. SEM, XRD, and EDS were performed to analyze morphology, crystallinity and element abundance in the smart materials. The fabricated smart material was proved to be a promising material for fabrication of photovoltaic device and nanooptoelectronic device [21].

Superabsorbent hydrogel are capable of absorbing water 1000 times than its own weight. It undergoes swelling forming a stable colloid. The impact of different factors on swellability of hydrogel was also analyzed. Polyacrylate starch based superabsorbing hydrogel infused with polyaniline was prepared. The morphology of the super absorbent was studied by SEM and thermal analysis was done by TGA. The electrical conductivity of the synthesized polymer was also analyzed [22]. Field emission electron source is considered to be more expedient than thermionic cathodes because there is no loss of energy in the vacuum. It can also endure high radiation. Silicon-based microstructures were fabricated with the help of isotropic as well as anisotropic dry and wet etching method. Silicon wafers were doped with boron. It was analyzed that the anisotropy of reactive ionic etching was influenced by power as well as pressure. Power was dependent upon pressure. Due to increased anisotropy silicon wafer was determined to be more homogenous. It was justified from the results that the homogenous cathode can bring more advancement in order to develop microelectronic devices [23].

4.5 Application in field of medical

Skin is the utmost significant sense organ present in our body which is capable of sensing different kinds of stimulus. Skin is featured with remarkable characteristics such as stretching, elastic nature and ability of being self-healed. Therefore, electronic skins were designed using smart materials, sensing the stimulation functions as well as mechanical properties of skin. The designing of electronic skin was considered to be the biggest challenge due to narrow responsiveness and integration of sensors to obtain multifunctioning. In order to obtain sense functioning as well as elasticity, polymer substrate like polyvinyl alcohol, polyamide, and polyethylene terephthalate were merged with active material like silver, gold, platinum and nanowires of silicone as well as zinc oxide [24]. Oxidation as well as reduction reaction in the conducting polymeric film capable of substituting anionic entity as well as solvent were analyzed through voltmetric studies in presence of different solvent and salt solutions. Charge, energy, and potential at different steps during the reaction were scrutinized. The conduction polymers behaved as a three-dimensional gel reactor, through which no large anion can infiltrate from the solution into the polymeric film mimicking in form of intracellular medium reaction [25]. Natural or artificial food contaminant may cause a disastrous effect on human health. It may be cancerous or may cause hormonal imbalance. Present-day analysis technology is quite sensitive and cannot be applied to the food samples. Moreover, this method required some pretreatments. The drawback of the technique is that it is very expensive and do not shows suitability for the harsh environment. In order to combat this problem a smart material is required. Molecular imprinted solid phase extraction technique is significantly used for analysis of food contaminants for extracting herbicides, drug residues as well as toxic contaminants [26]. Fig. 3.4 is indicating use of smart material in aircraft.

FIGURE 3.4   Use of smart material in aircraft.

4.6 Application of smart material in the field of aerospace

Smart materials and some structures have also been employed in morphing the aircraft. Outmoded hydraulic as well as mechanical are not well-thought-out techniques to morph aircraft. Smart materials are more advantageous due to higher energy density, easy to control, and its endurance to strains. This idea leads the research workers to design and morph aircraft [27].

Designing and characterization of biomolecular material has enabled the utilization of biomolecules in form of transducer for the smart material. It has been utilized for developing sensors and energy-saving devices [28]. Cellulose is hydrophilic, biodegradable as well as biocompatible in its nature. It is the most abundant material in our biosphere. Smart materials have been reported to be synthesized from cellulose by chemical alteration, incorporation, and blending processes. As cellulose is possessed by the hydroxyl groups they can undergo different reactions like etherification, esterification as well as oxidation reactions. This stimulus responsive natural polymer exhibits properties of smart material. Smart materials prepared from cellulose in form of gel, film, membrane as well as aggregates are of great importance. They are highly responsive to change in the external conditions. The smart material produced from it have been used for wastewater treatment [29]. The influence of improvisation of polymeric matrix with shape memory alloys were studied in order to enhance stored energy. Smart materials have been altered by modifying some parameters such as temperature, stress and electric as well as magnetic fields. Shape memory alloy is a class of smart materials which have a property to memorize and returning back to its original conformation on subjecting it to elevated temperature. Smart materials have also been employed to design aeronautical structures in order to mollify bird-striking requirements [30]. Shape memory alloy-based material were overviewed in different fields of engineering especially for aerodynamic performance of vehicles through modification of spoiler shape as well as shape of uppermost panel [31]. Fig. 3.5 is indicating use of smart material as biosenser device.

4.7 Application of smart material in the field of biosensors

Biosensors are the excellent analytical tool, which have been used in different fields such as food industry, agriculture, medical field, and army. Biosensors are considered to be more advantageous in comparison to other analytical methods because of its portable nature and compact size. Moreover, analysis can be performed with the minimum amount of samples [32]. Biosensors have been employed to analyze organic as well as inorganic pollution causing agents in the environment. It is also employed to detect heavy metal contaminants in the environment. Phenol as well as its derivative is considered to be highly toxic in its nature. Such pollutants present in waste water shows interaction with deoxy-ribonucleic acid, causing severe effects in humans.

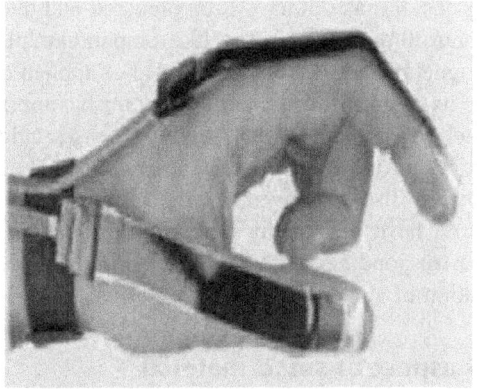

FIGURE 3.5   Use of smart material as biosensor device.

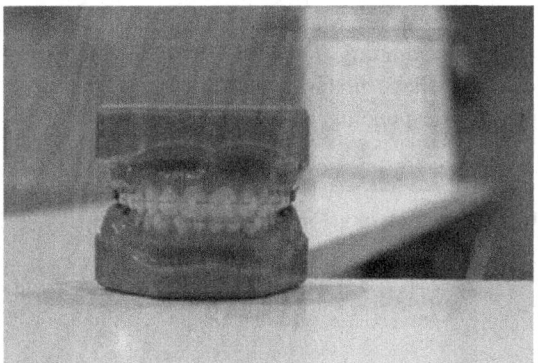

FIGURE 3.6   Applicability of smart material in dentistry.

Interaction amid DNA as well as contaminant have been used as electrochemical deoxy-ribonucleic acid biosensor [33]. Fig. 3.6 is indicating applicability of smart material in dentistry.

4.8 Application of smart material in the field of dentistry

Smart materials are considered as a promising, reliable and long termed efficient materials in the field of dentistry. Intelligent sutures comprised of plastic or silk thread coated with sensors capable of detecting infection was developed. They possess shape memory property as well as biodegradability. When the temperature is elevated suture get shrinked and tightened. Pheromones guided smart material was developed which was capable of eliminating pathogen such as streptococci mutant responsible for causing dental caries [34].

Biomimicking dental material have been prepared and they can be further transformed by controlling various factors like temperature, pH, stress, electric field as well as magnetic field. But on the removal of applied change the smart material returns to its original form. Enamel provarnish comprising amorphous calcium phosphate was prepared which slowly releases calcium as well as phosphate in ionic form from its gel thereby neutralizing pH. Moreover, casein phosphopeptides were also employed for the remineralization in dentistry. Such smart material have the quality of executing definite functions in an intelligent manner, responding to change in environment [35]. Table 3.2 is indicating application of smart material.

5. Impact and aspect of smart material

The smart materials are multidisciplinary in its nature. They are also referred to as an intelligent system proficient in sensing and actuation. It is characterized by adaption of structural features as well as having an active control in a way that its responses as well as its characteristics can be changed under influence of change in stimulus. They find its application in different fields such as piezoelectric, shape memory devices, photovoltaic cells, photomechanical, temperature responding, self-healing devices, etc. Smart materials

TABLE 3.2 Applications of smart materials.

S.No	Applicable fields
1.	Smart fabrics
2.	Morphing aircraft
3.	Sports good
4.	Microcrack healing material
5.	Fabrication of robots
6.	Security devices
7.	Microelectronic devices
8.	Optoelectronic devices
9.	Magnetorheological-devices
10.	Medical surgeries
11.	Dentistry
12.	Biosensors
13.	Analysis for food contamination

can take the researchers to the multiple levels giving excellent results as it is not bound to a single function. The role of such material is highly significant and can be applied to different fields. The practical studies reveals that the use as well as development of such system brings improvement in the growth of sustainable development. The smart materials have reached their full potential and have become very common. Their adaptive designing and performance is more beneficial over its cost make, making its bigger step toward sustainable development. Moreover, they develop as well as integrate more in their applications.

The negative impact of smart material is that the level of adaption is dependent upon complexity of the smart material. It becomes very important to know about the developmental stages of the material with time and how its qualities have flourished. Smart materials may be detrimental with respect that, it may lead to dropping out labor which may further lead to global crisis, high cost, recognition problems, difficulty in arrangements and more dependability on the conventional materials. It is expensive to produce. It may cause environmental pollution. Moreover, the lifetime effects of smart material are also not known [36]. Fig. 3.7 is indicating aspects of smart material.

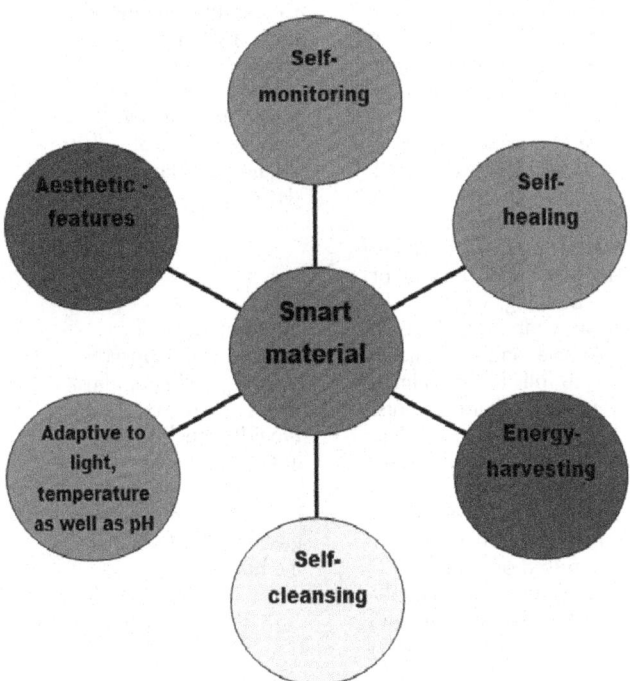

FIGURE 3.7 Aspects of smart material.

6. Impact of smart material on the sustainability

To achieve the environment sustainability, innovative smart material with improved as well as upgraded functions is the greatest solution to long-lasted complications, which have a deteriorating effect on the environment. Such an intelligent material offers protection to our environment, provides the stability to the standing structures as well as provides renovations. Such materials acts as an extensions for the conventionally utilized materials. They have the capability of altering themselves, adapting its surroundings.

They are referred as smart as they get adjusted to the changed environment. Smart material generates the responses for variable environment. Being self-activated they are capable of stimulating itself. The response exhibited by the smart materials is direct, expected as well as distinct. Renewable resources have been utilized to develop smart materials through sustainable methods to decrease the over utilization of the fossil resources.

7. Comparative analysis for the evaluation of smart material in sustainable environment

S.No	Smart material	Conventional material	References
1.	Smart bricks are capable of monitoring vibrations and alterations in temperature.	Ordinary bricks requires sensors-wireless link to warn the natural calamities.	[36]
2.	Smart concrete help in detection of stress as well as deformation. They also exhibits enhanced electrical resistance. It also analysis the occupancy of the building.	Concrete material requires additional devices.	[37]
3.	Smart wrap is composed of substrate lamination layers which rolls to form film capable of altering colors. It provides a refreshed look to the buildings and also provides rigidity.	Ordinary warp material is comprised of large number of film layers and do not provides much rigidity to the structures.	[38]
4.	Smart paints have extra capabilities. They are comprised of binder as well as pigment material which extract out energy from light source, resulting into glowing lights.	Ordinary paints just gives the finishing to the buildings.	[39]

5.	Lignin-based renewable smart material are the high performing materials which have benchmarked their performance replacing plastics.	Synthetic plastics.	[40]
6.	Biosensors are modified for detection of biospecies. Lignin-based biosensors are in great considerations as they can detect signals from mixture of species.	Ordinary sensors cannot detect signal from mixtures of biospecies.	[41]
7.	Graphene oxide is highly dispersible in water because of presence of carboxylic as well as hydroxyl group. It absorbs particles forming stable complex. It has been magnetically functionalized and efficiently used for water purification as well as employed in the field of drug delivery.	Chitosan or cellulose-based material.	[42]
8.	Light emission diode: an energy-saving device is a smart technology determining the visible response of an eye. It is used for designing screens as well as display devices these days.	Conventional lights.	[43—49]
9.	Thin film smart transistors are gaining large attentions for past few years due to its high performing level in various fields. They are applicable in displays, electronic papers, identification of radio frequencies, controlling humidity, determination of pH analysis of biological species, measuring cell growth as well as in the field of tissue regeneration. They have also been utilized to develop highly specialized sensing devices.	Ordinary transistors are not much specialized.	[50—60]

7.1 Future challenges and benefits

As we know that our population is growing day by day, so it is our duty to work on future challenge as soon as possible. Population growth is directly proportional to sustainable development. It always requires an objective that goes afar of individual scientific disciplines and even the subject of the research, so as to better grasp its possibility and penalty [61–67].

The most common challenge happens when the hugely dissimilar approaches from physical and social sciences need to be combined, for example, in studying the remediation plan of a contaminated urban area. In this case, we require to combine proficiency in soil remediation and chemical speciation of the contaminants, environmental fate, human health effects, ecotoxicological impacts, socioeconomic perspectives, urban planning, and so on. The broad variety of expert who are needed to do this will not share the same training and perspective to address the same issues, nor will they perceive or even define the problem from the same angles or focus on the same subsets of milestones and potential solutions.

In spite of the very difficult situation in which numerous developing countries at present find themselves in, sustainable development is achievable; however, it would require a lot of concentrated and coordinated effort. Stipulation suitable supply side policies, such as education and vocational programmers, were to be implemented, illiteracy rates would drop and people would be made more aware about the environment surrounding them, which would contribute significantly to an increase in environmental awareness.

8. Conclusion

The 21st century is facing a harmful threat to the environment. To combat this problem a sustainable fabricated smart material can offer a long-term solution to improve the qualities of life, thereby conserving the resources. Fabricated smart material can upgrade the relations in between the material as well as consumer. There is a need to lead innovative ideas by strengthening the relationship amid technical awareness as well as practice to eradicate the negative impacts. Smart materials are new generation material having adaptability to external stimuli with intrinsic intelligence. The use of smart material science technology has resulted in the application in different fields. The motive of the author is to provide the knowledge, promoting field of technology, and inspire the research workers, so that may contribute to this field. The purpose of the research is to understand the intelligent materials, which is very important for the development of novel smart material as well as to develop an advanced material with enhanced proficiency and reliability.

The objective of the researcher is to work endlessly for developing a system as well as a great smart material which can be controlled, guided, and is adaptive toward the environment also having very negligible negative impacts. Efforts must be directed to generate renewable smart materials. Functionality of such smart materials can be enhanced by blending the natural polymers with some metal oxides, fabricating it into some hybrid smart materials. Modified material so obtained exhibits biocompatibility, sustainability, biodegradability, higher degree of modification accompanied by excellent mechanical strength. Large struggles are required to develop an intelligent smart material.

Appendix A. Supplementary data

Supplementary data to this article can be found online at https://doi.org/10.1016/B978-0-12-824038-0.00002-X.

References

[1] Parihar S, Khandagale K, Jivrag P. Smart materials. J Mech Civil Eng 2016;13:28−32.

[2] Hubballi S, Harkude M, Jadav P. A review on introduction, classification and application of smart materials. Int J Manag Technol Eng 2018;8:2249−7455.

[3] Tembely M, Musa S. Smart materials: a primer. Int J Adv Res Comput Sci Software Eng 2017;7:43−4.

[4] Mishra J. Smart material- type SS and their application: a review. Int J Mech Prod Eng 2017;5. 2320-2092.

[5] Kamila S, Gopal V. Acoustics and thermal studies of conventional heat transfer fluids mixed with ZnO nano flakes at different temperatures. Heliyon 2019;5:02445.

[6] Tai J, He T, Pan Q, Zhang D, Wang X. A fast beam forming method to localize an acoustic emission source under unknown wave speed. Materials 2019;12:1−11.

[7] He L, Wang Y, Gao J, Wang J, Zhao T, He Z, Zhong Z, Zhang X, Zhong L. Enhancing the energy density of tricritical ferroelectrics for energy storage applications. Materials 2019;12:611.

[8] Sorbo G, Truda G, Bifulco A, Passaro J, Petrone G, Vitolo B, Ausanio G, Vergara A, Marulo F, Branda F. Non monotonous effects of noncovalently functionalized graphene addition on the structure and sound absorption properties of polyvinylpyrrolidone (1300 kDa) electrospun mats. Materials 2019;12:108.

[9] Ee B, Konovalov R, Konovalov S, Kuzmenko A, Tsaplev V. On the shaping of a short signal at the output of the receiving piezoelectric transducer in the radiation-reception system. Materials 2018;11:974.

[10] Kumar S, Goud S. Durability and performance analysis of smart materials for space applications- a review. Int J Res Appl Sci Eng Technol 2017;5:2321−9653.

[11] Luan N, Vuong L, Chuong T, Tho N. Structure and physical properties of PZT-PMnN-PSN ceramics near the morphological phase boundary. Adv Mater Sci Eng 2014;2014:821404.

[12] Ren G, Jhang K. Application of macrofiber composite for smart transducer of lamb wave inspection. Adv Mater Sci Eng 2013;2013:281575.

[13] Futsch R, Mejejri I, Rakotozafy H, Rougier A. PEDOT: PSS-V2O5 hybrid for color adjustment in electrochromic system. Front Mater 2020;7:78.

[14] Farrigia C, Borg R, Ferrara L, Buhagiar J. Application of *Lysinibacillus sphaericus* for surface treatment and crack healing in mortar. Front Built Environ 2019;5:62.

[15] Ersan Y, Palin D, Tesdamir S, Tesdamir K, Jonkers H, Boon N, Bilie N. Volume fraction, thickness and permeability of the sealing layer in microbial self- healing concrete containing biogranules. Front Built Environ 2018;4:70.

[16] Mageswari S, Palanivel B. Influence of Al, Ta doped ZnO seed layer on the structure, morphology and optical properties of ZnO nanorods. Current Smart Mater 2019;4:45−58.

[17] Yang J, Sun S, Zhang S, Li W. Review of structural control technologies using magneto rheological elastomers. Current Smart Mater 2019;4:22−8.

[18] Hassan A, Mahmud H, Jumaat Z, Subari B, Abdulla A. Effect of magnesium sulphate on self-compacting concrete containing supplementary cementitious materials. Adv Mater Sci Eng 2013;2013:232371.

[19] Zhang B, Wang S, Li X, Zhang X, Yang G, Qiu M. Crack width monitoring of concrete structures based on smart film. Smart Mater Struct 2014;23:045031.

[20] Cho S, Jung H, Lee I. Smart passive system based on magneto rheological damper. Smart Mater Struct 2005;14:707−14.

[21] Prabhakar R, Kumar D. Studies on polyacrylate-starch/polyaniline conducting hydrogel. Curr Smart Mater 2019;4:36−44.

[22] Sharma R, Sharma P, Malviya R. Polysaccharide-based scaffolds for bone marrow regeneration: recent work and commercial utility (Patent). Curr Smart mater 2019;4:29−35.

[23] Lawrowski R, Berger C, Langer C, Dams F, Schreiner R. Improvement of homogeneity and aspect ratio of silicon tips for field emission by reactive-ion etching. Adv Mater Sci Eng 2014;2014:948708.

[24] Almansoori M, Xuan Li X, Zheng L. A brief review on e-skin and its multifunctional sensing applications. Curr Smart Mater 2019;4:3−14.

[25] Otero T, Alfero M, Martinez V, Parez M, Martinez J. Biomimetic structural chemistry from conducting polymer: processes, charges and energies. Coulovoltammetric results from films on metal revisited. Adv Funct Mater 2013;23:3929−40.

[26] Baggiani C, Anfossi L, Giovannoli C. Solid phase extraction of food contaminants using molecular imprinted polymers. Anal Chim Acta 2007;591:29−39.

[27] Sun J, Guan Q, Liu Y, Leng J. Morphing aircraft based on smart materials and structures: a state-of-the-art review. J Intell Mater Syst Struct 2016;2016:1−24.

[28] Sarles S, Leo D. Membrane-based biomolecular smart materials. Smart Mater Struct 2011;20:094018.

[29] Qiu X, Hu S. "Smart" materials based on cellulose: a review of the preparations, properties and applications. Materials 2013;6:738−81.

[30] Guida M, Sellitto A, Marulo F, Riccio A. Analysis of the impact dynamics of shape memory alloy hybrid composites for advanced applications. Materials 2019;12:153.

[31] Sellitto A, Riccio A. Overview and future advanced engineering applications for morphing surfaces by shape memory alloy materials. Materials 2019;12:708.

[32] Mossberg B, Buchner M, Rishpon J. Electrochemical biosensors for pollutants in the environment. Electroanalysis 2007;19:2015−28.

[33] Silva L, Melo A, Salgado A. Biosensors for environmental applications. Environ Biosens 2015;4:283085176.

[34] Gupta V. Smart materials in dentistry: a review. Int J Adv Res Dev 2016;3:89−96.

[35] Padmawar N, Pawar N, Joshi S, Mopagar V, Pendyala G, Vadvadgi V. Biosmart dental materials: a new era in dentistry biosmart dental materials. Int J Oral Health Med Res 2016;3:2395−7387.

[36] Nihalani S, Joshi U, Meeruty A. Smart materials for sustainable and smart infrastructure. Mater Sci Forum 2019;969:278−83.

[37] Mahmoudian M, Sharifikheirabadi P. Uses of new/smart material in the green building with sustainability concern. Int Trans J Eng Manag Appl Sci Technol 2020;11:2228−9860.

[38] Morenoa A, Sipponen M. Lignin-based smart materials: a roadmap to processing and synthesis for current and future applications. Mater Horiz 2020;10:1039.

[39] Koduru J, Karri R, Mubarak N. Smart materials, magnetic graphene oxide-based nanocomposites for sustainable water purification. Sustain Polym Compos Nanocompos 2020;10:1007.

[40] Addington M. Smart material and sustainability. Strat Technol 2017;2:12.

[41] Das T, Prusty S. Biopolymer composite in field effect transistors. Biopolym Compos Electron 2017;10:1016.

[42] Mittal H, Kumar V, Saruchi RSS. Adsorption of methyl violet from aqueous solution using gum xanthan/Fe_3O_4 based nanocomposite hydrogel. Int J Biol Macromol 2016;89:1−11.

[43] Saruchi KV, Mittal H, Alhassan SM. Biodegradable hydrogels of tragacanth gum polysaccharide to improve water retention capacity of soil and environment-friendly controlled release of agrochemicals. Int J Biol Macromol 2019;132:1252−61.

[44] Saruchi KBS, Kumar V, Jindal R. Biodegradation study of enzymatically catalyzed interpenetrating polymer network: evaluation of agrochemical release and impact on soil fertility. Biotechnol Rep 2016;9:74−81.

[45] Kaith BS, Saruchi JR, Bhatti MS. Screening and RSM optimization for synthesis of a gum tragacanth−acrylic acid based device for in situ controlled cetirizine dihydrochloride release. Soft Matter 2012;8:2286−93.

[46] Mittal H, Alili AA, Alhassan SM. Solid polymer desiccants based on poly(acrylic acid-co-acrylamide) and laponite RD: adsorption isotherm and kinetics studies. Colloids Surf A 2020;599:124813.

[47] Saruchi KV. Separation of crude oil from water using chitosan based hydrogel. Cellulose 2019;26:6229−39.

[48] Mittal H, Morajkar PP, Alili AA, Alhassan SM. In-situ synthesis of ZnO nanoparticles using gum Arabic based hydrogels as a self-template for efective malachite green dye adsorption. J Polym Environ 2020;28:1637−53.

[49] Saruchi KV, Kaith BS. Synthesis of hybrid ion exchanger for rhodamine B dye removal: equilibrium, kinetic and thermodynamic studies. I & EC Res 2016;55(39):10492−9.

[50] Naushad M, Mittal A, Rathore M, Gupta V. Ion-exchange kinetic studies for Cd(II), Co(II), Cu(II), and Pb(II) metal ions over a composite cation exchanger. Desalin Water Treat 2015;54:2883−90. https://doi.org/10.1080/19443994.2014.904823.

[51] Saruchi KV. Adsorption kinetics and isotherms for the removal of rhodamine B dye and Pb^{+2} ions from aqueous solutions by a hybrid ion-exchanger. Arab J Chem 2019;12:316−29.

[52] Mittal H, Kumar V, Alhassan SM, Ray SS. Modification of gum ghatti via grafting with acrylamide and analysis of its flocculation, adsorption and biodegradation properties. Int J Biol Macromol 2018;114:283−94.

[53] Naushad M, ALOthman ZA. Separation of toxic Pb^{2+} metal from aqueous solution using strongly acidic cation-exchange resin: analytical applications for the removal of metal ions from pharmaceutical formulation. Desalin Water Treat 2015;53:2158−66. https://doi.org/10.1080/19443994.2013.862744.

[54] Mittal H, Alili AA, Alhassan SM. High efficiency removal of methylene blue dye using carrageenan-poly(acrylamide-co-methacrylic acid)/AQSOA-Z05 zeolite hydrogel composites. Cellulose 2020;27:8269−85.

[55] Saruchi TP, Kumar V. Kinetics and thermodynamic studies for removal of methylene blue dye by biosynthesize copper oxide nanoparticles and its antibacterial activity. J Environ Health Sci Eng 2019;2019. https://doi.org/10.1007/s40201-019-00354-1.

[56] Faisal AAH, Al-Wakel SFA, Assi HA, Naji LA, Naushad M. Waterworks sludge-filter sand permeable reactive barrier for removal of toxic lead ions from contaminated groundwater. J Water Proc Eng 2020;33:101112. https://doi.org/10.1016/j.jwpe.2019.101112.

[57] Sethi S, Kaith BS, Saruchi KV. Fabrication and characterization of microwave assisted carboxymethyl cellulose-gelatin silver nanoparticles imbibed hydrogel: its evaluation as dye degradation. React Funct Polym 2019;142:134−46.

[58] Naushad M. Surfactant assisted nano-composite cation exchanger: development, characterization and applications for the removal of toxic Pb^{2+} from aqueous medium. Chem Eng J 2014;235:100−8. https://doi.org/10.1016/J.CEJ.2013.09.013.

[59] Saruchi, Verma R, Kumar V, et al. Comparison between removal of Ethidium bromide and eosin by synthesized manganese (II) doped zinc (II) sulphide nanoparticles: kinetic, isotherms and thermodynamic studies. J Environ Health Sci Engineer 2020. https://doi.org/10.1007/s40201-020-00536-2.

[60] Naushad M, Vasudevan S, Sharma G, Kumar A, Alothman ZA. Adsorption kinetics, isotherms, and thermodynamic studies for Hg^{2+} adsorption from aqueous medium using alizarin red-S-loaded amberlite IRA-400 resin. Desalination Water Treat 2016;57(39):18551−9. https://doi.org/10.1080/19443994.2015.1090914.

[61] Saruchi SM, Hatshan MR, Kumar V, Rana A. Sequestration of eosin dye by magnesium (II)-Doped zinc oxide nanoparticles: its kinetic, isotherm, and thermodynamic studies. J Chem Eng Data 2020. https://doi.org/10.1021/acs.jced.0c00810.

[62] Naushad M, Alqadami AA, AlOthman ZA, Alsohaimi IH, Algamdi MS, Aldawsari AM. Adsorption kinetics, isotherm and reusability studies for the removal of cationic dye from aqueous medium using arginine modified activated carbon. J Mol Liq 2019;293:111442. https://doi.org/10.1016/j.molliq.2019.111442.

[63] Dhiman B J, Kaith S. Fabrication of high performance biodegradable *Holarrhena antidysenterica* fiber based adsorption devices. Arab J Chem 2020. https://doi.org/10.1016/j.arabjc.2020.10.004.

[64] Kaith BS, Dhiman J, Bhatia JK. Synthesis and characterization of MHa-g-poly(HEMA) $PO_4^{2-}2H^+$ cation exchanger-effective removal of methylene blue from waste water. RSC Adv 2015;5(50):39771−84. https://doi.org/10.1039/c5ra00670h.

[65] Kaith BS, Dhiman J, Kaur Bhatia J. Preparation and application of grafted *Holarrhena antidycentrica* fiber as cation exchanger for adsorption of dye from aqueous solution. J Environ Chem Eng 2015;3(2):1038−46. https://doi.org/10.1016/j.jece.2015.03.001.

[66] Bhatia JK, Kaith BS, Dhiman J. RSM optimized soy protein fibre as a sorbent material for treatment of water contaminated with petroleum products. Desalin Water Treat 2016;57(9):4245−54. https://doi.org/10.1080/19443994.2014.993720.

[67] Bhatia JK, Kaith BS, Dhiman J. Synthesis and optimization of soy protein fiber based graft copolymer through response surface methodology for removal of oil spillage. Polym Bull 2013;70(11):3155−69. https://doi.org/10.1007/s00289-013-1014-0.

Chapter 4

Application of nanotechnology in pesticides adsorption with statistical optimization and modeling

Kamalesh Sen

Environmental Chemistry Lab, Department of Environmental Science, The University of Burdwan, Bardhaman, West Bengal, India

1. Introduction

Agricultural water pollution plays about 70% of freshwater, which is detrimental to the health of human beings, plants, flora and fauna as well as water scarcity, is one of the major environmental problems in the world [1−4]. It has major demands for removal of agrochemical as sustainable ways. There are many application in purpose of wastewater recycling, including agricultural land, aquaculture, production costs, recreational and environmental practices, and recharge of artificial groundwater [5]. Anthropogenic practices and industrial abuses leads to increase the emerging contamination of environment, issues become the public health concern all over world [4,5]. Corresponding pesticides become a major threat of natural outlook, for instance, domestic, fish, wildlife, birds, and ecosystem; regarding chemicals separated into chemical groups [6,7].

The chemical groups have segregated with organochlorines, organophosphates, and carbamates [8]. Pesticides meant that controlling pests regarding pesticides, herbicides, insecticides, termiticides, nematicides, molluscides, avicides, rodenticides, bactericides, insect repllents, antimicrobials, and fungicides [8,9]. In the present report, the pesticides used in the world near about 2 million tons, annually; where China becomes major participant, according to the United States, Argentina, and India, and rising rapidly. Corresponding of this 47.55% of herbicides, 29.51% of insecticides, 17.57% of fungicides, and 5.5% of other pesticides are attributed [4,10]. By the year of 2020, the predictable increases up to 3.5 million tons [4]. In this context, a serious

Cognitive Data Models for Sustainable Environment. https://doi.org/10.1016/B978-0-12-824038-0.00005-5
65

consequence may pose a huge amount of pesticides, because of their persisting in the environment.

The toxicity in few reports, chronic induces the herbicides like glyphosate act as a nonselective herbicide to resist the plant growth and arrested the enzyme, helping dying the plants regarding no-selective types [11,12]. Exhausts of pesticides can cause the environmental problems, and essentials to removing unnecessary residues. Various type of pollutants are usually dealing to protocols in sustainable ways, such as sedimentation, photocatalysis, coagulation, electroplating, membrane filtrations, microbial remediation, and adsorption [13,14]. Herewith, focus only the adsorption due to its application are carried out in last few decades with having more sophistication, less pollutant generated, and very high-energy efficiency [15,16].

Adsorption study deals with surface adhesion of adsorbent, so that the large surface could be increased in the adsorption capacity [17–19]. However, the adsorption utilizes two major process: batch adsorption and column adsorption [17,20]. The applicability of adsorbents in terms of biological substances as cellulose and lignin basis materials, zeolites, soils, biochar's, activated carbon from biological materials, nanomaterials, nanobiocomposite, nano-nanocomposite, ceramic substances, chitosan, and nanodoped materials [19,21–23]. Therefore, the applicability of nanoparticle-based adsorbents have more demanded regarding increasing adsorption efficacy. Moreover, in the context of adsorption have more emphasized with nanobase adsorbent onto pesticides separation, mechanism, characterization, and statistical optimization based on adsorbent preparation and adsorption study [24].

Nanoparticles is a unique high-tech material that have unique properties [25], i.e., catalysts, factionalized large surfaces area, coating, high reactivity, sensitivity and small size, that surface has permissible to more peasant of adsorption, considering recent years the nanomaterials utilized on organic and inorganic molecules, even toxic elements [26–28]. It has irregular or uniformed surface, deals with following aspect crystalline, amorphous, crystal solids, and agglomerated. Synthesized protocols deals with biological, physical, chemical; corresponding as bottom-up method, sol-gel, spinning, chemical vapor deposition (CVD), pyrolysis, biosynthesis, mechanical milling, nanolithography, thermal decomposition, and sputtering [27,29]. However, this sustainable to conducted preparation of nanoparticles may increase to production. The optimization study regarding adsorbent/nanoparticle synthesis and adsorption study has to be apply as response surface methodology (RSM) and artificial intelligence (AI) [14,30]. Kinetics and isotherm models are applied for adsorption mechanisms according to monolayer, multilayer, chemisorption, or physisorption [31,32].

The main objective of the study is to review of applied nanotechnology on the pesticide adsorption from recent literature, focused mainly on nanomaterial preparation techniques and mechanism, with adsorption modeling and statistical optimization.

2. Different techniques of pesticide removal

Pesticide pollution has become a global issue that relies heavier chemical substance like as carbamate, organochlorine, organophosphate, etc. These chemicals need to remediate as chemically, physically, and biologically from water and soil; it is essential to investigate which methods are more sustainable or environmentally friendly. The biological process had taken the place of pesticides removal through phytoremediation and microbial remediation, and organic pesticides are removed with bioaugmentation and biostimulation [33−35]. Different biopurification systems were applied previously to follow as remediation of fungicides through a biomixture augmented with ligninolytic fungi toward used bioaugmented activated sludge, which was produced in vitro condition using microbial consortium, pressurized activated sludge, anaerobic-aerobic biological treatment, and membrane bioreactor [36,37]. The physical process as treatment of physical ways, such as clay materials, polymeric substance, activated carbon, zeolites, and others. The clay has a low-cost availability, the minerals interact with hydrophilic and negative charges, resulting in the easy retention of cationic pesticides [38]. Pesticides cationic nature can obtain from aliphatic and aromatic amines from interlayer cations, which are generally adsorbed by the clay anionic minerals [38,39]. On the other hand, pesticides contain large hydrophobic moieties, which affect the surface to adsorb the interactions of van der Waals with surface oxygen atoms [38]. Membrane filtration is one of the techniques that are widely used in pesticides removal, the filtration is largely reliable for microfiltration and nanofiltration to cut off sized deal with contamination by water percolation mechanism [40]. Chemical wastewater treatment regarding chemical to help hydrolyzing contaminants into safer chemicals [41]. The main utility of chemical treatments usually photocatalysis and/or membrane techniques, iron-enhanced sand filters, chlorination, advanced oxidation processes, and adsorption [42]. The physicochemical techniques among these methods like as electrocoagulation, adsorption, here basically more opportunity to dealing the adsorption to low-cost and easy methods [43]. Adsorption is well-known techniques leads to equilibrium separation process, which are verified in terms of flexibility, less harm, and sensitivity toward the utility of adsorbents.

Adsorbents types are segregated, regarding techniques as carbon-base, cellulose-base, mineral-base, waste-base adsorbent, polymer-base adsorbent, chitosan-base adsorbent, and nanoparticle-base adsorbents [44,45]. It is more suitable of nanobase adsorbents to leads more sufficient adsorption compare regarding others. This finding could benefit nanobase carbon and activated carbon, the highest showed in pesticides adsorption in nanoparticles-based, applied pesticides as organophosphorus, organochlorinated, carbomate, and others [23,46]. The mechanisms are dealing with area of surface and porous contains, compared with uptake capacity, has a greater subset of pesticide adsorption.

3. Nanomaterial synthesis and characterization

The synthesis of nanoparticles is mainly associated with the physical, biological, and chemical; therefore, it is a very challenging task, constructed with physical methods to vapor deposition, mechanical pressure, high-energy radiation, thermal energy, or electrical energy to support nanoparticle synthesis [25,28,47]. Corresponding methods are applied previously, following as high-energy ball milling, laser ablation, electrospraying, inert gas condensation, physical vapor deposition, laser pyrolysis, flash spray pyrolysis, and melt mixing are some of the most regularly used physical methods to generate nanoparticles [25,48]. The chemical process to synthesis of nanoparticles carried out following as sol-gel method, microemulsion technique, hydrothermal synthesis, polyol synthesis, and chemical vapor with plasma deposition [49–51]. Biological process as one of the sophisticated process that the nanoparticles are formed to associate with biomolecules from plants, act as a capping agent, helping to reduce of metal ions. This is established that phytochemical and biomolecules are considered the reducing and capping agent, corresponding molecules are NADH-dependent reductase, terpenoids, sugars, alkaloids, flavonoids, phenols, tannins, and proteins [52]. Nanoparticle synthesis processes are applied in many fields, and it is necessary to know about the nanoparticle size that controls in the process.

The nanoparticle research has another tool of characterization: it can be derived regarding formation or synthesis, structural pattern, elemental composition, reducing agent, capping agent, and then utilized. Existing oriented research to describe nanoparticle characterization is focused on the most advanced tools, which include spectroscopy analysis as UV-visible (UV-vis), X-ray diffraction (XRD), X-ray photoelectron (XPS) (quantitative analysis chemical state), and energy dispersive (EDS) and Fourier Transform Infrared (FTIR), morphology analysis as Zeta potential, dynamic light potential (DLS), field emission scanning electron microscopy (FESEM), transmission electron microscope (TEM), and selected area electron diffraction (SAED). The synthesis nanoparticle has its own properties that ensure nanoparticle formed to configure the regulatory properties [29,48,53,54]. UV-Vis could apply for the color of absorbance to introduce particular surface plasmon resonance. This resonance conformed to band gaps between absorbance. However, it is a very effective characterization for conforming to the nanoparticles. XRD pattern has to be provided for the nature of nanoparticle, either crystalline or amorphous nature. The Zeta potential deals with bulk fluorescence in particle dispersion according to layers containing charged ions associated with the nanoparticles surface [54]. DLS is a deterministic method to deal with the size distribution profile in nanoparticles by light scattering [55]. The FTIR analysis to characterized functional groups in constituent related with nanomaterials. The morphological structure of nanoparticles have been observed using FESEM and TEM, and these studies specified structures, either spherical, oval, irregular, triangular, or hexagonal shapes; the EDS were analyzed for elemental quantifications that have existing precursors [32].

4. Nanoadsorbent materials and properties

Nanoadsorbency is a unique characterization by their particles size, which ensure the containing charge. Adsorption, increasing or decreasing, could depend on the surface area, adsorbate attachment sites, and contained material. Nanomaterials are discussed based on the removal of pesticides like carbon-based nanoadsorbents (CNT, graphene, fullerene, and carbon composite materials), miscellaneous nanoparticles (including metal, ceramic, calcium base, doped, etc.), nanocomposites (containing different type such as nanoparticle composites, bionanocomposites, etc.), and nanoclay (Fig. 4.1).

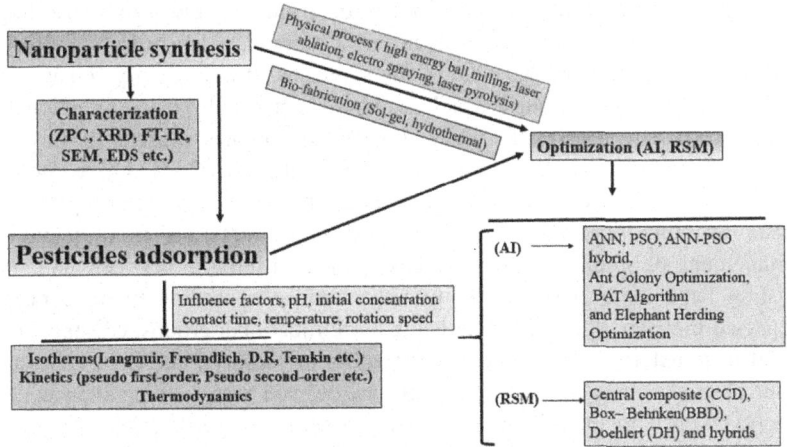

FIGURE 4.1   The graphical information about adsorption, nanoparticle synthesis, optimization and modeling.

4.1 Carbon-based nanoadsorbents

A new class of carbon-based nanoparticles such as carbon nanotube (CNT), graphite/graphene, fullerene, and carbon composite materials have different properties. These belong to sp^2-bonded carbon atoms by predominant structures, which are classified as a number of dimension-spacing such as zero-dimension (fullerene), carbon nanotubes (one-dimensional), and graphene (two-dimensional) [47]. The CNT exists either as a single-wall or multiwall, or as a one-dimensional macromolecular structure (1−100 nm) [56]. The nanoparticles have some application in adsorption toward various types of pesticides. The adsorption related with carbon-based molecules have the capability to adsorb remarkable amounts due to the existence of a large number of functional groups regarding modification. The structural properties depend on their preparation or modification process. The CNT involves a different mechanism according to hydrophobic, covalent bonding, π-π interactions, hydrogen bonding, and electrostatic interactions. The single-wall CNT have branching of graphene sheet, regarding diameter 1 nm, rolled up to concentric cylinders layer by spacing at 0.3−0.4 nm [57]. The multiwall CNT adsorption mechanism derived as organophosphorus pesticides (malathion); the pH played a significant role of adsorption, and achieved a maximum high pH value, regarding mechanism interacted electrostatic interaction related with chemisorption [58]. Adsorption may play a role between adsorbate and adsorbent with −COOH, −OH, −NH$_2$ functional group with organic substances. Fenuron pesticide was adsorbed by MWCNTs and possible mechanism described as hydrophobic interacting patterns of π-σ (1), π-π stacked shape (7), and π-π (1) T-shaped and π-alkyl (3) [18].

Graphene oxide (GO) can be synthesized from graphite flasks regarding intercalation and oxidation using Hummer's method, a key role by sulfuric acid and potassium permanganate of the synthesis. It accordingly consists of epoxide and hydroxyl groups, which are strongly hydrophilic and affectionate in relation to hydrogen and water molecules. GO has present in surface functional group regarding hydroxyl (-OH), alkoxy (C−O−C), carbonyl (C=O), carboxylic acid (-COOH) with other species of oxygen groups. Graphene quantum dots (GQD) is a new class of materials from quantum confinement of graphene sheet. It has unique difference between GO for existing band gaps, have benefit to large surface according to mass ratio, dispersed in water, due to edges existing more functional groups. GO and GO-based nanomaterials was described according to carbaryl, catechol, and fluridone pesticides their interacting patterns are derived π-π stacking and van der Waals, taking a major role between GO and pesticides [59]. Further application GO bases silica coated magnetic nanoparticles functionalized with 2-phenylethylamine (Fe3O4@SiO2@GO) as adsorption of chlorpyrifos, parathion, and malathion from aqueous solution, due to improvement efficacy of GO toward adsorption. Synthesis of adsorbent, GO to Fe$_3$O$_4$@SiO$_2$@GO nanoparticles by covalent coupling of graphene oxide, regarding mechanism carried out between graphene oxide nanosheets and Fe$_3$O$_4$@SiO$_2$−NH$_2$, covalently bonded with the nanosheets produced through amide bond. Supporting FTIR peak as amide carbonyl group at 1632 and 1425 cm^{-1}, rename as −CONH amide band, and C−N stretch of amide, respectively [19]. On the other word, the Fe$_3$O$_4$@SiO$_2$@GO interacted with pesticides as hydrogen bonding and π-π interaction of the electronegative atoms (P, N, and S). As the current area of GO, improve the efficacy derive with other nanocomposites, because of that GO easily bonded the molecules negative and positive together. In addition, ability of GO as pesticides adsorption by molecular interacted possibilities as H-bonded and electrostatic, due to edged molecules are strongly hydrophilic to make polar bridge as pesticides negative atoms [60].

Fullerenes (C60) belongs to the large family, spherical structure, and consists of 60 carbon atoms, with a molecular weight of 720 g/mol and diameter about 1 nm. Fullerenes are a type of carbon molecule that is not graphite or diamond. They have a spherical, elliptical or cylindrical arrangement of millions of carbon atoms [61]. Adsorbent behavior higher than activated carbon, the adsorption progresses to the physisorption method mainly through the scattered interaction force. As a comparison between fullerene and graphitized carbon black for adsorption, the addition of C60 to metallic oxides such as silica, alumina, and iron may enhance the adsorption activity [62,63]. A report of fullerene C_{60} are applied on glyphosate (organophosphorous herbicide) adsorption in aqueous medium, regarding C_{60} fullerene continues to show a neutral net charge, pH influences the impressment of organic functional groups to be able to enhance the negative dissociation of glyphosate adsorption through covalent bonds (functionalization) [64]. Fullerene functionalization deals with polar OH groups as polyhydroxylated fullerene $C_{60}(OH)_{24}$ to ensure better facilities to enhancement adsorption regarding metallic oxide, and properties of catalytic influences not only single but improvement when conjugated with nanotitanium dioxide [65]. Therefore, the fullerene is one of the best adsorbent to pesticides adsorbent, integration for adsorption when metallic nanoparticle conjugated.

4.2 Miscellaneous nanomaterials

Metal nanoparticles (MN) are nanoscale subjected to made of pure metals (e.g., gold, platinum, silver, titanium, zinc, cerium, iron, and thallium) localized of the surface plasmon resonance property derives broad adsorbent band gaps by optoelectrical formation. The formation and preparing ability from the respective reducing agents at the biochemical processes. The physical and mechanical process for made over ball milling and CVD methods. MN was enabled to increase the surface properties to effectiveness as well over their size. Adsorption properties dealing their surface area, metal nanoparticle could enhance band gaps according to size, shape, doped, and ionic basis [15]. Nanoscale zero-valent iron (nZVI) have strong efficiency to removal of pesticides by reduction mechanism. The sodium citrate-activated zero-valent iron nanoparticles enhance adsorption 2,4-D dechlorination. When dechlorination of 2,4-D, the nZVI sheared electrons by protonation and reduced the 2-4D by deprotonation, then deoxidized surface of iron easily bounded by metal chelation [66]. Magnetic nanoparticle have more demanding currently applied to some researchers, it has unique properties surface of magnetic core increases adsorption efficiency. The organophosphorus pesticides adsorbed by magnetic nanoparticles modified with poly (p-phenylenediamine-co-thiophene) [67]. Another application of magnetic with MWCNTs based on organic framework ZIF-8 to adsorption organophosphorus pesticides [68]. Ceramics materials are inorganic solid to produce nanoparticle by different techniques, applied the pesticides adsorption. It consists in silica or alumina to produce in amorphous, polycrystalline, well-dense, porous, and nanosize. Ceramic nanoparticles have large surface area with positive charged and enabled to enhance adsorption in a negative charge of adsorbate. Acicular mullite ceramic (AMC, $3Al_2O_3 \cdot 2SiO_2$) as an aluminosilicate mineral have applicable regarding adsorption with iron-oxide embedded in contaminants, increasing adsorption of positive state of pollutants [69]. Calcium-based nanoparticles has more effectiveness of adsorption study, including oxide, carbonate, and sulfate. It has a positive surface with porous, interacted with adsorbate molecules through electrostatic interaction. Calcium peroxide nanoparticles interacted with organic pollutants to electrostatic mechanism [31]. Nanodoped adsorbents is currently applied to increasing the surface energy and band gaps, actually lower energy atom conjugating with higher energy atoms to stabilization band gaps. It has not only produced the effective surfactant but also helps to photocatalytic degradation of organic pollutant. A gadolinium-doped cobalt ferrite nanoparticle applicability of adsorption onto organic pollutants capacity increases at 263.2 mg/g, which is better than undoped ferrite nanoparticles [70]. Another application of nitrogen doped with magnetic carbon nanoparticles enhanced uptake capability and surface properties compared with undoped magnetic carbon nanoparticles [71]. Semiconductor materials hold properties according to metals and nonmetals, so they have many applications because of these properties. Especially increasing electronic properties and band gaps could be conducted to produce surface-binding sites and energy [51]. Organic polymer-based nanoparticles leads to a particle matrix as enhancing surface functional groups, i.e., $-OH$, CHO, $=C = O$ etc., and large interactions encouraged due to surface availability, nanoparticles are able to increase vacant sites [72]. Polymer-based nanoparticle has insoluble structure with high surface properties, could combines with high dense function groups, are able to enrich the binding properties of adsorption.

4.3 Nanocomposite

A composite produced by nanomaterial to develop for increase the properties. It makes to improve physical and chemical properties. It displays to lengthen the adsorption properties according to the surface area based on nanoparticle functionalized. Nanocomposite not only nanoparticle-nanoparticle composite but also nanoparticle-biological composite are more effectiveness of adsorption study. It has current demand to increase surface area and colloidal properties. Nanocomposite are classified with categorized based on structural and matrix differences, such as metal-based nanocomposite, polymer-based nanocomposite, ceramic-based nanocomposite and carbon-based nanocomposite. Metal-based nanocomposites are Fe_3O_4/CNT, Fe/MgO, $Al/Mg/Ca$, $Fe/$hydroxyl apatite, etc., capable incensement charge by disordered grain surface. The ceramic-based nanocomposite utility in adsorption chemistry, deals to impregnate with cellulose fibers, or metallic nanoparticle combined with ceramic matrices, showed high adsorptive capacity along with ceramic precursors, such as Al_2O_3/TiO_2 and SiO_2/Ni. The biopolymer based the composite as chitosan—ZnO nanoparticle for removal of permethrin pesticides, showed uptake ability up to 99% [23]. Zr with MOFs of UiO-67 composite are enhanced to removal regarding organophosphorus pesticides [73]. Fenarimol adsorption by Fe_2O_3/attapulgite nanocomposite are applied for adsorption efficiency to improve; mechanism between such minerals as polar interactions either hydrogen or $\pi-\pi$ bonding on outer face of adsorbent surfaces, due to little dimension ($0.89 \times 0.97 \times 1.01$ nm) may restricted in inner surface adsorption [74]. Triazines is a class of nitrogen-containing heterocyclic pesticide, which was removed by nanocomposite of Fe_3O_4@graphene nanocomposite, mechanism by FTIR, and clearly showed stretching and bending corresponding as N—H, C—H, and C—N of aliphatic and aromatic groups. Interaction between triazines and Fe_3O_4@graphene of electrostatic by ($\equiv N^+ -$) groups of pesticide and oxygen containing of Fe_3O_4@graphene. Hydrophobic nature of Fe_3O_4@graphene could strongly interacted with the heterocyclic pesticides molecule by $\pi - \pi$ interactions [22]. Composite of nanoparticles with activated char/activated carbon/cellulose-based bionanocomposite used more applicability, due to surface integration of composite. Zinc oxide nanoparticles impregnated with pea peels to enhance the chlorpyrifos adsorption leads to the surface area (m^2/g) as 305.50 and the adsorption capacity as 47.846 mg/g [75].

4.4 Nanoclay

Nanoclay is an adsorbent that contains clay minerals. It can produce a variety of minerals, also constructs with geometrical shape as tetrahedral and octahedral sheets. The physicochemical arranged to configure according to geometrical by electrical force, could be constructed the structure. The recognizing minerals produced the polymetric construct as $O^{2+,}$ Si^{4+}, Al^{3+}, Fe^{2+}, Mg^{2+}, Ca^{2+}, Na^{2+}, and K^+, the main family of phyllosilicates which containing the silicon layer with other metallic composition. It was applied to pesticides adsorption due to low production of cost with nontoxic, regarding silicon clay majorly applied in adsorption such as bentonite, kaolinite, illite, sepiolite, montmorillonite, and pyrophyllite. The nanoclays have more surface areas and well interacted with ion-exchange by available minerals during adsorption. The nanoclays need modification over nanocomposite materials. The greatest potentiality is seen using polymer nanocomposite layer silicate, i.e., poly(methylmethacrylate) activation montmorillonite was applied for organochlorine pesticides, surface are improved modified montmorillonite (18.7 m^2/g) from polymererztion montmorillonite (246.80 m^2/g) [76]. Improvement efficiency onto surfactant modified montmorillonite applied in diazinon insecticide, adsorption capacity achieved at 1428.5 mg/g [77]. Now, current research could be emphasized with nanoimpregnated and modified with clay minerals due to enhancing the efficiency of adsorption and the available minerals not sufficient to the pesticides adsorbed.

5. Nanomaterials application toward pesticides adsorption and modeling

The adsorption study referred to the batch and the column methods, which was based on surface phenomena according to surface interaction between adsorbate and adsorbents. It has to influence the factors such as pH, contact time, dose, temperature, particle motion, and adsorbate amount [19,78]. The pH played a vital role for adsorption, which produced the adsorbent surface where interacted either acidic or alkaline region of pH. Thus, ZPC takes a role to describe the adsorbate interaction through deprotonated or protonate. The contact time and temperature are described to kinetic and isotherm modeling by adsorbate initial concentration. The adsorbent dose to help adsorption, dose increases the adsorption at a certain level. When the adsorption reaches the equilibrium condition, dose and adsorbate cannot interact due to surface availability decrease [79]. The adsorption capacity increased when the surface area enhanced, nanoparticle become more potential to improve the surface properties due to size, intermolecular forces and functionalized substances, i.e., amino functionalized silicon nanoparticle [13]. The equilibrium of the adsorption has to be described with isotherm and kinetic model regarding with adsorption capacity (q_e, mg/g), calculated by the following equation:

$$q_e = \frac{V(C_0 - C_e)}{m} \qquad (4.1)$$

Where V, m, C_0, and C_e are adsorbate volume, adsorbent amounts, initial, and final strength of adsorbate. Isotherm models have to be demonstrated to adsorb amount per unit weight of adsorbate, according to bulk adsorbate concentration at various temperatures. Here review reveals with some pesticide with isotherms and kinetics model regarding nanoparticle (Table 4.1). Whereas applied different type of pesticides with various types of nanomaterials, the pseudosecond-order model was fitted in maximum kinetic model.

Isotherm could assume the following representing model as Langmuir and Freundlich, Temkin, Dubinin—Radushkevich [80,81]. It not only perceives linear concepts but also interacts with nonlinear phenomena. Model efficacy has to be estimated by correlation coefficient (R^2). However, it is essential to describe the adsorption pattern, it could vary into adsorbate to adsorbate or adsorbent to adsorbent. Langmuir isotherm have to capable describe the monolayer adsorption regarding homogenous surface phenomena, it could be extracted of saturation coverage according to occupancy adsorbate sites. The linear (Eq. 4.2) and nonlinear model (Eq. 4.3) given below shows C_e is the initial strength of equilibrium, and Q_{max} and K_L are related to monolayer capacity of empirical and affinity of adsorbent as Langmuir constants, respectively.

$$\frac{C_e}{q_e} = \frac{1}{Q_{max}K_L} + \left(\frac{1}{Q_{max}}\right)C_e \qquad (4.2)$$

TABLE 4.1 Descriptive study of pesticides adsorption with different nanoadsorbents.

Groups of pesticides	Target pesticides	Adsorbent	Description	Adsorbent capacity	References
Neonicotinoid	Thiacloprid	$Fe_4O_3-GO-\beta-CD$	Model fitted Langmuir monolayer, Freundlich and pseudo-second-order kinetics model	2.88 mg/g	[99]
	Thiamethoxam	$Fe_4O_3-GO-\beta-CD$	Model fitted Langmuir monolayer, Freundlich and pseudo-second-order kinetics model	2.88 mg/g	[99]
	Acetamiprid	$Fe_4O_3-GO-\beta-CD$	Model fitted Langmuir monolayer, Freundlich and pseudo-second-order kinetics model	3.11 mg/g	[99]
Carbomate	Carbaryl	$\beta-CD@POSS@Fe_3O_4$	Mechanism at van der Waals; hydrogen bond; hydrophobic interactions Maximum adsorbed at pH 3.0 to 6.0 Initial concentration 5.0–60.0 mg/L Contact time at 25 min	27.22 mg/g	[100]
	Carbofuran	$\beta-CD@POSS@Fe_3O_4$	Contact time = 20 min pH range 3.0–6.0 Initial concentration (5.0–60.0 μg/mL)	19.53 mg/g	[100]

Continued

TABLE 4.1 Descriptive study of pesticides adsorption with different nanoadsorbents.—cont'd

Groups of pesticides	Target pesticides	Adsorbent	Description	Adsorbent capacity	References
Organo-Cholorine	Dichlorodiphenythreechloroethen (DDT)	Shell-tunable mesoporous Fe_3O_4@HMS	Fast adsorption rate; magnetic property with mesoporous material; convenient and highly efficient	7.5 mg/g	[101]
	Dicamba	Biopolymer based clay composite	Langmuir model and pseudo-second-order kinetic model showed a higher correlation; pH = 2.0	251.9 mg/g	[102]
	Diuron	α-γ-Fe_2O_3-Sh.	pH = 6.2 Pseudo-second-order kinetics model and the Langmuir model fitted; T = 45°C. Endothermic adsorption	30.290 mg/g	[103]
	Fenarimol	Fe_2O_3–palygorskite nanoparticles	Freundlich model fitted; heterogeneous surface; Pseudo-second order	344 mg/g	[74]

Pesticide	Nanomaterial	Details	Capacity	Ref.
4-Chloro-2-methylphenoxyacetic acid	Nano Cu–Fe–NO$_3$ with layered double hydroxide	Langmuir and Freundlich, pseudo-second-order model fitted; initial concentration = 50 mg/L; T = 298 K, pH = 6.0	1099 mg/g	[104]
Chlorpyrifos	Zinc Oxide@Pea Peels	Temkin isotherm (R^2 = 0.99)and pseudo-second-order model best fitted; pH = 2.0	47.846 mg/g	[75]
Chlorpyrifos	Fe$_3$O$_4$@SiO$_2$@GO-PEA	Pseudo second order kinetics fitted by non-linear fitted; Langmuir, Freundlich, Temkin, Dubinin–Radushkevich, Sips, and Redlich and Peterson models fitted Dose = 1.5 g; pH = 7.0; T = 25°C	25.6 mg/g	[19]
Permethrin	Chitosan@zinc oxide nanocomposite	Adsorbent dose = 0.5 g; room temperature; pH = 7, remove efficiency = 99%; initial concentration = 25 mL volume, 0.1 mg/L	—	[23]

Continued

TABLE 4.1 Descriptive study of pesticides adsorption with different nanoadsorbents.—cont'd

Groups of pesticides	Target pesticides	Adsorbent	Description	Adsorbent capacity	References
Organo Phosphate	Glyphosate	Biochar@ Fe(0)	Pseudo-second order and Langmuir best fitted; pH = 4; T = 298 K, Contact time = 1600 min, initial concentration 0.5 −100 mg/L, solid/ solution 0.015 g/25 mL	80 mg/g	[32]
	Dichlorvos	Zinc-silver bimetallic nanoparticles embedded in montmorillonite-biopolymer nanobiocomposite	Langmuir; Freundlich; and Dubinin −Radushkevich Pseudo-first order; pseudo-second order; intraparticle diffusion Initial concentration = 60 mg/L pH = 9.0	250 mg/g	[105]
	Malathion	Fe$_3$O$_4$@SiO$_2$@GO-PEA	Pseudo second order kinetics and Langmuir, Freundlich, Temkin, Dubinin−Radushkevich, Sips, and Redlich and Peterson isotherm; Dose = 1.5 g; pH = 7.0; T = 25°C	61.9 mg/g	[19]

	Parathion	Fe$_3$O$_4$@SiO$_2$@GO-PEA	Pseudo second order kinetics and Langmuir, Freundlich, Temkin, Dubinin–Radushkevich, Sips, and Redlich and Peterson isotherm; Dose = 1.5 g; pH = 7.0; T = 25°C	135 mg/g	[19]
	Diazinon	Surfactant-modified montmorillonites	Pseudo-second order kinetic and Langmuir isotherm models fitted; pH = 3	1428.5 mg/g	[77]
	Dimethoate	Gold Nanospheres	Langmuir model; initial concentration = 100 mg/L; T = 25° C	456 mg/g	[15]
	Dimethoate	Gold nanorods	Langmuir model; initial concentration = 100 mg/L; T = 25° C	57.1 mg/g	[15]
Alycyclic	Atrazine	MWCNTs	Polanyi–Manes isotherm model; spontaneous and exothermic	100.43 mg/g	[57]
	Triazines	Cellulose/Graphene Composite	Langmuir and second order kinetic pH = 9, temp = 308 K	9.5877 mg/g	[106]
	Ametryn	Functionalized iron nano particles	Pseudo-second-order and liquid film diffusion mechanisms; Langmuir and D-R isotherm fitted; contact time = 30 min; dose = 2.5 g/L; pH = 7.0 and T = 20°C	14.29 µg/g	[107]

$$q_e = \frac{Q_{max}K_L C_e}{1 + K_L C_e} \tag{4.3}$$

$$R_L = 1/(1 + K_L C_0) \tag{4.4}$$

Langmuir model can be verified of dimensionless constants referring to separating factors, R_L, which expresses by Eq. (4.4). However, R_L indicated that adsorption to be unfavorable ($R_L > 1$), linear ($R_L = 1$), irreversible ($R_L = 0$) and favorable ($0 < R_L < 1$). Freundlich isotherm have predicted to multilayer adsorption with heterogeneous surface. Nonlinear and linear models are describing of energy distribution in terms of exponential function ($1/n$) varied between adsorptive sites, expressed as

$$q_e = K_F C_e^{\frac{1}{n}} \tag{4.5}$$

$$\log q_e = \log K_F + \frac{\log C_e}{n} \tag{4.6}$$

Where, Freundlich constants, K_F (Freundlich constant, L/mg) and n are calculated from intercept and the slope, n indicates that model as favoring to multilayer process, which is generally termed as 1 to 10. If $1/n$ is below the one, run monolayer to be normal adsorption phenomena [22,73]. Temkin isotherm applied in adsorption, according to estimation of factors affecting binding energy calculate of the heat adsorption mechanism [81]. The heat increments to layer of adsorption decreases linearity in terms of surface interaction of adsorbate and adsorbent, which is expressed below as linear and nonlinear term. As parameters are calculated by Eqs. (4.7) and (4.8), which are heat of sorption constant ($RT/b_T = B$ J/mol), and equilibrium binding ($A_T = $ L/g)

$$q_e = (RT / b_T)\ln(A_T C_e) \tag{4.7}$$

$$q_e = \frac{RT}{b_T}(\ln A_T + \ln C_e) \tag{4.8}$$

D–R isotherm is an empirical isotherm model, which also applied for known about Gaussian energy distribution as mechanism with heterogeneous surfaces [82]. The linear and nonlinear based mechanism postulated, adsorption is that pore filling instead of layer-by-layer surface coverage, that encourage of van der Waals forces according to the physisorption mechanism.

$$\ln q_e = \ln q_{max} - \beta E^2 \tag{4.9}$$

$$q_e = q_{max}\exp(-K_{DR}\varepsilon^2) \tag{4.10}$$

$$\varepsilon = RT\ln(1 + 1 / C_e)$$

Whereas, Polanyi potential (ε), D–R constant (β), the mean free energy of adsorption ($K_{DR} = $ mol^2/kJ^2) with internal energy ($E = 1 / \sqrt{2B}$) are used. The

Hill isotherm is one of the model define by three parameters to described the binding on homogeneous substrate [81]. It assumed that the adsorption affects by a binding sites, produces multiadsorption at the same adsorbate molecules, which are expressed as

$$q_e = \frac{q_H C_e^{nH}}{K_h + C_e^{nH}} \tag{4.11}$$

$$\log \frac{q_e}{q_H - q_e} = n_H \log(C_e) - \log(K_h) \tag{4.12}$$

Where, q_H, n_H, and K_h denotes as equilibrium capacity from Hill model, number of binding sites of multiadsorption, and Hill constant of energy potential, respectively [81,83]. Isotherm for pesticides with nanoparticle has more effective area in terms of adsorption applicability. Previously applied pesticides adsorption by different precursors, such as glyphosate onto resin D301 [84], and fenuron onto MWCNTs, mechanism applied using Temkin, Freundlich, Langmuir, and DR isotherms [18]. The organophosphorus pesticides was applied by $Fe_3O_4@SiO_2@GO$ nanoparticles, regarding linear fitted as well with D−R isotherm, nonlinear in terms of parathion and chlorpyrifos showed highest ($R^2 = 0.9$), but the marathon did not take place. The Temkin model fitting as well of organophosphorus, energy-derived physical adsorption due to adsorption energy less 20 kJ/mol, herein parathion (1.352 kJ/mol), malathion (1.562 kJ/mol), and chlorpyrifos (2.139 kJ/mol), interaction pattern π-π and hydrogen bonding [19]. Model fitting are very essential, some case obtained interactive in terms of species composition, the kinetics can have an effect over a certain period of time to be uptake. The kinetics are generally applied in pesticides adsorption for rate control mechanism, which are pseudofirst order (pFO), pseudosecond order (pS), and intraparticle diffusion mechanism (iD) (Eqs. 4.13−4.15). The adsorption mechanism assumed that the rate controlled by particle diffusion (external rate > internal transport), film diffusion (external transport < internal transport), and liquid film barrier (external rate > internal transport) [13]. The pFO model as known as Lagergren model, assume rate involved to uptake at time consume. The pS model are derived the adsorption rate proportionate with adsorbent and adsorbate, understanding the driving force with chemisorption mechanism. The iD mechanism are widely used to understanding the rate limiting step, involve film diffusion, surface diffusion and pore diffusion mechanism [85]. The corresponding model showed equation simultaneously, q_t as adsorption capacity at equilibrium time, t as time, C as iD capacity, q_e as equilibrium adsorption capacity regarding kinetic model and k_1, k_2 and k_{dif} are pFO, pS, and iD, respectively.

$$\log(q_e - q_t) = \log q_e - \frac{k_1}{2.303} t \tag{4.13}$$

$$\frac{t}{q_t} = \frac{1}{k_2 q_e^2} + \frac{t}{q_e} \tag{4.14}$$

$$q_t = k_{dif} t^{1/2} + C \tag{4.15}$$

These kinetics are applied in organophosphorus pesticide adsorption by amino functionalized silicon, showed highest fitted with pS kinetics, achieving at 30−40 mg/L glyphosate concentrations, where deals with film diffusion with multilayer phenomena [13].

6. Statistical optimization-related synthesis of precursors and adsorption

Statistical optimization as stable tools for adsorption study and material synthesis. Experimental design produces, the efficient related with experimental subset factors, are traditional theoretical method of changing only one variable [86]. All applicable techniques are population based, initial term into randomly assigned as an input parameters within specified ranges. If the ranges are longer to difficulty recoding of models, then sophisticated as well as fitness-based models demand more for further optimization of the objective functionality. Applicability of optimization based on adsorption in some recent paper, maximum showed on RSM and artificial intelligence (AI) [30,86]. The RSM-based optimization are full and fractional factorial, central composite, Box−Behnken, Doehlert and hybrids. On the applicability of AI-based techniques are generally established, such as genetic algorithm (GA), particle swarm optimization (PSO), ant colony optimization (ACO), BAT algorithm (BA), elephant herding optimization (EHO), and hybrids based on artificial neural network [87]. All strategies have some drawbacks of optimization, and it has some advantages that are reviewed (Table 4.2).

6.1 RSM-based optimization

RSM as one of the predicted best unit to adsorption with variables. This is collected of mathematical and statistical ways based on polynomial fitting onto experimental data, which derive the behaviors of variables tendency to statistical making previsions. The most frequently applied design to achieving goals by following methods as full and fractional factorial, central composite (CCD), Box−Behnken (BBD), Doehlert (DH), and hybrids [80,86]. The so-called designs can be applied by understanding simple response surfaces of linear forms for all investigators, but are usually used to determine which experimental components are most important for the investigation and which factors do not significantly influence experimental results. It is elaborated here as multivariate investigation in terms of linear model at two-factor case

$$\hat{y} = b_0 + b_1 x_1 + b_2 x_2 + b_{12} x_1 x_2 \tag{4.16}$$

TABLE 4.2 Descriptive study of statistical design regarding experimental number.

Statistical design	Description	Limitation	Experimental number	References
Randomized Complete Block Design (RCB)	Application based on blocking analysis (reduces variation) by one or more specific factors. No restriction on the treatment numbers or replicates, missing plots are easily estimated.	Error for large number of treatments because the blocks become too large and the whole block contains considerable variability.	$N = \Pi_{i=1}^{k} L_i; k = $ factors $L = $ level	[108]
Latin Square Experimental Design (LSE)	Requires less samples; each block in single experiment; requires some conditions (i.g., pH, temperature) that helps to model estimation.	Equal number of levels and blocking of factors, no interaction each variables.	$N = L^2$ $L = $ level	[109,110]
Full Factorial Design (FF)	Technique applied for meta models; constructed by two or three level basis. All possible combination are chosen from input factors of the sample.	Difficulty of experimenting with more than two factors, or many levels and needs to be planned carefully, as any one of the level defects or general operation will jeopardize the bulk of the work.	$N = L^k$	[111]
Fractional Factorial Design (FrF)	This can reduce total experiments/simulations to half, one-third, or one-fourth of the sample size; select the samples are balanced combination.	Error increasing when sample size increases.	$N = L^{k-p}$	[112]

Continued

TABLE 4.2 Descriptive study of statistical design regarding experimental number.—cont'd

Statistical design	Description	Limitation	Experimental number	References
Star Experimental Design (SE)	It subjected two axial points on the axis of each factor, aspect of physical distance between center and center each point from all the factors which calculative mean value.	Physical distance applied only from the center point, multivariate is not generated.	$N = 2k + 1$	[113]
Central Composite Design (CCD)	It combination between two-level of FF and SE, the design is able to estimate the curvature of the space. Predicted in quadric level of three factorial design.	Limitation to describe according to 3^k, 4^k ... n^k factorial. This design considers experiments with all factors at a negative or positive level, which is a drawback because runs conducted in extreme situations can induce unsatisfactory results.	$N = k^2 + 2k + 1$	[114,115]
Box–Behnken Experimental Design (BBD)	It could help on incomplete three-level of FF, reduce the sample size; predicted to degree of coefficients by least squares polynomial, in nonlinear relation.	In this design has not point at vertex of the cubic area created by the upper and lower limits for each variable, which can lead to unsatisfactory results.	$N = 2k(k-1) + C$ $C = constant$	[14,114]
Doehlert matrix	Filling of identical spaces; likelihood of sorting; helps to reuse of experiments when the boundaries were not justified at the beginning, it can be provided that the old and the new considered adjacent boundaries.	Level of k equal to 3, the factors can be studied with three, five and seven levels, it not estimated at fractional and negative or positive level.	$N = k^2 + k + C$	[114]

Design	Description	Notes	N	Ref.
Plackett–Burman Experimental Design (PBE)	Look for experimental designs to investigate the dependence of some measured quantity on several individual variables, level at 3, 4, 5, or 7 as integer, applied only when the power of N as multiple 4 not 2.	Limited number of experiments the idea was to find smaller designs; Plaquecat - Burman designs with original effects, deals designs do not allow to recognized between certain effects and specific interactions, it's very confusing.	$N = k + 1$	[116]
Taguchi Experimental Design (TED)	Minimization of sensitivity of the problem by controllable and noise factors; when uncontrollable factors, subscripts and refer between inner and outer factors.	The main disadvantage of the Taguchi method is that the results obtained are only relative and do not indicate precisely which parameter has the greatest impact on the characteristic quality of performance.	$N = L^{k_1 + k_0}$	[117,118]
Random Experimental Design (RE)	Sample are place in independently and techniques are relied to fill the space of uniform design, factorial level do not need to be discretized.	Relatively low accuracy due to lack of limitations which permits environmental variant to enter experimental error	$N = 1$	[119]
Latin Hypercube Design (LHD)	Helps to unfound result of Monte Carlo simulation range at 0 to 1 at first position; after that each interval randomly select.	Large magnitude is not detected, magnitude is least.	$N = 1/k$	[120]

In the context of optimization, the theoretical model aspect of controllable variable regarding output response, either nonlinear or linear relation, which made complexity to the experimental data fitting as prediction. RSM is a sophisticated analysis of statistical optimization that can detect predictions in 3^K factorial based on experimental data. It is significant to fit this mathematical model for the approximate relationship between feedback and individual variables, and resulting to the optimal conditions for these variables as a result of the maximum response [88]. Since the theorem is commonly applied to RSM, the first order and second order models are followed

$$Y = \beta_0 \sum_{i=1}^{k} \beta_i X_i + \varepsilon \qquad (4.17)$$

$$Y = \beta_0 \sum_{i=1}^{k} \beta_i X_i + \sum_{i=1}^{k} \beta_{ii} X_i^2 + \sum_{i=1}^{k} \sum_{i \neq j=1}^{k} \beta_{ij} X_i X_j + \varepsilon \qquad (4.18)$$

Where Y is output of the prediction as response, β_0 as constant of the model, β_i as linear effect of X_i, β_{ii} related with quadratic effect of X_{ii}, β_{ij} as the parameter interaction of $X_i \& X_j$, and ε subjected output residual. From the first-order model has represented while inadequate strong relationship between independent variables, the second-order model is benefited in term of resulting as complex interaction, high constructed, flexible, and varied to identify the best point [89]. Moreover, the most frequently applied with polynomial model of the RSM through the second-order model. It is suitable to the design of a second-order model to factorial influence of 3^k, BBD, CCD, and DH [90]. As the literature have applied of those model for nanoparticle synthesis and adsorption study, which are particularly benefits respect of symmetrical differ as experimental, variable number, selecting block, and runs [14,90]. The 3^k factorial design most popularly used for combinations of the level as variable with three level each, low, medium, and high [83]. This model requires large number of variable factors while experimental runs, and loss of efficiency at quadratic functions. If the quadratic functions trial since more efficiency to needs the number of factors less than five. Thus, the 3^k factorial design have optimized using two or three variability, the efficiency varied from lower to higher ranges. Recent application of this model for 3^4 variability of adsorption which are temperature, pH, adsorbents, and activating agents, with significant merged from P-value which less than 0.05, and correlation coefficient subjected below 0.9 [78], because factors intake at four. The Doehlert has selected for heterogeneous levels of variables, where beneficial for variables are restricted such as cost and/or material constraints, allows when any variables at the critical or secondary level [91,92]. This model has unique properties as dealing intervals which intake distribution as uniformed. This design has hexagonal architecture that represents two variable at central point with surrounded six interacting vectors. If three variable, the design converted to octahedron, it has the ability to change under different experimental metrics. Although its metrics cannot be orthogonal or rotated, it does offer some advantages, such as the need for a smaller number of experimental points for its utilization and high efficiency. The Doehlert modeling of chitosan nanoparticles production regarding factors as stirring rate, temperature, acetic acid concentration, chitosan, and tripolyphosphate volume ratio and pH adjustment with response as size of the nanoparticles [93], whereas the model was significantly fitted ($P < .001$), and highly predicted efficiency regarding experimental ($R^2_{adjsted} = 0.9920$, $R^2_{predicted} = 0.9720$). The BBD provide at three level design (-1, 0, $+1$) for several variable regarding equal spacing. Also, this design has no dots at the vertices of the cube formed by the upper and lower boundaries for several variable. Therefore, BBD can apply to find the best experimental conditions, resulting in optimal efficiency of various processes. The BBD applied for carbaryl insecticide adsorption using eggshell powder with operating factors such as initial carbaryl concentration, pH, adsorbent dose, and contact time; results were well described with a quadric model ($R^2 = 0.9985$ and P-value$<.001$) [80]. The central composite design is the most widely used design fitting over the second-order models, has received a great deal of attention for theoretical development of its features, such as its adsorption study. It consists of two level of full factorial design aspect, at least one point in the center regarding the experimental region, and restricted minimum three variables. Optimization of the adsorption of malathion using MWCNTs performed of CCD modeling, variable as independent as adsorbent doses, malathion concentration, contact time and pH with removal percentage as responses, as well fitted of second order CCD model at 26 experimental trials [6].

6.2 AI-based optimization

AI-based optimization has no limit in this area for development of new model, here justifying problem solved very concisely [94]. In the previous literature commonly used AI-based models, program-based algorithms were specified as hybrids with GA, PSO, ACO, random forest, BA, and EHO, artificial neural networks (ANN) [30], i.e., PSO-ANN hybrid, and ACO-ANN. Those models are the best solution inspired by the behavior of natural entities, have the ability to provide optimal of adsorption or nanoparticle preparation. Previous literature has been applied to some ANN, quantum-layer perceptron, and PSO, but some lack of existing research according to AI-based fields for adsorption and synthesized base [95]. The applicability of ANN-based optimization is one of the generic algorithms that face of model interaction arias parameter variable and behaviors of probability. A feed-forward ANN has the ability to give adequacy with an arbitrary producing of a sufficient number of neuron by one hidden layer [87]. The number of hidden layers can provide the movement factors regarding testing values, the results adjusted by require adequacy. It is only applied when inputs parameter are restricted with nonlinear phenomena. Training of the network can process with suitable term of weights and bias regarding machine learning methods as input-output data orientation. Normalization of inputs variable and optimized output date are performed according to the ranges between 0 and 1, corresponding training network [87].

$$y = \frac{x_i - x_{\min}}{x_{\max} - x_{\min}} \tag{4.19}$$

Where y is normalized testing value, $x_{\max}$ and the $x_{\min}$ as the maximum and minimum of x_i, respectively. Application PSO is a heuristic based on swarm intelligence, and develop from swarm intelligence regarding bird and fish flock movement behavior. It is mostly application on adsorption in basis of swarm algorithm. It consists as n particle and D-dimensional space at the position of each particles. The particles movement such condition during optimized as parameters keeps inertia, change of optimal position according to parametric condition, and optimized in maximum output condition swarm position [96]. The position of the individual particle aspect of the swarm interacted through the optimist position during training movement. Also, it has individual benefits such as no chances of overlapping and research speeds, which are very fast with easily optimized nonlinear prediction. Moving the particles based on the previous position randomly produces inertia weights, such as position and motion as repetition incorporating all elements [95,97]. The weight of inertia follows in Eq. (4.20)

$$W_i = W_{\min} + (W_{\max} - W_{\min}) \times \frac{iter_{\max} - iter}{iter_{\max}} \tag{4.20}$$

Whereas, $W_{\min}$ and $W_{\max}$ are weight of inertia vectors in terms of minimum and maximum respectively on $iter$ and $iter_{\max}$ number of iterations. The back-propagation could be integrated in hybrid with particle swarm to achieve maximum goal, rectify regarding better estimation of weight by ANN-PSO [97]. The efficiency increment of experimental matrix, date set perform 50 number experimental with maximum achieved in meta-heuristic optimization by ANN embedded PSO, well depicted as testing with R^2 equal to 0.9106 and 0.9279, respectively. It has better performing as compare with RSM-CCD optimization. Increasing the efficiency of the experimental matrix, the data set ANN embedded PSO model performed the experiment of 50 with the highest score in meta-heuristic optimization ($R^2 = 0.9106$ and 0.9279, respectively, as training and testing). It has performed even better compared to RSM-CCD optimization [98]. Although this review is more significant in terms of adsorption, removal can be a beneficial aspect, and as reported in Table 4.3, it may help ANN, RSM-based optimization during experimental runs.

TABLE 4.3 Application of statistical modeling regarding with study of pesticides adsorption and nanoparticles synthesis.

Purposes	Description	Optimization	References
Adsorption study malathion by MWCNTs	Optimized by RSM fitted in linear ($P > .001$), ($R^2 = 0.709$ and 0.612 as adjusted and predicted); Experimental trial $= 26$. Three-factor RSM methodology remove almost 100% of malathion; optimized conditions of initial concentration $= 6$ mg/L; initial MWCNTs concentration $= 0.5$ g/L; contact time $= 30$ min	Central Composite Design	[58]
Synthesis gentamicin-loaded calcium carbonate nanoparticles	Three level of factorial design (3 factor); $CaCl_2$: Na_2CO_3 (X_1); concentration of drug (X_2); speed of homogenization (X_3), optimized X_1–X_3 levels, prediction for particle size of 80.23 nm with 30.80% efficiency.	Box–Behnken design	[121]
Silver nanoparticles synthesized by bio fabricated method	Process modeling of ANN weights, biases, neuron numbers and activation functions (one hidden layer through PSO optimization): Ag NPs size prediction; variables as $AgNO_3$:Opium; Feed rate (mL/min); Agitation speed (rpm) and pH/temperature; Function as Tansig-Purelin; neuron $= 25$–1; $R^2 = 0.9972$	Hybrid artificial neural network-particle swarm optimization algorithm	[122]
Silver nanoparticles prepared by montmorillonite interlayer space	Hidden layers $= 2$ and3 with the different candidate of neurons in each layer is examined (4 neurons and 1 hidden layer produced); model fitted, $R^2 = 0.985$ for ANN modeling of nanocomposites behaviors by an-sigmoid function to transfer and linear	Feed-forward ANN network	[123]

Continued

TABLE 4.3 Application of statistical modeling regarding with study of pesticides adsorption and nanoparticles synthesis.—cont'd

Purposes	Description	Optimization	References
Glyphosate adsorption by amino-functionalized silicon	Variable for initial concentration (5–50 mg/L); adsorption dose (0.01–0.5) g; pH (2–12); contact time (5–120 min); temperature (293–373 K); 22 hidden layers for 100 weights of neuron; optimized concentration = 39.99 mg/L; pH = 12.66; dose = 0.299 g; contact time = 60 min; temperature = 100.00°C; removal = 98.99%; desired = 0.677.	ANN-multilayer perceptron	[13]

7. Conclusion and future perspectives

Over the past few decades, agricultural pollutants have increased, causing serious health risks and environmental threats. This chapter reviewed pesticide adsorption based on nanoparticles. Current demands are toward improving effective production of super adsorbents through nanotechnology because of its high adsorptive capacity. As a future perspective, the super absorbent would not only induce a single pesticide for adsorption, but it will be effective in a variety of pesticides. Because of that, agricultural releases not only insecticides, but it also discharges different types of pesticides; reusable adsorbents will be more efficient and sustainable. As compared to previous reports for adsorption study, statistical optimization did not apply to higher efficacy nanomaterial, lacking this part, basically, adsorption induces more efficiency when experimental data increases, which could help to optimize and predict actual adsorption value. Most of the adsorption studies were used as RSM based optimizations, few AI-based, but the current demand for nonlinear modeling, which is more accurately explained by AI-based than RSM based optimization. Non-linear based optimization depends on the error function, the error and adequacy are inversely proportional, various AI optimization models could increase the adequacy by self-regenerating and reducing errors. Statistical optimization is essential for short-term experiments, with maximum adsorption achieved by minimum experimental runs. Mechanisms of adsorption are assumed by molecular binding energy which occurred from XPS, XPS analysis has more essential to know the proper adsorption. Nanoparticle-based fertilizers have won ability to recovery to plant growth, as more essential to synthesis nanofertilizer based adsorbent, i.e., zinc-iron composites and hydroxyapatite. Nanoparticle-based materials primarily used as adsorbents, when adsorbents efficiency decreases then disposed of dumping area; further could apply the adsorbents in the agricultural field as nano fertilizers.

Nanotechnology is subjected to superior performance of adsorption. All adsorptive models like kinetics and isotherm are essential and are used for equilibrium study. This model linear and nonlinear are most effective for drive adsorptive mechanism, i.e., multilayer, monolayer, chemisorption, physical adsorption, binding energy, free energy, and so forth. Graphene-based nano-adsorbents are more effective with alicyclic and aromatic pesticides due to their hydrophobic nature. Metal nanoparticle have more demands for pesticides adsorption, but there is a lack of research according to uranium-based nanoparticles, silver nanoparticles, copper nanoparticles, iron-oxide nano-particles, manganese oxide nanoparticles with biochar, activated carbon, and row wastage material impregnation. Statistical optimization is best on AI with hybrid types, and RSM-based optimization is more efficient of Doehlert matrix than BBD and CCD. This review article describes pesticide adsorption based on nanotechnology with statistical optimization. Finally, this study is concluded that the nanomaterials have the ability to adsorb pesticides when surface functional groups increase to enhance adsorptive properties. The statistical optimization of ANN is more suitable than RSM.

Appendix A. Supplementary data

Supplementary data to this article can be found online at https://doi.org/10.1016/B978-0-12-824038-0.00005-5.

References

[1] Kanianska R. Agriculture and its impact on land-use, environment, and ecosystem services. Lands Ecol − Influ Land Use Anthropog Impacts Lands Creat 2016. https://doi.org/10.5772/63719.

[2] Abhilash PC, Singh N. Pesticide use and application: an Indian scenario. J Hazard Mater 2009;165:1−12. https://doi.org/10.1016/j.jhazmat.2008.10.061.

[3] S.O.I. Ph.D. Effects of agricultural pesticides on humans, animals, and higher plants in developing countries. Arch Environ Health 1991;46:218−24. https://doi.org/10.1080/00039896.1991.9937452.

[4] Sharma A, Kumar V, Shahzad B, Tanveer M, Sidhu GPS, Handa N, Kohli SK, Yadav P, Bali AS, Parihar RD, Dar OI, Singh K, Jasrotia S, Bakshi P, Ramakrishnan M, Kumar S, Bhardwaj R, Thukral AK. Worldwide pesticide usage and its impacts on ecosystem. SN Appl Sci 2019;1:1446. https://doi.org/10.1007/s42452-019-1485-1.

[5] Khatri N, Tyagi S. Influences of natural and anthropogenic factors on surface and groundwater quality in rural and urban areas. Front Life Sci 2015;8:23−39. https://doi.org/10.1080/21553769.2014.933716.

[6] García-Galán MJ, Monllor-Alcaraz LS, Postigo C, Uggetti E, López de Alda M, Díez-Montero R, García J. Microalgae-based bioremediation of water contaminated by pesticides in peri-urban agricultural areas. Environ Pollut 2020;265:114579. https://doi.org/10.1016/j.envpol.2020.114579.

[7] Daniel O, Meier MS, Schlatter J, Frischknecht P. Selected phenolic compounds in cultivated plants: ecologic functions, health implications, and modulation by pesticides. Environ Health Perspect 1999;107:109−14. https://doi.org/10.1289/ehp.99107s1109.

[8] Jayaraj R, Megha P, Sreedev P. Organochlorine pesticides, their toxic effects on living organisms and their fate in the environment. Interdiscipl Toxicol 2016;9:90−100. https://doi.org/10.1515/intox-2016-0012.

[9] Mamta RJR, Wani KA. Status of organochlorine and organophosphorus pesticides in wetlands and its impact on aquatic organisms. Environ Claims J 2019;31:44−78. https://doi.org/10.1080/10406026.2018.1519315.

[10] De A, Bose R, Kumar A, Mozumdar S. Targeted delivery of pesticides using biodegradable polymeric nanoparticles. Springer India; 2014. https://doi.org/10.1007/978-81-322-1689-6.

[11] Bai SH, Ogbourne SM. Glyphosate: environmental contamination, toxicity and potential risks to human health via food contamination. Environ Sci Pollut Res Int 2016;23:18988−9001. https://doi.org/10.1007/s11356-016-7425-3.

[12] Kostopoulou S, Ntatsi G, Arapis G, Aliferis KA. Assessment of the effects of metribuzin, glyphosate, and their mixtures on the metabolism of the model plant *Lemna minor* L. applying metabolomics. Chemosphere 2020;239:124582. https://doi.org/10.1016/j.chemosphere.2019.124582.

[13] Sen K, Mondal NK. Facile fabrication of amino-functionalized silicon flakes for removal of organophosphorus herbicide: in silico optimization. Water Conserv Sci Eng 2020;5:67−80. https://doi.org/10.1007/s41101-020-00085-7.

[14] Sen K, Datta JK, Mondal NK. Glyphosate adsorption by *Eucalyptus camaldulensis* bark-mediated char and optimization through response surface modeling. Appl Water Sci 2019;9:162. https://doi.org/10.1007/s13201-019-1036-3.

[15] Momić T, Pašti TL, Bogdanović U, Vodnik V, Mraković A, Rakočević Z, Pavlović VB, Vasić V. Adsorption of organophosphate pesticide dimethoate on gold nanospheres and nanorods. J Nanomater 2016. https://doi.org/10.1155/2016/8910271.

[16] Sen K, Mondal NK, Chattoraj S, Datta JK. Statistical optimization study of adsorption parameters for the removal of glyphosate on forest soil using the response surface methodology. Environ Earth Sci 2016;76:22. https://doi.org/10.1007/s12665-016-6333-7.

[17] Feng Z, Zhu Z, Sun T. Batch and fixed-bed column adsorption of tetrabromobisphenol A onto metal organic resin: equilibrium, kinetic and mechanism studies. New J Chem 2020;44:12771−8. https://doi.org/10.1039/D0NJ02389B.

[18] Ali I, Alharbi OML, ALOthman ZA, Al-Mohaimeed AM, Alwarthan A. Modeling of fenuron pesticide adsorption on CNTs for mechanistic insight and removal in water. Environ Res 2019;170:389−97. https://doi.org/10.1016/j.envres.2018.12.066.

[19] Wanjeri VWO, Sheppard CJ, Prinsloo ARE, Ngila JC, Ndungu PG. Isotherm and kinetic investigations on the adsorption of organophosphorus pesticides on graphene oxide based silica coated magnetic nanoparticles functionalized with 2-phenylethylamine. J Environ Chem Eng 2018;6:1333−46. https://doi.org/10.1016/j.jece.2018.01.064.

[20] Bayat M, Alighardashi A, Sadeghasadi A. Fixed-bed column and batch reactors performance in removal of diazinon pesticide from aqueous solutions by using walnut shell-modified activated carbon. Environ Technol Innov 2018;12:148−59. https://doi.org/10.1016/j.eti.2018.08.008.

[21] Awasthi A, Jadhao P, Kumari K. Clay nano-adsorbent: structures, applications and mechanism for water treatment. SN Appl Sci 2019;1:1076. https://doi.org/10.1007/s42452-019-0858-9.

[22] Boruah PK, Sharma B, Hussain N, Das MR. Magnetically recoverable Fe3O4/graphene nanocomposite towards efficient removal of triazine pesticides from aqueous solution: investigation of the adsorption phenomenon and specific ion effect. Chemosphere 2017;168:1058–67. https://doi.org/10.1016/j.chemosphere.2016.10.103.

[23] Moradi Dehaghi S, Rahmanifar B, Moradi AM, Azar PA. Removal of permethrin pesticide from water by chitosan–zinc oxide nanoparticles composite as an adsorbent. J Saudi Chem Soc 2014;18:348–55. https://doi.org/10.1016/j.jscs.2014.01.004.

[24] Ghaedi M, Khafri HZ, Asfaram A, Goudarzi A. Response surface methodology approach for optimization of adsorption of Janus Green B from aqueous solution onto ZnO/Zn(OH)$_2$-NP-AC: kinetic and isotherm study. Spectrochim Acta Mol Biomol Spectrosc 2016;152:233–40. https://doi.org/10.1016/j.saa.2015.06.128.

[25] Ealia SAM, Saravanakumar MP. A review on the classification, characterisation, synthesis of nanoparticles and their application. IOP Conf Ser: Mater Sci Eng 2017;263:032019. https://doi.org/10.1088/1757-899X/263/3/032019.

[26] Hassan M, Naidu R, Du J, Liu Y, Qi F. Critical review of magnetic biosorbents: their preparation, application, and regeneration for wastewater treatment. Sci Total Environ 2020;702:134893. https://doi.org/10.1016/j.scitotenv.2019.134893.

[27] Sadegh Mazloom M, Hemmati-Sarapardeh A, Husein MM, Shokrollahzadeh Behbahani H, Zendehboudi S. Application of nanoparticles for asphaltenes adsorption and oxidation: a critical review of challenges and recent progress. Fuel 2020;279:117763. https://doi.org/10.1016/j.fuel.2020.117763.

[28] Aguilar-Pérez KM, Avilés-Castrillo JI, Ruiz-Pulido G. Nano-sorbent materials for pharmaceutical-based wastewater effluents – an overview. Case Stud Chem Environ Eng 2020:100028. https://doi.org/10.1016/j.cscee.2020.100028.

[29] Mondal P, Anweshan A, Purkait MK. Green synthesis and environmental application of iron-based nanomaterials and nanocomposite: a review. Chemosphere 2020;259:127509. https://doi.org/10.1016/j.chemosphere.2020.127509.

[30] Khezri V, Yasari E, Panahi M, Khosravi A. Hybrid artificial neural network–genetic algorithm-based technique to optimize a steady-state gas-to-liquids plant. Ind Eng Chem Res 2020;59:8674–87. https://doi.org/10.1021/acs.iecr.9b06477.

[31] Madan SS, Wasewar KL, Ravi Kumar C. Adsorption kinetics, thermodynamics, and equilibrium of α-toluic acid onto calcium peroxide nanoparticles. Adv Powder Technol 2016;27:2112–20. https://doi.org/10.1016/j.apt.2016.07.024.

[32] Jiang X, Ouyang Z, Zhang Z, Yang C, Li X, Dang Z, Wu P. Mechanism of glyphosate removal by biochar supported nano-zero-valent iron in aqueous solutions. Colloid Surface Physicochem Eng Aspect 2018;547:64–72. https://doi.org/10.1016/j.colsurfa.2018.03.041.

[33] Eevers N, White JC, Vangronsveld J, Weyens N. Chapter seven – bio- and phytor-emediation of pesticide-contaminated environments: a review. In: Cuypers A, Vangronsveld J, editors. Advances in botanical research. Academic Press; 2017. p. 277–318. https://doi.org/10.1016/bs.abr.2017.01.001.

[34] Rajmohan KS, Chandrasekaran R, Varjani S. A review on occurrence of pesticides in environment and current technologies for their remediation and management. Indian J Microbiol 2020;60:125–38. https://doi.org/10.1007/s12088-019-00841-x.

[35] Zhang H, Yuan X, Xiong T, Wang H, Jiang L. Bioremediation of co-contaminated soil with heavy metals and pesticides: influence factors, mechanisms and evaluation methods. Chem Eng J 2020;398:125657. https://doi.org/10.1016/j.cej.2020.125657.

[36] García-Peña I, Hernández S, Auria R, Revah S. Correlation of biological activity and reactor performance in biofiltration of toluene with the fungus Paecilomyces variotii CBS115145. Appl Environ Microbiol 2005;71:4280–5. https://doi.org/10.1128/AEM.71.8.4280-4285.2005.

[37] Cambronero-Heinrichs JC, Masís-Mora M, Quirós-Fournier JP, Lizano-Fallas V, Mata-Araya I, Rodríguez-Rodríguez CE. Removal of herbicides in a biopurification system is not negatively affected by oxytetracycline or fungally pretreated oxytetracycline. Chemosphere 2018;198:198–203. https://doi.org/10.1016/j.chemosphere.2018.01.122.

[38] Lagaly G. Pesticide–clay interactions and formulations. Appl Clay Sci 2001;18:205–9. https://doi.org/10.1016/S0169-1317(01)00043-6.

[39] Yariv S, Cross H. Colloid geochemistry of clay minerals. In: Yariv S, Cross H, editors. Geochemistry of colloid systems: for earth scientists. Berlin, Heidelberg: Springer; 1979. p. 287–333. https://doi.org/10.1007/978-3-642-67041-1_8.

[40] Obotey Ezugbe E, Rathilal S. Membrane technologies in wastewater treatment: a review. Membranes 2020;10:89. https://doi.org/10.3390/membranes10050089.

[41] Sharma S, Bhattacharya A. Drinking water contamination and treatment techniques. Appl Water Sci 2017;7:1043–67. https://doi.org/10.1007/s13201-016-0455-7.

[42] Gerba CP. Chapter 25 – drinking water treatment. In: Maier RM, Pepper IL, Gerba CP, editors. Environmental microbiology. 2nd ed. San Diego: Academic Press; 2009. p. 531–8. https://doi.org/10.1016/B978-0-12-370519-8.00025-0.

[43] De Gisi S, Lofrano G, Grassi M, Notarnicola M. Characteristics and adsorption capacities of low-cost sorbents for wastewater treatment: a review. Sustain Mater Technol 2016;9:10–40. https://doi.org/10.1016/j.susmat.2016.06.002.

[44] Singh NB, Nagpal G, Agrawal S, Rachna. Water purification by using adsorbents: a review. Environ Technol Innov 2018;11:187–240. https://doi.org/10.1016/j.eti.2018.05.006.

[45] Kyzas GZ, Kostoglou M. Green adsorbents for wastewaters: a critical review. Materials 2014;7:333–64. https://doi.org/10.3390/ma7010333.

[46] Ahmad T, Rafatullah M, Ghazali A, Sulaiman O, Hashim R, Ahmad A. Removal of pesticides from water and wastewater by different adsorbents: a review. J Environ Sci Health Part C 2010;28:231–71. https://doi.org/10.1080/10590501.2010.525782.

[47] Testa C, Zammataro A, Pappalardo A, Sfrazzetto GT. Catalysis with carbon nanoparticles. RSC Adv 2019;9:27659–64. https://doi.org/10.1039/C9RA05689K.

[48] Roostaee M, Sheikhshoaie I. Magnetic nanoparticles; synthesis, properties and electro-chemical application: a review. Curr Biochem Eng 2020;6:91–102. https://doi.org/10.2174/2212711906666200316163207.

[49] Patrón-Romero L, Luque PA, Soto-Robles CA, Nava O, Vilchis-Nestor AR, Barajas-Carrillo VW, Martínez-Ramírez CE, Chávez Méndez JR, Alvelais Palacios JA, Leal Ávila MÁ, Almanza-Reyes H. Synthesis, characterization and cytotoxicity of zinc oxide nanoparticles by green synthesis method. J Drug Deliv Sci Technol 2020;60:101925. https://doi.org/10.1016/j.jddst.2020.101925.

[50] Polonskyi O, Ahadi AM, Peter T, Fujioka K, Abraham JW, Vasiliauskaite E, Hinz A, Strunskus T, Wolf S, Bonitz M, Kersten H, Faupel F. Plasma based formation and deposition of metal and metal oxide nanoparticles using a gas aggregation source. Eur Phys J D 2018;72:93. https://doi.org/10.1140/epjd/e2017-80419-8.

[51] Semiconductor nanoparticles. In: Wang ZL, Liu Y, Zhang Z, editors. Handbook of nanophase and nanostructured materials. Boston, MA: Springer US; 2002. p. 813–48. https://doi.org/10.1007/0-387-23814-X_23.

[52] Iravani S, Korbekandi H, Mirmohammadi SV, Zolfaghari B. Synthesis of silver nanoparticles: chemical, physical and biological methods. Res Pharm Sci 2014;9:385–406.

[53] Titus D, James Jebaseelan Samuel E, Roopan SM. Chapter 12 – nanoparticle characterization techniques. In: Shukla AK, Iravani S, editors. Green synthesis, characterization and applications of nanoparticles. Elsevier; 2019. p. 303–19. https://doi.org/10.1016/B978-0-08-102579-6.00012-5.

[54] Mourdikoudis S, Pallares RM, Thanh NTK. Characterization techniques for nanoparticles: comparison and complementarity upon studying nanoparticle properties. Nanoscale 2018;10:12871−934. https://doi.org/10.1039/C8NR02278J.

[55] Lim J, Yeap SP, Che HX, Low SC. Characterization of magnetic nanoparticle by dynamic light scattering. Nanoscale Res Lett 2013;8:381. https://doi.org/10.1186/1556-276X-8-381.

[56] Kukovecz Á, Kozma G, Kónya Z. Multi-walled carbon nanotubes. In: Vajtai R, editor. Springer handbook of nanomaterials. Berlin, Heidelberg: Springer; 2013. p. 147−88. https://doi.org/10.1007/978-3-642-20595-8_5.

[57] Yan XM, Shi BY, Lu JJ, Feng CH, Wang DS, Tang HX. Adsorption and desorption of atrazine on carbon nanotubes. J Colloid Interface Sci 2008;321:30−8. https://doi.org/10.1016/j.jcis.2008.01.047.

[58] Dehghani MH, Niasar ZS, Mehrnia MR, Shayeghi M, Al-Ghouti MA, Heibati B, McKay G, Yetilmezsoy K. Optimizing the removal of organophosphorus pesticide malathion from water using multi-walled carbon nanotubes. Chem Eng J 2017;310:22−32. https://doi.org/10.1016/j.cej.2016.10.057.

[59] Wang H, Hu B, Gao Z, Zhang F, Wang J. Emerging role of graphene oxide as sorbent for pesticides adsorption: experimental observations analyzed by molecular modeling. J Mater Sci Technol 2020. https://doi.org/10.1016/j.jmst.2020.02.033.

[60] Zhou Q, Fang Z. Graphene-modified TiO_2 nanotube arrays as an adsorbent in micro-solid phase extraction for determination of carbamate pesticides in water samples. Anal Chim Acta 2015;869:43−9. https://doi.org/10.1016/j.aca.2015.02.019.

[61] Prato M. Fullerene materials. In: Hirsch A, editor. Fullerenes and related structures. Berlin, Heidelberg: Springer; 1999. p. 173−87. https://doi.org/10.1007/3-540-68117-5_5.

[62] Chandrakumar KRS, Ghosh SK. Alkali-metal-induced enhancement of hydrogen adsorption in C60 Fullerene: an ab initio study. Nano Lett 2008;8:13−9. https://doi.org/10.1021/nl071456i.

[63] Davydov VY, Kalashnikova EV, Karnatsevich VL, Kirillov AI. Adsorption properties of multi-wall carbon nanotubes. Fullerenes Nanotub Carbon Nanostruct 2005;12:513−8. https://doi.org/10.1081/FST-120027215.

[64] Lima JDM, Gomes DS, Frazão NF, Soares DJB, Sarmento RG. Glyphosate adsorption on C60 fullerene in aqueous medium for water reservoir depollution. J Mol Model 2020;26:110. https://doi.org/10.1007/s00894-020-04366-9.

[65] Djordjevic A, Šojić Merkulov D, Lazarević M, Borišev I, Medić P, Pavlović V, Miljević B, Abramović B. Enhancement of nano titanium dioxide coatings by fullerene and polyhydroxy fullerene in the photocatalytic degradation of the herbicide mesotrione. Chemosphere 2018;196:145−52. https://doi.org/10.1016/j.chemosphere.2017.12.160.

[66] Zhou H, Zhao Y, Xiang J, Huang N, Baig SA, Zeng S. Mechanism and influence factors of 2,4-D dechlorination by sodium citrate-activated bimetallic palladium-zero valent iron nanoparticles. Appl Organomet Chem 2020;34:e5324. https://doi.org/10.1002/aoc.5324.

[67] Targhoo A, Amiri A, Baghayeri M. Magnetic nanoparticles coated with poly(p-phenylenediamine-co-thiophene) as a sorbent for preconcentration of organophosphorus pesticides. Microchim Acta 2017;185:15. https://doi.org/10.1007/s00604-017-2560-1.

[68] Liu G, Li L, Huang X, Zheng S, Xu X, Liu Z, Zhang Y, Wang J, Lin H, Xu D. Adsorption and removal of organophosphorus pesticides from environmental water and soil samples by using magnetic multi-walled carbon nanotubes @ organic framework ZIF-8. J Mater Sci 2018;53:10772−83. https://doi.org/10.1007/s10853-018-2352-y.

[69] Zhou Q, Long T, He J, Guo J, Gao J. Cadmium removal from water by enhanced adsorption on iron-embedded granular acicular mullite ceramic network. J Taiwan Inst Chem Eng 2020;106:92−8. https://doi.org/10.1016/j.jtice.2019.10.003.

[70] Zhao X, Wang W, Zhang Y, Wu S, Li F, Liu JP. Synthesis and characterization of gadolinium doped cobalt ferrite nanoparticles with enhanced adsorption capability for Congo Red. Chem Eng J 2014;250:164−74. https://doi.org/10.1016/j.cej.2014.03.113.

[71] Shin K-Y, Hong J-Y, Jang J. Heavy metal ion adsorption behavior in nitrogen-doped magnetic carbon nanoparticles: isotherms and kinetic study. J Hazard Mater 2011;190:36−44. https://doi.org/10.1016/j.jhazmat.2010.12.102.

[72] Mahmud HNME, Huq AKO, Yahya R. Polymer-based adsorbent for heavy metals removal from aqueous solution. IOP Conf Ser: Mater Sci Eng 2017;206:012100. https://doi.org/10.1088/1757-899X/206/1/012100.

[73] Zhu X, Li B, Yang J, Li Y, Zhao W, Shi J, Gu J. Effective adsorption and enhanced removal of organophosphorus pesticides from aqueous solution by Zr-based MOFs of UiO-67. ACS Appl Mater Interfaces 2015;7:223−31. https://doi.org/10.1021/am5059074.

[74] Ouali A, Belaroui LS, Bengueddach A, Galindo AL, Peña A. Fe$_2$O$_3$−palygorskite nanoparticles, efficient adsorbates for pesticide removal. Appl Clay Sci 2015;115:67−75. https://doi.org/10.1016/j.clay.2015.07.026.

[75] ul Haq A, Saeed M, Usman M, Naqvi SAR, Bokhari TH, Maqbool T, Ghaus H, Tahir T, Khalid H. Sorption of chlorpyrifos onto zinc oxide nanoparticles impregnated Pea peels (*Pisum sativum* L): equilibrium, kinetic and thermodynamic studies. Environ Technol Innov 2020;17:100516. https://doi.org/10.1016/j.eti.2019.100516.

[76] Youssef AM, Malhat FM, Abdel Hakim AA, Dekany I. Synthesis and utilization of poly (methylmethacrylate) nanocomposites based on modified montmorillonite. Arab J Chem 2017;10:631−42. https://doi.org/10.1016/j.arabjc.2015.02.017.

[77] Hassani A, Khataee A, Karaca S, Shirzad-Siboni M. Surfactant-modified montmorillonite as a nanosized adsorbent for removal of an insecticide: kinetic and isotherm studies. Null 2015;36:3125−35. https://doi.org/10.1080/09593330.2015.1054319.

[78] Tezcan Un U, Ates F, Erginel N, Ozcan O, Oduncu E. Adsorption of Disperse Orange 30 dye onto activated carbon derived from Holm Oak (Quercus ilex) acorns: a 3k factorial design and analysis. J Environ Manag 2015;155:89−96. https://doi.org/10.1016/j.jenvman.2015.03.004.

[79] Sen K, Datta JK, Mondal NK. Box−Behnken optimization of glyphosate adsorption on to biofabricated calcium hydroxyapatite: kinetic, isotherm, thermodynamic studies. Appl Nanosci 2020. https://doi.org/10.1007/s13204-020-01612-7.

[80] Chattoraj S, Mondal NK, Sen K. Removal of carbaryl insecticide from aqueous solution using eggshell powder: a modeling study. Appl Water Sci 2018;8:163. https://doi.org/10.1007/s13201-018-0808-5.

[81] Al-Ghouti MA, Da'ana DA. Guidelines for the use and interpretation of adsorption isotherm models: a review. J Hazard Mater 2020;393:122383. https://doi.org/10.1016/j.jhazmat.2020.122383.

[82] Condon JB. Equivalency of the Dubinin−Polanyi equations and the QM based sorption isotherm equation. B. Simulations of heterogeneous surfaces. Microporous Mesoporous Mater 2000;38:377−83. https://doi.org/10.1016/S1387-1811(00)00158-X.

[83] Ganesamoorthi B, Kalaivanan S, Dinesh R, kumar TN, Anand K. Optimization technique using response surface method for USMW process. Procedia Soc Behav Sci 2015;189:169−74. https://doi.org/10.1016/j.sbspro.2015.03.211.

[84] Chen F, Zhou C, Li G, Peng F. Thermodynamics and kinetics of glyphosate adsorption on resin D301. Arab J Chem 2016;9:S1665−9. https://doi.org/10.1016/j.arabjc.2012.04.014.

[85] Kajjumba GW, Emik S, Öngen A, Özcan HK, Aydın S. Modelling of adsorption kinetic processes—errors, theory and application, advanced sorption process applications. 2018. https://doi.org/10.5772/intechopen.80495.

[86] Ahmadi Azqhandi MH, Ghaedi M, Yousefi F, Jamshidi M. Application of random forest, radial basis function neural networks and central composite design for modeling and/or optimization of the ultrasonic assisted adsorption of brilliant green on ZnS-NP-AC. J Colloid Interface Sci 2017;505:278−92. https://doi.org/10.1016/j.jcis.2017.05.098.

[87] Ghaedi AM, Vafaei A. Applications of artificial neural networks for adsorption removal of dyes from aqueous solution: a review. Adv Colloid Interface Sci 2017;245:20−39. https://doi.org/10.1016/j.cis.2017.04.015.

[88] Herath GAD, Poh LS, Ng WJ. Statistical optimization of glyphosate adsorption by biochar and activated carbon with response surface methodology. Chemosphere 2019;227:533−40.

[89] Bezerra MA, Santelli RE, Oliveira EP, Villar LS, Escaleira LA. Response surface methodology (RSM) as a tool for optimization in analytical chemistry. Talanta 2008;76:965−77. https://doi.org/10.1016/j.talanta.2008.05.019.

[90] de Oliveira LG, de Paiva AP, Balestrassi PP, Ferreira JR, da Costa SC, da Silva Campos PH. Response surface methodology for advanced manufacturing technology optimization: theoretical fundamentals, practical guidelines, and survey literature review. Int J Adv Manuf Technol 2019;104:1785−837. https://doi.org/10.1007/s00170-019-03809-9.

[91] Caldas LFS, de Paula CER, Brum DM, Cassella RJ. Application of a four-variables Doehlert design for the multivariate optimization of copper determination in petroleum-derived insulating oils by GFAAS employing the dilute-and-shot approach. Fuel 2013;105:503−11. https://doi.org/10.1016/j.fuel.2012.10.026.

[92] Zolgharnein J, Shariatmanesh T, Asanjarani N, Zolanvari A. Doehlert design as optimization approach for the removal of Pb(II) from aqueous solution by Catalpa Speciosa tree leaves: adsorption characterization. Desalination Water Treat 2015;53:430−45. https://doi.org/10.1080/19443994.2013.853625.

[93] Abdel-Hafez SM, Hathout RM, Sammour OA. Towards better modeling of chitosan nanoparticles production: screening different factors and comparing two experimental designs. Int J Biol Macromol 2014;64:334−40. https://doi.org/10.1016/j.ijbiomac.2013.11.041.

[94] Swarnkar A, Swarnkar A. Artificial intelligence based optimization techniques: a review. In: Kalam A, Niazi KR, Soni A, Siddiqui SA, Mundra A, editors. Intelligent computing techniques for smart energy systems. Singapore: Springer; 2020. p. 95−103. https://doi.org/10.1007/978-981-15-0214-9_12.

[95] Fan M, Hu J, Cao R, Xiong K, Wei X. Modeling and prediction of copper removal from aqueous solutions by nZVI/rGO magnetic nanocomposites using ANN-GA and ANN-PSO. Sci Rep 2017;7:18040. https://doi.org/10.1038/s41598-017-18223-y.

[96] Fang W, Sun J, Ding Y, Wu X, Xu W. A review of quantum-behaved particle swarm optimization. IETE Tech Rev 2010;27:336−48. https://doi.org/10.4103/0256-4602.64601.

[97] Mahmoodi-Babolan N, Heydari A, Nematollahzadeh A. Removal of methylene blue via bioinspired catecholamine/starch superadsorbent and the efficiency prediction by response surface methodology and artificial neural network-particle swarm optimization. Bioresour Technol 2019;294:122084. https://doi.org/10.1016/j.biortech.2019.122084.

[98]   Karri RR, Sahu JN. Modeling and optimization by particle swarm embedded neural network for adsorption of zinc (II) by palm kernel shell based activated carbon from aqueous environment. J Environ Manag 2018;206:178−91. https://doi.org/10.1016/j.jenvman.2017.10.026.

[99]   Liu G, Li L, Xu D, Huang X, Xu X, Zheng S, Zhang Y, Lin H. Metal−organic framework preparation using magnetic graphene oxide−β-cyclodextrin for neonicotinoid pesticide adsorption and removal. Carbohydr Polym 2017;175:584−91. https://doi.org/10.1016/j.carbpol.2017.06.074.

[100]   Ma Q, Liu X, Zhang Y, Chen L, Dang X, Ai Y, Chen H. Fe_3O_4 nanoparticles coated with polyhedral oligomeric silsesquioxanes and β-cyclodextrin for magnetic solid-phase extraction of carbaryl and carbofuran. J Separ Sci 2020;43:1514−22. https://doi.org/10.1002/jssc.201900896.

[101]   Tian H, Li J, Shen Q, Wang H, Hao Z, Zou L, Hu Q. Using shell-tunable mesoporous Fe_3O_4@HMS and magnetic separation to remove DDT from aqueous media. J Hazard Mater 2009;171:459−64. https://doi.org/10.1016/j.jhazmat.2009.06.029.

[102]   Pinto M de CE, Gonçalves RGL, dos Santos RMM, Araújo EA, Perotti GF, Macedo R dos S, Bizeto MA, Constantino VRL, Pinto FG, Tronto J. Mesoporous carbon derived from a biopolymer and a clay: preparation, characterization and application for an organochlorine pesticide adsorption. Microporous Mesoporous Mater 2016;225:342−54. https://doi.org/10.1016/j.micromeso.2016.01.012.

[103]   Barbosa de Andrade M, Sestito Guerra AC, Tonial dos Santos TR, Cusioli LF, de Souza Antônio R, Bergamasco R. Simplified synthesis of new GO-α-γ-Fe_2O_3-Sh adsorbent material composed of graphene oxide decorated with iron oxide nanoparticles applied for removing diuron from aqueous medium. J Environ Chem Eng 2020;8:103903. https://doi.org/10.1016/j.jece.2020.103903.

[104]   Nejati K, Davari S, Rezvani Z, Dadashzadeh M. Adsorption of 4-Chloro-2-methylphenoxy acetic acid (MCPA) from aqueous solution onto Cu-Fe-NO_3 layered double hydroxide nanoparticles. J Chin Chem Soc 2015;62:371−9. https://doi.org/10.1002/jccs.201400403.

[105]   Sahithya K, Das N. Enhanced removal of dichlorvos from aqueous solution using zinc-silver bimetallic nanoparticles embedded in montmorillonite-biopolymer nano-biocomposites: equilibrium, kinetics and thermodynamic studies. Res J Pharm Technol 2017;10:1105−14. https://doi.org/10.5958/0974-360X.2017.00200.1.

[106]   Zhang C, Zhang RZ, Ma YQ, Guan WB, Wu XL, Liu X, Li H, Du YL, Pan CP. Preparation of cellulose/graphene composite and its applications for triazine pesticides adsorption from water. ACS Sustain Chem Eng 2015;3:396−405. https://doi.org/10.1021/sc500738k.

[107]   Ali I, AL-Othman ZA, Alwarthan A. Green synthesis of functionalized iron nano particles and molecular liquid phase adsorption of ametryn from water. J Mol Liq 2016;221:1168−74. https://doi.org/10.1016/j.molliq.2016.06.089.

[108]   Liu L, Berger VW. Randomized block design: nonparametric analyses. In: Wiley StatsRef: statistics reference online. American Cancer Society; 2014. https://doi.org/10.1002/9781118445112.stat06564.

[109]   Yu F, Qiu F, Meza J. 12 − Design and statistical analysis of mass-spectrometry-based quantitative proteomics data. In: Ciborowski P, Silberring J, editors. Proteomic profiling and analytical chemistry. 2nd ed. Boston: Elsevier; 2016. p. 211−37. https://doi.org/10.1016/B978-0-444-63688-1.00012-4.

[110]   Kim BG, Stein HH. A spreadsheet program for making a balanced Latin Square design. Revista Colombiana de Ciencias Pecuarias 2009;22:591−6.

[111] Hamdi L, Toumi LB, Salem Z, Allia K. Full factorial experimental design applied to methylene blue adsorption onto Alfa stems. Desalination Water Treat 2016;57:6098−105. https://doi.org/10.1080/19443994.2015.1029003.

[112] Chapter 23 Fractional factorial designs. In: Massart DL, Vandeginste BGM, Buydens LMC, De Jong S, Lewi PJ, Smeyers-Verbeke J, editors. Data handling in science and technology. Elsevier; 1998. p. 683−99. https://doi.org/10.1016/S0922-3487(97)80053-6.

[113] Hariri-Ardebili MA, Seyed-Kolbadi SM, Noori M. Response surface method for material uncertainty quantification of infrastructures, shock and vibration, vol. 2018; 2018. p. e1784203. https://doi.org/10.1155/2018/1784203.

[114] Zolgharnein J, Shahmoradi A, Ghasemi JB. Comparative study of Box−Behnken, central composite, and Doehlert matrix for multivariate optimization of Pb (II) adsorption onto Robinia tree leaves. J Chemometr 2013;27:12−20. https://doi.org/10.1002/cem.2487.

[115] Sadhukhan B, Mondal NK, Chattoraj S. Optimisation using central composite design (CCD) and the desirability function for sorption of methylene blue from aqueous solution onto Lemna major. Karbala Int J Mod Sc 2016;2:145−55.

[116] Guzun AS, Stroescu M, Jinga SI, Voicu G, Grumezescu AM, Holban AM. Plackett−Burman experimental design for bacterial cellulose−silica composites synthesis. Mater Sci Eng C 2014;42:280−8. https://doi.org/10.1016/j.msec.2014.05.031.

[117] Sadat-Shojai M, Khorasani M-T, Jamshidi A. Hydrothermal processing of hydroxyapatite nanoparticles—a Taguchi experimental design approach. J Cryst Growth 2012;361:73−84. https://doi.org/10.1016/j.jcrysgro.2012.09.010.

[118] Ghosh SB, Mondal NK. Application of Taguchi method for optimizing the process parameters for the removal of fluoride by Al-impregnated Eucalyptus bark ash. Environ Nanotechnol Monit Manag 2019;11:100206. https://doi.org/10.1016/j.enmm.2018.100206.

[119] Hurrion RD, Birgil S. A comparison of factorial and random experimental design methods for the development of regression and neural network simulation metamodels. Null 1999;50:1018−33. https://doi.org/10.1057/palgrave.jors.2600812.

[120] Petelet M, Iooss B, Asserin O, Loredo A. Latin hypercube sampling with inequality constraints. AStA Adv Stat Anal 2010;94:325−39. https://doi.org/10.1007/s10182-010-0144-z.

[121] Dizaj SM, Lotfipour F, Barzegar-Jalali M, Zarrintan M-H, Adibkia K. Application of Box−Behnken design to prepare gentamicin-loaded calcium carbonate nanoparticles, Artificial Cells. Nanomed Biotechnol 2016;44:1475−81. https://doi.org/10.3109/21691401.2015.1042108.

[122] shafaei A, Khayati GR. A predictive model on size of silver nanoparticles prepared by green synthesis method using hybrid artificial neural network-particle swarm optimization algorithm. Measurement 2020;151:107199. https://doi.org/10.1016/j.measurement.2019.107199.

[123] Shabanzadeh P, Senu N, Shameli K, Tabar MM. Artificial intelligence in numerical modeling of silver nanoparticles prepared in montmorillonite interlayer space. J Chem 2013;2013:e305713. https://doi.org/10.1155/2013/305713.

Chapter 5

Sustainability issues in upcoming wastewater treatment plants at Patna

Nityanand Singh Maurya[1], Sulagna Roy[2], Astha Kumari[1]
[1]*Department of Civil Engineering, National Institute of Technology Patna, Patna, Bihar, India;*
[2]*L&T Construction, Chennai, Tamil Nadu, India*

1. Introduction

In this era of rapid urbanization, the concept of sustainability has come to play a major role in all aspects of life. Sustainability in any sphere of life refers to maintaining a balance in the ecosystem. The consumption of a resource may be deemed as sustainable when the level of consumption just touches the optimum mark, thereby ensuring sufficient availability of the resource for the future generations. Sustainability of a resource is based on the basic principle of creating a balance between the three dimensions, namely, environmental, economic, and social [1]. Sustainable development can only occur when these three dimensions coexist harmoniously.

Water is often considered to be the most mispriced and misused natural resource that finds extensive use in industrial, domestic, and agricultural sectors [2]. The increase in population, adoption of high living standards, and urbanization tend to increase the pressure on the resource consumption, which results in increased cost of water supply and demands better practices related to water management [3]. It is generally assumed that 80% of the water demand contributes to the generation of wastewater in the system [4]. Over the years, these wastewaters have been found to be highly contaminated and polluted with known or unknown toxic substances that can be directly or indirectly regarded as the effect of the urbanization and industrialization. Thus, employing a suitable degree of treatment for the generated wastewater has become utmost important in order to ensure an overall protection of human health and ecosystem.

Wastewater treatment plants (WWTPs) equipped with activated sludge process and aerated lagoons have been in use since a long time. In the recent

Cognitive Data Models for Sustainable Environment. https://doi.org/10.1016/B978-0-12-824038-0.00012-2
101

years, the volume of wastewater generated has increased tremendously, together with a steady decline in the quality of wastewater. Due to this, the discharge of treated wastewater into the surface water bodies no longer remains a simple issue and the increasing concern over the matter has led to the idea of "Sustainable WWTPs."

As quoted earlier, the idea of adopting sustainable WWTPs shall take into account all the three major dimensions of sustainability, which means that the treatment scheme adopted shall not only be environment friendly, but should also be economically feasible and socially acceptable. A sustainable wastewater treatment plant refers to the type of plant that shall fulfill the present needs without affecting the ability and efficiency of the future generations to meet their own needs [5].

Out of the three key indicators of sustainability, the first indicator that holds immense significance is the environmental sustainability. Environmental sustainability generally involves providing long-term support in the development of environment by protecting natural resources and reducing emissions [1]. It is often observed that the discharge from a WWTP may constitute low concentration of chemicals but the concentration of nutrients in the discharge may be significantly high, thereby creating an imbalance in the receiving water bodies and nutrient flux. This imbalance eventually disrupts the hydrological as well as the ecological regimes [6]. Excessive discharge of nutrients (like nitrogen and phosphorus) into the water bodies may give rise to a phenomenon called eutrophication, which is highly detrimental to the ecosystem [1]. Moreover, a WWTP also consumes considerable amount of energy during its operation, which amounts to a significant contribution to the carbon dioxide emissions [6]. Thus, while designing a WWTP, the basic motto of keeping the environment safe has to be mandatorily borne in mind.

The second key indicator that defines the existence of a wastewater treatment plant, or rather any infrastructure is the economic sustainability. Upon the conceptualization of a WWTP, the first and the foremost thing that comes to a designer's mind is whether the proposed scheme is economically feasible and financially viable. Economic sustainability involves carrying out a cost-benefit analysis, such that the cost incurred shall never exceed the benefits, thereby ensuring a proper balance between the two aspects [1]. The overall cost of constructing and operating a treatment plant can be broadly divided into two parts, the capital cost (CAPEX) and the operating cost (OPEX). While the funds for initial construction of the treatment facilities can be obtained as one-time grant from a central government or soft loan from various international funding agencies like Asian Development Bank (ADB) and World Bank, the funds required for operation and maintenance continue to remain a challenging issue. In many cases, the bifurcation of funds for CAPEX and OPEX are not proper, due to which the operation and maintenance of the plants suffer to a greater extent. Thus, it is mandatory to provide a reality check on the economic aspect of the treatment plant, so that the cost can be controlled in both phases of construction and operation, without affecting the benefits to be reaped during the course of its operation.

The third key indicator is the social sustainability, which can be viewed from numerous aspects. One aspect of attaining social sustainability can be fulfilled if the people benefiting from the treatment scheme realize its importance, understand the challenges, feel responsible, and accept the consequences of their actions. This can be achieved by educating people and spreading public awareness about the technologies involved [1]. Another aspect of the social indicator being adopting a technology which is culturally accepted that shall promote public participation; and also be beneficial to the people in terms of increasing the job opportunities, providing quality education, and improvement in the local environment [6].

India, a country accommodating over 1.3 billion people, has seen an overwhelming increase in the volume of wastewater generation over the past couple of decades. However, the corresponding capacity of wastewater treatment plants have been found to be either insufficient or the quality of operation and maintenance in the existing treatment plants have declined over time. As per the report of Central Pollution Control Board in 2009, the estimated quantity of sewage generated from Class I cities and Class II towns has been 38,254.82 million litres per day, wherein only 11,787.38 million litres per day (31%) is treated (Advisory Note on Septage Management in Urban India, 2013) [7]. This gap in wastewater generation and wastewater treatment is a matter of great concern, as it has led to the failure of the objective of wastewater treatment plants. Thus, the idea of adopting sustainable WWTPs in the country has come into picture in order to meet the increasing demands of the huge populace, without compromising on the needs of the future generation.

However, the concept of sustainable wastewater treatment plants is not an easy feat to achieve. A well structured and executable plan at every stage has been the core requirement behind the fulfillment of this idea. This prompted the government to come up with National Mission for Clean Ganga as an initiative toward attaining the goal of establishing sustainable wastewater treatment plants as one of the major issues in the rejuvenation of Holy River Ganga in the country.

In this chapter, the city of Patna (in India), has been considered as a case study, which can give a thorough insight into the journey of wastewater treatment plants in India, starting from the conceptualization to the commissioning and maintenance of the same. The subsequent sections will further elaborate on the existing treatment scenario in Patna, highlight the reasons associated with the failure of the old wastewater treatment plants and suggest certain sustainable measures which may help in overcoming the encountered problems with a brief about the status of national mission for clean ganga (NMCG) program in the city.

2. Scenario of existing wastewater treatment plants in Patna

Patna, also known as Pataliputra, is the capital city of Bihar. It is one of the oldest cities in India whose history sprawls across centuries. Situated on the banks of the holy river Ganges, it is surrounded by three major perennial rivers, namely, Ganga (in the North), Sone (in the West), and Punpun (in the South-East). Geographically, Patna is located at 25 degrees36′40″ N and 85 degrees08′38″ E, with an average elevation of 53 m above the MSL. The average annual rainfall in the city is 1100 mm, with a temperature variation of 46°C in summer to 10°C in winter (City Development Plan of Patna, 2010−2030) [8]. The city had four WWTPs with installed capacity of 109 MLD based on activated sludge process, aerated lagoons/pond systems at four locations (Table 5.1). Due to poor collection systems coupled with nonfunctioning sewage pumping stations, the WWTPs of the city never operated on design capacity. The following sections describe the existing sanitation and wastewater collection system, and WWTPs of the city.

2.1 Sanitation and wastewater collection system

The city of Patna practices a wide variety of sanitary options, which ranges from households with water closet latrines to open defecation. Fig. 5.1 depicts the percentages of different sanitation options that are being practised in the city. It can be directly inferred from Fig. 5.1 that only 17% of the total households are connected to the sewers, which is quite an alarming figure stating the underutilization of the wastewater treatment plants installed in the city.

The estimated major sewer length in the city runs for 27.4 km. In the year 2011, the wastewater generation was estimated around 224.6 MLD, which is

TABLE 5.1 Scenario of underutilization for individual WWTPs.

Sl No.	Name of STP	Installed capacity (MLD)	Operating capacity (MLD)	Percentage underutilized (%)
1.	Saidpur	45	27	60
2.	Pahari	25	13	52
3.	Beur	35	18	51.4
4.	Karmali Chak	4	2	50
Total		109	60	55.1

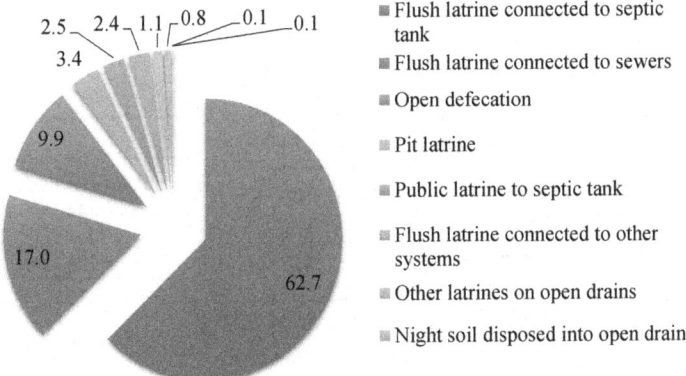

■ Flush latrine connected to septic tank

■ Flush latrine connected to sewers

■ Open defecation

▦ Pit latrine

■ Public latrine to septic tank

▦ Flush latrine connected to other systems

▦ Other latrines on open drains

▦ Night soil disposed into open drain

FIGURE 5.1 Sanitation options in Patna (per Census of India, 2011).

expected to rise up to 551 MLD by the end of 2030 [8]. It has been observed that the 17% of households connected to the sewers actually cover the old residential colonies of Rajendra Nagar, Kankarbagh, Kadam Kuan, and the government staff colony in Gardani Bagh. Apart from these areas, the rest of the city does not have a proper sewerage system. This lack of proper sewers is responsible for the discharge of untreated wastewater directly into the open or closed drains, which further connect to the major storm water drains that feed the discharge into the surface waters of River Ganga or River Punpun. This inadequacy of sewer network also leads to spillage and accumulation of wastewater in the low lying areas thereby affecting the water bodies and the natural aquifers.

2.2 Wastewater treatment plants

The history of wastewater treatment system in Patna goes way back to 1936 when the first wastewater treatment plant was installed at Saidpur with a capacity of 4.5 MLD, employing the activated sludge process for treating the wastewater collected from Rajendra Nagar Colony [8]. This was followed by the installation of Beur sewage treatment plant (STP), using trickling filters for treating the wastewaters from Gardani Bagh government staff quarters. Later on, in the year 1968, the capacity of Saidpur STP was revamped to 28 MLD, while retaining the old capacity of 4.5 MLD [8]. Moreover, in the same year, the old Beur STP using trickling filter technology was demolished and a new treatment plant with a capacity of 15 MLD was installed in its place. The two WWTPs of Saidpur and Beur were subsequently augmented to 45 MLD and 35 MLD respectively, in order to fulfill the requirement of the Central and Western Zone of Patna under the Ganga Action Plan (Phase 1) [9]. As per the

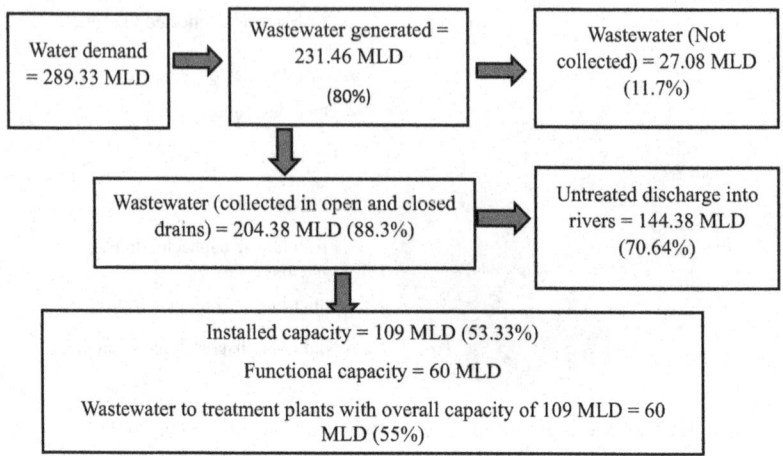

FIGURE 5.2   Fate of wastewater in Patna (year 2016).

data furnished in the City Development Plan of Patna (2010–2030), the city is equipped with four WWTPs based on activated sludge process and aerated lagoons/pond systems at Saidpur (45 MLD), Pahari (25 MLD), Beur (35 MLD) and Karmali Chak (4 MLD). The WWTPs at Pahari and Karmali Chak have been installed with the intention of catering to the requirement of the Southern and Eastern Zone of Patna [8]. Although the overall installed capacity of the treatment plants is quite high, the amount of sewage that reaches the plants for undergoing the desired treatment is quite low.

Fig. 5.2 depicts the fate of wastewater generated in the households of Patna through a flow diagram, which emphasizes on the underutilization of the individual WWTPs. Considering per capita demand of 135 lpcd and the projected population of Patna in the year 2016 as 2,143,216 (Based on the Census, 2011), the water demand can be calculated as 289.33 MLD [10]. As stated in the previous section that 80% of water amounts to the quantity of wastewater generated, the wastewater demand arrived is 231.46 MLD.

As stated above, the underutilization of the existing WWTPs can be clearly deduced from Fig. 5.2. In spite of the overall installed capacity of the existing WWTPs being 109 MLD, it has been observed that only 60 MLD is being treated before discharge into the surface water bodies (Table 5.2). This emphasizes on the fact that the overall treatment scheme remains unutilized to the extent of 44.9% of the total installed capacity. Thus, the population growth and the corresponding increase in the generation of wastewater with inefficiency in the existing treatment scheme turns out to be of major concern in the city of Patna.

TABLE 5.2 Sewage infrastructure projects in Patna under "Namami Gange" program.

SI No.	Project name	WWTP capacity (in MLD)	Sewer network length (in km)	Percentage completion (physical progress)	Remarks
1	Beur WWTP	43	—	100%	In O&M phase.
2	Beur sewerage network	—	179.74	74%	134.373 km of network completed.
3	Saidpur WWTP and adjacent network	60	55.10	96.95%	53.803 km of pipeline laid, casting of SBR Basin, sludge dewatering building, etc. completed.
4	Saidpur sewerage network	—	172.50	64%	84.2 km of pipeline laid, construction of 2830 nos. of manholes, and 1796 nos. of house connections established.
5	Karmalichak WWTP	37	—	100%	In O&M phase.
6	Karmalichak sewerage network	—	96.54	62.62%	69.709 km of sewer network completed with the construction of 2523 nos. of manholes and 9133 nos. of house connections established.
7	Pahari WWTP	60	—	70%	Construction of SBR, equalization tank completed and construction of sludge dewatering building, sludge sump under progress.
8	Pahari zone IV a sewerage scheme	—	87.69	86.01%	In O&M phase.
9	Pahari zone V sewerage scheme	—	115.93	88.01%	134.373 km of network completed with the construction of 4721 nos. of manholes.

Continued

TABLE 5.2 Sewage infrastructure projects in Patna under "Namami Gange" program.—cont'd

Sl No.	Project name	WWTP capacity (in MLD)	Sewer network length (in km)	Percentage completion (physical progress)	Remarks
10	Digha WWTP and sewer network (outskirts of Patna)	100	288	0.78%	Survey for sewer network has been completed, approval of basic engineering package (BEP) and other design documents and drawings for WWTP obtained from BUIDCo
11	Kankarbagh WWTP and sewer network	50	150	—	Soil testing is under progress, while certain design documents and drawings for WWTP have been approved by BUIDCo.
12	Barh WWTP (outskirts of Patna)	11	—	42.43%	Construction of SBR tank completed upto 93.5% and construction of Grit Chamber completed upto 58%.
13	Mokama WWTP (outskirts of Patna)	8	—	37%	Survey and corresponding design is complete and construction of SBR is completed upto 81%.
14	Maner WWTP and adjacent network (outskirts of Patna)	6.5	3	—	At present, land is not available for WWTP and official proceedings for land procurement are under progress.
15	Danapur WWTP (outskirts of Patna)	25	—	—	In bid evaluation stage
16	Phulwarisariff WWTP (outskirts of Patna)	13	—	—	In bid evaluation stage

2.3 Reasons for failure of the existing wastewater treatment plants (WWTPs)

There may be several known or unknown reasons behind the failure of the existing wastewater treatment plants in the city of Patna. Yet, it has been observed that the following causes play a significant role in this process, namely,

  i. Poor operation and maintenance
  ii. Inadequate supply of power
  iii. Lack of trained manpower
  iv. Poor wastewater collection system
  v. Lack of coordination and good governance

2.3.1 Poor operation and maintenance

The operation and maintenance cost for an STP may include continuous charges in the form of energy and chemical procurement, followed by labor charges and the replacement of equipment [6,11]. The cost estimate for Beur STP in Patna highlights the fact that the OPEX for the STP can be estimated @5% of the capital cost per annum (Detailed Project Report - Package 1 (Beur STP), 2013) [12]. Similar cost estimates can be observed as well for the WWTPs at Pahari (Detailed Project Report: Package 3; Pahari: STP, 2013) [13] and Karmali Chak (Detailed Project Report: Package 4; Karmali Chak Sewer Network and STP, 2013) [14]. This immense expenditure can only be met with an uninterrupted supply of funds. It has been observed that in most cases, the Central and State Governments fall short of funds to equip and maintain the WWTPs in a proper way. One such way of mitigating this problem is revenue collection. Revenue collection for WWTPs plays a major role in providing the necessary fund for OPEX. In the Detailed Project Report of Beur STP in 2013 [12], the revenue collection has been estimated based on the assumption that 100% of the households and commercial establishments shall connect to the sewers by the end of 2019. Yet, the financial analysis shows that the total revenue generation will not be able to surpass the expenditure before 2025. Similar scenarios can be observed in the financial analysis statements of WWTPs at Pahari and Karmali Chak [13,14]. The failure to manage a sustainable revenue system can be studied in the research conducted by Murray and Drechsel in 2011 on the sanitation sector in Ghana, wherein the lack of a sustainable revenue stream has proven to be one of the causes behind the failure of WWTPs in Ghana [15].

2.3.2 Inadequate supply of power

The second important cause responsible for the failure of WWTPs is the inadequate supply of power. As stated in the Guidelines for Siting of Sewage

Treatment Plant (Urban Development and Housing Development, government of Bihar), the proper functioning of WWTPs can only be guaranteed when there is a continuous power supply for carrying out the smooth operation of the electromechanical equipment such as pumps, aeration tanks, and so forth [16]. This is due to the fact that the reliance of a technology on electricity is a critical factor in determining the fragility of the system [15]. Irregular power supply often disrupts the treatment mechanism which results in untreated sludge being discharged into the surface water bodies. In the study conducted on the sanitation sector in Ghana, it has been observed that the rate of failure of the electricity dependent technologies was very high without the availability of a generator [15]. It has been observed that cities like Patna experience frequent power cuts, which can be dealt with the help of alternate power sources such as generators.

According to the STP Guide (Design, Operation, and Maintenance) published by Karnataka State Pollution Control Board, the minimum requirement of power backup is equal to the combined power for all units with an additional margin of 20% [17]. However, very few WWTPs make use of alternate power sources and hence, continue to be affected from the erratic power supply. The Detailed Project Report - Package 1 (Beur STP) published in 2013, clearly stated that the four WWTPs in Patna, namely, Saidpur, Beur, Pahari and Karmali Chak, do not operate continuously due to power shortage, thereby highlighting the problem to a large extent [12]. In order to deal with this crisis, certain WWTPs in India have taken up the initiative of generating power from the treated sewage, such as the K&C Valley STP of capacity 60 MLD under Bangalore Water Supply and Sewerage Board (BWSSB), which uses the technology of activated sludge process with power generation [18].

According to an evaluation of the operation and maintenance costs on Gebze wastewater treatment plant by Turkmenler et al., it has been inferred that the energy costs contribute almost 1.5% of the annual operation and maintenance cost of the plant [19]. Fig. 5.3 denotes the power consumption (in kWh/m^3) over the period July 2013–June 2014 in Gebze WWTP, Turkey [19]. Thus, the provision of generating additional power from treated sludge is highly recommended so that it can be sold off to the respective authorities, which has the potential of giving a financial stability in the operation of the treatment plants. This approach is yet to be adopted in the city of Patna.

2.3.3 Lack of trained manpower

One more factor that contributes to a great extent toward the smooth functioning of a wastewater treatment plant is the availability of trained/skilled manpower. As per the research conducted by Khursheed et al. in 2014 [20] on the challenges faced in the operation and maintenance of sewerage systems in India, it has been reported that the chief representatives of various state water supply and sewerage boards consider the unavailability of trained manpower/ staff as one of the reasons that hinder the operation and maintenance of the

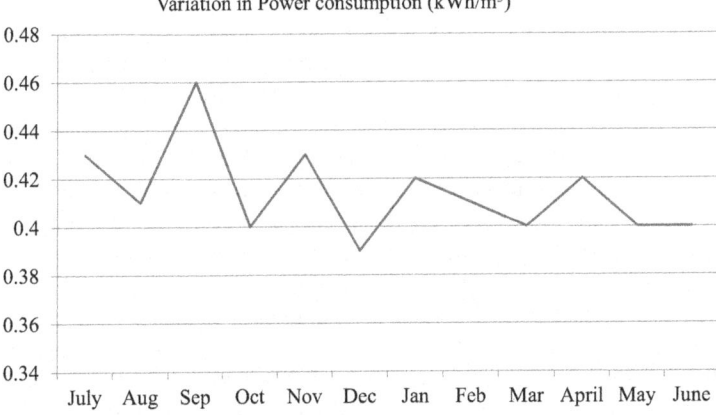

FIGURE 5.3 Variation in the power consumption at Gebze wastewater treatment plant, Turkey over the period of July 2013—June 2014.

plants. It has been observed in the study that the private contractors in charge of O&M generally depute under qualified staffs which contributes to the poor functioning of the wastewater treatment plants [20]. The importance of trained/skilled manpower has also been emphasized in the report prepared by Central Pollution Control Board (CPCB), Zonal Office—Bhopal on the Performance Evaluation of Sewage Treatment Plants in Central Zone. According to the report, the operation and maintenance of the plant at Mata Mandir, Bhopal, has been found to be satisfactory, where the plant is being operated only with skilled manpower. Similarly, the plants at Badwai, Gondarmau, and Mohali Damkheda in Bhopal (to name a few) also exhibit satisfactory performance in terms of operation and maintenance with the availability of skilled manpower [21]. Moreover, in the Performance Evaluation of Sewage Treatment Plants under National River Conservation Directorate (NRCD), the availability of skilled manpower has been identified as one of the three major factors on which the operation and maintenance of the plant depends [22]. Hence, it is quite clear that the advanced and conventional treatment technologies require skilled professionals for their proper functioning as one would need to understand the technology involved in each and every treatment unit for ensuring the proper effluent quality at the end.

2.3.4 Poor wastewater collection system

It has been observed that the quantity of wastewater reaching the wastewater treatment plants plays an important role in determining its efficiency. Based on the amount of wastewater treated, it can be determined whether the capacity of the wastewater treatment plant is underutilized. Many cities lack a proper sewerage network for the collection of wastes, as a result of which it is discharged directly or indirectly in open drains, thereby polluting the surface water bodies [21]. As per the Census of India 2011, the number of urban households not connected to any sewer system comes around 30 million [23]. As stated in the Guidelines for Septage Management in Maharashtra, published in February 2016, around 70% of the households in Maharashtra are equipped with individual toilets, out of which only 53% are connected to a proper sewer network, leaving the remaining 47% to be connected to septic tanks, pits and other collection systems [23]. Similar scenario can be perceived in the city of Patna, as discussed in the previous sections, wherein the percentage of households connected to the sewer network is very low. In the Detailed Project Report: Package 1 (Beur STP) published in 2013 [12], the gap between the installed capacity and the treatment capacity of the plants has been highlighted, where the lack of wastewater flow due to insufficient sewer length is identified as one of the causes behind the underutilization of the treatment capacity of the plants. Thus, from the data stated above, it can be inferred that proper collection and channeling of wastewater to the wastewater treatment plants is essential for the effective utilization of the treatment plants.

2.3.5 Lack of coordination and good governance

Last but not the least, improper governance within the WWTPs plays a significant role behind the failure of most of the treatment plants. As stated in the case study for the sanitation sector of Ghana, the plant operators, especially in the public sector, are seldom committed to their job and hence, fail to keep a proper check on the effluent characteristics [15]. The analysis indicates that there is a lack of proper incentive which is being paid to the plant operators, thereby triggering the reluctance in their behavior [15]. As stated in the report on the performance evaluation of WWTPs conducted by CPCB in August 2013, under the NRCP of Ministry of Environment and Forests [4], it has been observed that certain sewage treatment plants have reported issues such as the lack of lab facilities at site and hence, inconsistent sampling with no proper records; several nonoperational units leading to poor sewage quality at the outlet; contradictory log books for water meter, chlorination, energy meter, DG sets, etc. as maintained by the plant operators; improper maintenance of DG sets; frequency of sludge removal from different ponds varying from 18 to 20 months, etc. These problems, when compared with the scenario at the sanitation sector of Ghana, clearly show the lack of proper governance inside the WWTPs. In addition to the reason cited by Murray and Drechsel [15], there may be other reasons behind this lack of coordination and governance, the chief among them being appointment of unskilled personnel who lack the proper knowledge on the maintenance of various WWTP units. Moreover, it has also been observed in the public sector WWTPs that the trained personnel/operators are often transferred to other treatment facilities, resulting in an inconsistency in the system. Once a trained operator is replaced with a novice, it usually takes time for the newcomer to get adapted to the system, which in turn affects the operation of the treatment plant. Another reason which may be considered is the irregular working hours of the plant operators inside the treatment plants. The plant operators are usually entitled to work in 8-h shifts. Due to the lack of proper coordination inside the treatment plants, the working hours of a particular operator may extend thereby leading to fatigue and the inability of the person to deliver the desired output. These instances, although appear smaller in magnitude, can be detrimental to the life of a wastewater treatment plant.

3. Upcoming environmental infrastructure projects

The upcoming environmental infrastructure projects pertaining to the development of wastewater treatment plants and sewerage network along the Ganga Basin have been placed under "Namami Gange Program." "Namami Gange Program," which has been recognized as a "Flagship Program" in June 2014 by the Govt. of India under National Mission for Clean Ganga (NMCG), has taken up the overall rejuvenation of River Ganga as its primary objective, with an estimated budget of INR 20,000 crores [9]. The initiatives considered under this program have been tuned to focus on controlling pollution along River Ganga, followed by conservation of biodiversity and subsequent rejuvenation. Certain key initiatives of this program that have been designed to meet its goal include the development of sewage treatment infrastructure and river front, cleaning of surface water, conservation of biodiversity, enhancing afforestation, monitoring the discharge of industrial effluent and spreading general awareness among public. The adoption of Ganga Gram that focuses on the development of sanitation sector across 1674 g panchayats, is another initiative under this program. The implementation of all the initiatives have been categorized under three different headings, namely, entry-level activities (which shall focus on immediate impact), medium-term activities (which shall be implemented within a span of five years) and long-term activities (which shall be implemented within a span of 10 years) [9].

3.1 Status of projects pertaining to sewage treatment infrastructure in India

As stated earlier, the development of sewage treatment infrastructure, being one of the key initiatives under this program primarily focuses on the development of wastewater treatment plants and proper sewerage network. According to the Project Progress Report of July 2020 [24], the total number of projects to be executed in India under the development of sewage treatment infrastructure is 151, out of which 54 projects have been completed, 72 projects are under progress, and 25 projects are in the bid evaluation stage. Fig. 5.4 gives an overview on the status of the sewage treatment infrastructure projects in India.

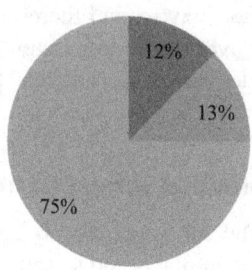

FIGURE 5.4 Status of the sewage infrastructure projects in India.

3.2 Overview on sewage treatment infrastructure projects undertaken in Bihar and Patna

All the sewage infrastructure projects under "Namami Gange" have been spread over a total of eight states, wherein Bihar bags a total of 30 projects, with a sanctioned cost of INR 5328.6 crores [25]. Patna, the capital city of Bihar, has 16 (out of 30 sewerage projects that include the construction of wastewater treatment plants and the development of sewer network) under its scope, out of which two projects have been completed, two projects are in the bid evaluation stage, while 12 projects are under progress [26,27]. Fig. 5.5 depicts the status of the sewage treatment infrastructure projects in Patna and its outskirts. As per the status reported on December 2020, the construction of WWTPs at Beur (with a treatment capacity of 43 MLD) and Karmali Chak (with a treatment capacity of 37 MLD) stand completed and these projects

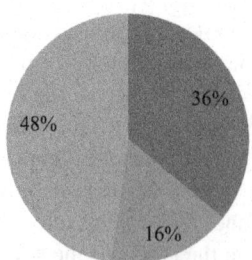

FIGURE 5.5 Status of the sewage infrastructure projects in Patna and its outskirts.

have duly entered the O&M phase. The report also depicts that the construction of WWTP at Saidpur with an overall treatment capacity of 60 MLD is over 95% complete and it shows that the overall target is to achieve a wastewater treatment capacity of 350 MLD and establishing sewer network of ~1145 km in the main parts of Patna. Table 5.2 highlights the status of the sewage infrastructure projects undertaken in Patna and its outskirts [27].

The funds for the major WWTPs in Patna have been granted by World Bank and the corresponding projects have been identified as Externally Aided Project (EAP). This way NMCG has ensured that there is no shortage of funds during the construction of the sewage infrastructure. Moreover, the adoption of the Hybrid Annuity Model in certain projects shall also help in overcoming the shortage of funds faced during the operation and maintenance period. This is because under this model, only 40% of the capital cost shall be paid to the contractor during the construction phase, whereas the balance 60% shall be payable during the operation and maintenance period of the plant in the form of annuities based on the performance evaluation of the treatment plant [28]. This technique shall not only ensure the financial viability of the plant, but will also help in maintaining the desired performance of the treatment units. In order to meet the desired performance, the executing agencies shall definitely take care of the governance inside the treatment plants by appointing trained, capable and efficient personnel, thereby ensuring proper coordination inside the WWTPs [22]. Thus, the projects enlisted in Table 5.2 shall be operated and maintained by the respective contractors for a period of 15 years, with a structured allocation of manpower. Moreover, in order to ensure proper power backup for the operation of the treatment plants, all projects have been equipped with diesel powered electrical generation sets of desired capacity. This way the problems associated with power failure shall be mitigated.

As indicated in the official website of NMCG, the Ganga Committee is present both at the state as well as the district levels and constitutes of District Magistrates (DMs), rural, and urban local bodies (ULBs). Moreover, the major policy initiatives undertaken by NMCG also include third-party inspections, establishment of city-level monitoring centers, public involvement, and so forth [24]. This administrative structure in line with the major policies shall not only help in project execution but shall also contribute toward the sustainability of the projects, in terms of ensuring proper governance in the system and also keeping up the performance of the system even after the transference of the projects to the ULBs.

4. Sustainability measures taken in upcoming projects

Upon a thorough analysis of the causes responsible for the failure of the existing wastewater treatment plants, the following sustainable measures can be suggested for the upcoming projects.

4.1 Appropriate cost analysis of the projects

The overall cost analysis of the projects should be appropriately carried out at the initial stage, wherein the estimation of capital cost and operational cost should be given equal importance. It is often observed that the capital cost is thoroughly estimated at the initial stage without paying much heed to the operational cost of the plant as the capital cost tends to be on the higher side. However, this little negligence often turns out to be detrimental to the life of the treatment plant. This is owing to the fact that many operational problems crop up during O&M, which require sufficient fund to be resolved. Hence, proper bifurcation and allocation of capital cost and operational cost is mandatory for ensuring the viability of a WWTP. As emphasized in the previous section, the adoption of a sustainable revenue collection system for a specific period can turn out to be a good preventive measure in overcoming the fund shortage for operation and maintenance.

4.2 Adoption of alternate power generation sources

The necessity of maintaining uninterrupted power supply in a wastewater treatment plant is immensely high. As explained in the previous section, the performance of electromechanical equipment is highly dependent on power. During project planning, the provision for additional power backup (preferably in the form of generators) should be taken into account. This will ensure that when the plant becomes fully operational, the treatment units face less hindrance while running and are capable of delivering the desired output quality. This can act as a stable preventive measure toward ensuring the smooth operation of wastewater treatment plants and thereby ensuring its sustainability in future.

4.3 Appointment of skilled and trained operators

As explained earlier, the presence of skilled plant operators is highly essential in ensuring the smooth functioning of the WWTPs. It is mandatory for a plant operator to understand the treatment scheme and the function of each treatment unit, so that he is able to troubleshoot the system in case of a sudden breakdown. As cited in the previous section, many treatment plants fail to deliver the desired performance owing to the lack of trained operators. Thus, the best way to prevent the occurrence of such instances is to appoint skilled manpower during the commissioning of the treatment plant and provide frequent training in due course of time to update them with the latest developments in the adopted processes/technologies.

4.4 Simultaneous development of wastewater collection network

Any WWTP cannot operate at its designed capacity if it does not receive the required quantity of influent to be treated. It is evident from the instances cited earlier that most of the wastewater treatment plants fall short of running at its installed capacity, owing to the poor wastewater collection system. Therefore, during the inception of a project, an equivalent study shall be carried out on the existing deficit in the wastewater collection network and the possible ways of addressing the problems shall be figured out subsequently. This preventive measure may turn out to be beneficial in terms of development of wastewater collection network and subsequently help in solving the problem of under-utilization of wastewater treatment plants.

4.5 Development of a proper administrative structure

During the inception of a project, as a preventive measure, it should be made mandatory to devise a proper administrative structure and define the roles and responsibilities at different levels of governance. As cited in the previous section, proper governance inside the treatment plant as well as at the administrative level is essential for the sustainability of the project. As soon as the responsibility of the wastewater treatment plant is transferred from the implementing agency to the ULB, respective officials appointed at different levels shall be made aware of their roles and responsibilities and frequent evaluation or inspection of the system shall be carried out accordingly. While devising an appropriate administrative structure and ensuring good gover-nance, it is also essential to consider certain aspects like providing proper incentives to the plant operators and taking into consideration the overall working hours for every individual. These minute things, when arranged in a systematic manner may help in preventing the occurrences of irregularities in the treatment plant administration.

It is observed that most of the sustainable measures discussed above have been taken up in the wastewater treatment projects to be executed under "Namami Gange Program," which is discussed in the subsequent section.

4.6 Smooth transfer to urban local body

Upon further analysis, it is observed that the successful running of a WWTP also depends on the ULB or the municipal area in which it is located. The commissioning of a WWTP by an implementing agency is followed by the O&M period. Once the O&M period gets over for the implementing agency, the responsibility of operation and maintenance of the plant is transferred to the respective ULBs. As stated in the Memorandum of Agreement for Bai-dyabati Municipality signed in February 2010 in the state of West Bengal, the ULB is responsible for the operation and maintenance and is expected to submit a report regarding the same which would undergo thorough evaluation from time to time before the release of funds [29]. Moreover, the Detailed Project Report - Package 1 (Beur STP) published in 2013 [12], also states that Bihar Urban Infrastructure Development Finance Corporation (BUIDCo) would transfer the responsibility of operation and maintenance to the respective ULBs. This is also evident in the Detailed Project Report: Package 3 [13] (Pahari: STP) and the Detailed Project Report: Package 4 (Karmali Chak Sewer Network and STP) published in 2013 [14]. Considering this data, it can be concluded that the smooth transfer of the treatment plant from the implementing agency to the ULB holds great significance behind the successful functioning of the WWTPs. As stated in the Detailed Project Report on Sewerage for Digha of Patna, Bihar—Comprehensive Waste Water Management, Digha and Kankarbagh, published in 2017 [30], the concept of Hybrid Annuity Model has been adopted in the cost sharing of project wherein, the O&M cost is covered under the overall project cost, borne by the central government or the State government or the ULBs. However, in case of a cost overrun in the project, the share for the central government shall be limited to a definite amount sanctioned by the Cabinet Committee on Economic Affairs (CCEA). This mandates active participation of the ULBs in the overall planning of the project in order to facilitate a smooth transfer and ensure better operation and maintenance with a consistent supply of funds.

5. Sustainability analysis of the upcoming projects

5.1 Environmental sustainability

Environmental sustainability largely depends on the selection of appropriate treatment technology. As briefed in the earlier sections, the motto of attaining environmental sustainability depends on the adoption of methods that would cause least possible effect on the environment. In case of the upcoming projects under "Namami Gange," it has been observed that the treatment scheme selection is based on Sequential Batch Reactor (SBR). As indicated in the Detailed Project Report of Beur STP in 2013, the percentage of BOD removal envisaged is around 90% [12]. It is essential to meet the desired effluent standard pertaining to BOD and COD removal for discharging the same into the surface water bodies, as these water sources are being extensively used in municipal water treatment, and this efficiency of the treatment scheme has to be viable at all times. However, the capability of this technology to function at its desired efficiency after 15 years still remains in question. As previously indicated in Fig. 5.2, the list of technologies initially adopted in the WWTPs were activated sludge process, aerated lagoons, and so forth whose maintenance was far easier compared to the technologies like sequential batch reactors. But the WWTPs remained underutilized and the treatment technologies could not perform up to the desired level. This further strengthens the question on the capability and effective functioning of the SBR technology to deliver the desired effluents beyond the 15-year time frame. Moreover, discussion with NMCG officials has also thrown light on the fact that the sludge produced during the treatment process shall undergo sludge compression and will then be transported to the landfill sites. A more suitable approach would have been sludge digestion, which could help in the production of biogas and fertilizers, thus reducing the load on the environment. Moreover, with the increase in urbanization and infrastructure development, the load on wastewater treatment plants have increased which further contribute toward energy consumption and greenhouse gas (GHG) emissions [31]. The release of GHGs in the form of methane, carbon dioxide, and nitrous oxide is inevitable, so the only way out is to devise a strategy which would help in compensating the same. But such a plan has not been adopted yet in the upcoming projects. Therefore, these points raise concern on the long-term environmental sustainability of the projects.

5.2 Economic sustainability

As discussed in the previous section, the capital cost for the major WWTPs shall be borne by external agencies such as World Bank and the operation and maintenance cost shall be taken care on the basis of Hybrid Annuity Model. Moreover, upon discussion with NMCG officials, it has come to notice that the payback for the capital cost shall be taken care by central government which is indeed a great relief for the state government. However, the financial planning has not been carried out beyond 15 years and the maintenance of the plants after the time frame of 15 years still remains in question. As cited by Koul and John in the life cycle cost analysis for wastewater treatment plants based on different technologies, it has been observed that the capital cost and the total annual O&M cost for SBR is higher compared to other technologies such as Upflow Anaerobic Sludge Blanket Reactor (UASBR) and Moving Bed Biofilm Reactor (MBBR) [32]. Table 5.3 lists out the cost comparison between SBR, UASBR, and MBBR. In spite of that, an appropriate revenue collection model has not been devised in the Design Progress Report (DPR), which raises concern over the economic sustainability of the WWTPs after their

TABLE 5.3 Cost Comparison between wastewater treatment technologies [32].

Sl No.	Parameter	MBBR	UASBR	SBR
1	Capital cost (in crores/MLD)	0.21	0.26	0.7
2	Total annual O&M cost (in crores/MLD)	0.6	0.208	0.63

transference to the ULBs. Neither the transfer mechanism, nor the idea of revenue collection from the end users has been described in detail in the respective DPRs of the WWTPs. Moreover, the technology envisaged for these WWTPs being primarily SBRs, which itself is an energy intensive process, the consumption of power can be directly linked with the operational cost to be incurred in the maintenance of the WWTPs. Therefore, if proper planning is not carried out at the initial stage with respect to the maintenance of the WWTPs, then it would indeed be difficult to account for the economic sustainability of the projects.

5.3 Social sustainability

The idea of social sustainability can be achieved if there is scope for public involvement in the projects. Wherever the constructions of WWTPs have been envisaged, the concerned people residing in those locations could have been involved. However, the initiatives for the upcoming projects have been directly undertaken by the government, without involving the general public. Public participation has been found to be missing, starting from the inception to the execution of the projects. Therefore, it is quite difficult to gauge how far these projects can be evaluated on the parameter of social sustainability.

6. Provisions to be included in upcoming projects

6.1 Selection of treatment technology during the inception of the projects

Selection of treatment technology plays a vital role in determining the sustainability of a project. The selection of the treatment scheme should be such so that it touches the three important parameters of sustainability, viz., environmental, economic, and social. It is observed that the selection of treatment technology in the upcoming projects has been kept open for the bidders to decide. However, the initial costing is based on the adoption of SBR. This approach can be viewed as a major flaw in the system. The tendency of the bidders will always be to adopt such a treatment scheme that shall be cost effective, which in turn might also end up in the project being awarded to the bidder with low technical expertise. Also, this disparity between the initial costing being carried out based on one technology and the adoption of another technology during execution might also affect the overall viability of the project in the long run. Therefore, in order to keep the bidders at par and also to ensure that the best possible technological solution has been adopted, it is advisable to evaluate and fix the technology during the inception of the projects and include the same in the corresponding bid documents for the upcoming projects.

6.2 Resource recovery

Resource recovery is an important part of wastewater treatment process. It is also an important yardstick for assessing the environmental sustainability of a wastewater treatment plant. During the inception of a project, the technology selection and the corresponding sludge generation shall be evaluated simultaneously. And more focus should be given toward reduction in sludge generation by reusing the sludge for other purposes. The adoption of this approach in the upcoming projects is very much essential, owing to the fact that the projects have to touch upon the mark of environmental sustainability. Moreover, this will also help in the long-term viability of the projects as the valuable products generated from sludge (such as biogas, fertilizers, and so forth) will find their extensive use in other spheres of life, thereby ensuring a balance in the entire system.

6.3 Public participation

As already discussed in the previous section regarding the need of public participation in order to ensure social sustainability, this provision may be adopted in the upcoming projects. During the conceptualization of the WWTP, the demands and the opinions of the concerned end users may be taken into account. Also, the people in that region may be trained and converted to skilled personnel who would be able to fulfill the criteria of trained and skilled manpower, thereby contributing toward the generation of employment. This way the upcoming projects can be made to connect with the social cause as well.

7. Conclusion

Sustainability in wastewater treatment holds immense significance toward the development of a community and the notion holds good for the city of Patna as well. In the past, many WWTPs have been commissioned, yet they could not be sustained due to multiple factors, ranging from poor wastewater collection network to hindrance in successfully operating the treatment plants. This mandated the adoption of the concept of sustainable WWTPs in Patna. However, out of the three key indicators of sustainability, viz., environmental, economic, and social; economic sustainability turns out to be the key player in the arena. It can be inferred from the previous discussions that without assessing the financial viability of a project, which is dependent on multiple factors such as O&M cost, cost for ensuring uninterrupted power supply, skilled manpower, development of proper wastewater collection network, and so forth, it is indeed difficult to meet the goal of economic sustainability. While assessing the reasons for the failure of WWTPs in the past, it has been observed that most of the causes could have been mitigated had there been a proper study on the financial viability of the project. Thus, based on the study, it may be concluded that the preventive measures suggested in the respective section, that includes running an appropriate cost analysis of the projects, adoption of alternate power sources, appointment of skilled operators, development of a proper administrative structure, and simultaneous development of wastewater collection network may contribute toward the sustainability of the upcoming projects in Patna. The most notable aspect of this discussion is that the "Namami Gange Program" of government of India has adopted most of the measures stated earlier in order to ensure the execution of sustainable WWTPs. Starting with the overall cost estimation of the projects under "Namami Gange Program," it is observed that the cost sharing of the wastewater treatment plant projects in Patna is based on the Hybrid Annuity Model, which might prove to be helpful to some extent over a limited period of 15 years only. Yet the overall cost analysis has to be properly executed at the initial stage in order to avoid any cost overrun, which would otherwise have an impact on the smooth operation of the WWTPs. Although the cost for procuring alternate sources of power such as generators can be analyzed in the OPEX, yet NMCG has to focus on alternate strategies in terms of ensuring uninterrupted power supply in the WWTPs, such as generating power from the treatment processes. This approach shall not only help in curbing the power demand, but will also be beneficial in establishing self-sustained WWTPs. Another advantage of this approach is that it will ensure the environmental sustainability of the plant by taking care of the sludge generation. The presence of the Ganga Committee at the State and District Levels consisting of administrative officials can prove to be helpful in dispersing the roles and responsibilities in a proper way thus ensuring good governance and maintaining the operation of the wastewater treatment plants in an efficient way, thereby delivering the desired output.

However, achieving these milestones will not be an easy feat, and the sustainability analysis will have to be carried out consistently at each and every level, so as to ensure that the treatment plants always maintain a balance in between the three key indicators of sustainability.

Upon further analysis on the sustainability of the upcoming WWTPs, it has been observed that certain aspects related to overall sustainability are missing. For instance, there is no guarantee on the performance of the treatment technology after 15 years, coupled with no proper planning with respect to sludge disposal and GHG emissions that is a major concern in terms of environmental sustainability. Also, there is no proper revenue collection model that would take care of the functioning of the treatment plants beyond 15 years, which is the key point for ensuring economic sustainability. Moreover, the lack of public participation is the prime indicator that clearly points out the negligence in terms of ensuring social sustainability.

Hence, it can be inferred from the above discussions that in order to achieve the goal of sustainability in the upcoming wastewater treatment plants in Patna, a holistic approach has to be taken in resolving the respective issues, and provisions such as selection of treatment technology during the inception of the projects, resource recovery and public participation must be included such that the adopted treatment scheme not only produces the desired effluent, but is also capable of taking care of the operational costs, power generation and sludge generation, thereby touching the key aspects of sustainability.

Appendix A. Supplementary data

Supplementary data to this article can be found online at https://doi.org/10. 1016/B978-0-12-824038-0.00012-2.

References

[1] Dereszewska A, Cytawa S. Sustainability considerations in the operation of wastewater treatment plant 'swarzewo'. E3S Web Conf 2016;10:00014. https://doi.org/10.1051/e3sconf/2016000014.

[2] Yenkie KM. Integrating the three E's in wastewater treatment: efficient design, economic viability, and environmental sustainability. Cur Opin Che Eng 2019;26:131−8.

[3] Bdour AN, Hamdi MR, Tarawneh Z. Perspectives on sustainable wastewater treatment technologies and reuse options in the urban areas of the Mediterranean region. Desalination 2009;237:162−74.

[4] ENVIS Resource Partner on Control of Pollution Water, Air and Noise. Taken from: http://cpcbenvis.nic.in/cpcb_newsletter/sewagepollution.pdf.

[5] Iyer VG. Design and development of sustainable wastewater treatment plant for sustainable development management of wastewater. Econ World 2017;vol. 5:486−91. https://doi.org/10.17265/2328-7144/2017.05.011. Sep.−Oct.

[6] Muga HE, Mihelcic JR. Sustainability of wastewater treatment technologies. J Environ Manag 2007;88:437−47.

[7] Advisory Note on Septage Management in Urban India. Available on:http://cpheeo.gov.in/upload/uploadfiles/files/Advisory%20Note%20on%20Septage%20Management%20in%20Urban%20India.pdf.

[8] City Development Plan of Patna (2010−2030). http://urban.bih.nic.in/Docs/CDP/CDP-Patna.pdf.

[9] National Mission for Clean Ganga. Available on: https://nmcg.nic.in.

[10] Census. Available on: https://censusindia.gov.in/2011-prov-results/data_files/bihar/Provisional%20Population%20Totals%202011-Bihar.pdf, ; 2011.

[11] Jones CH, Meyer J, K Cornejo P, Hogrewe W, Seidel CJ, Cook SM. A new framework for small drinking water plant sustainability support and decision-making. Sci Total Environ 2019;695:1−10.

[12] Detailed project report - sewerage. Beur STP); 2013.

[13] Preparation of detailed project reports (DPRs) & bid documents tendering for execution; construction supervision & quality Control of sewerage projects of Patna under NGRBA, detailed project report − sewerage. Pahari STP); 2013.

[14] Detailed project report − sewerage system. Karmali Chak Network & STP); 2013.

[15] Murray A, Drechsel P. Why do some wastewater treatment facilities work when the majority fail? Case study from the sanitation sector in Ghana. Waterlines 2011;30:135−49.

[16] Siting of Sewage Treatment Plant (Urban Development and Housing Development, Govt. of Bihar). Available on: http://urban.bih.nic.in/Docs/STP-Guidelines.pdf.

[17] Design, operation and maintenance, published by Karnataka State Pollution Control Board. Available on: https://kspcb.gov.in/STP-Guide-web(Med).pdf.

[18] Bangalore Water Supply and Sewerage Board. Available from: https://www.bwssb.gov.in/com_content?page=3&info_for=4.

[19] Turkmenler H, Aslan M. An evaluation of operation and maintenance costs of wastewater treatment plants: Gebze wastewater treatment plant sample. Des Water Treat 2017;76:382–8.

[20] Khursheed A, Tyagi VK, Khan AA, Bhatia A, Gaur RZ, Ali M, Sharma M, Kazmi AA, Lo S. Operation and maintenance of sewerage systems: present challenges and possible solutions – an Indian experience. Desalination Water Treat 2014:1–16.

[21] Performance Evaluation of Sewage Treatment Plants in Central Zone. Available on: https://www.nqr.gov.in/sites/default/files/File_ETP.pdf.

[22] Performance Evaluation of Sewage Treatment Plants under NRCD. Available on: https://missionganga.thewaternetwork.com/_/wastetreatment/storage/TFX%5CDocumentBundle%5CEntity%5CDocumentaepIZ8b4WcmbOG4WQ8TXwg/6K2Gvuso8H51v07vpTZliA/file/NewItem_195_STP_REPORT.pdf.

[23] Guidelines for Septage Management in Maharashtra. Available on: https://www.cseindia.org/static/mount/recommended_readings_mount/25-Septage_Management_Guidelines_UDD_020216.pdf.

[24] District Magistrates, Rural & Urban Local Bodies Details, Ganga Committee, National Mission for Clean Ganga. Available on: https://nmcg.nic.in/DmUlbsDetails.aspx.

[25] Major Policy Initiatives under NGRBA, Ganga Basin, National Mission for Clean Ganga. Available on: https://nmcg.nic.in/major_policy.aspx.

[26] Preparation of detailed project reports (DPRs) & bid documents tendering for execution. Construction Supervision & Quality Control of Sewerage Projects of Patna under National Ganga River Basin Authority (NGRBA).

[27] Bihar NMCG Project Status as on 10.12.2020. https://nmcg.nic.in/writereaddata/fileupload/27_FinalupdatedMPR%20forMonthofNovemberr2020upload.pdf.

[28] Development of 50 MLD Sewage Treatment Plant at Varanasi under Hybrid Annuity based PPP mode. Available from: https://www.varanasihamstp.in/about-us.

[29] Memorandum of Agreement for Baidyabati Municipality. Available on: https://nmcg.nic.in/pdf/MOA_Baidyabati.pdf.

[30] Preparation of detailed project report on sewerage for Digha of Patna, Bihar-comprehensive waste water management, Digha and Kankarbagh. Bihar: BUIDCo; 2017.

[31] Chai C, Zhang D, Yu Y, Feng Y, Wong MS. Carbon footprint analysis of mainstream wastewater treatment technologies under different sludge treatment scenarios in China. Water 2015;7:918–38. https://doi.org/10.3390/w7030918.

[32] Koul A, John S. A life cycle cost approach for evaluation of sewage treatment plants. Int J Innov Res Adv Eng (IJIRAE) 2015;2(Issue 7):15–20.

Chapter 6

Community approach toward disaster resilience

Surbhi Sharma[1], Vaneet Kumar[2], Saruchi[3]

[1]*Department of Physics, Kanya MahaVidyalaya, Jalandhar, Punjab, India;* [2]*Department of Applied Sciences, CT Institute of Engineering, Management and Technology, CT Group of Institutions Jalandhar, Jalandhar, Punjab, India;* [3]*Department of Biotechnology, CT Institute of Pharmaceutical Sciences, CT Group of Institutions Jalandhar, Jalandhar, Punjab, India*

1. Introduction

Catastrophic natural disasters have undeniably posed threats to human lives as well as their subsistence all over the world. The last decade has shown the tremendous increase in the occurrence of natural hazards annually. Increasing impact of such hazards on lives, livelihood, economic situations, and the food security of vulnerable communities is the matter of concern. Researchers as well as scientists have come up with the common agreement that disasters neither occur accidently nor are *just an act of God.* Disaster is the convergence of vulnerability and hazardous conditions. Until 1970s terms natural hazard and natural disaster were used interchangeably. The intensity of disaster was measured by amplitude of the hazard caused. But later from the 1990s, with the development of research especially in social sciences and humanities, disaster began to be associated not only with the physical or structural impact but also with social and economic damage caused [63]. Natural hazards, namely, earthquakes, volcanoes, tornadoes, and so forth can be highly intense but not always regarded as a disaster. Any hazard is considered to be disaster if it directly or indirectly hits social and economic as well as environmental conditions of the community. Disaster is related to the resistance toward development where the probability of loss and damage characterizes vulnerability [62]. While it is a universal acceptance that disaster is the occurrence of disruption to the normal community life leading to the injury or death, damage to the property and deterioration of health and health-related services, which is beyond the capacity of authorities to cope with their own resources thus requiring special response and mobilization of resources from outside the affected community [14,47,59]. There is a consensus that magnitude of a disaster should be measured not only by extent of loss of lives or property but

Cognitive Data Models for Sustainable Environment. https://doi.org/10.1016/B978-0-12-824038-0.00003-1

also by development policies failure. Studies have overruled myths those considered negligible contribution of human endeavors over the occurrence of events like drought, floods, cyclones, and so forth. Acts like deforestation; desertification has triggered global warming leading to the absurd climatic patterns. Intergovernmental Panel on Climate Change [33] recognized adaptation to present climate scenario as disaster risk reduction activity, which can increase the community resilience to climatic change. Fig. 6.1 shows the elements of natural disaster.

United Nations Development Program [63] reveals that about 75% of world's total population has been exposed to at least one type of disaster, namely, earthquake, volcanoes, tornadoes, flood or drought and more than 184 deaths per day were reported between 1980 and 2000 [63]. Since the year 2000, these disasters have taken toll of more than 1.1 million of human lives whereas have affected more than 2.7 billion of people [21]. With the prime objective to identify ways of establishing disaster resilience among nations, a World Conference on Disaster Reduction was held in January 2005 in Kobe, Hyogo, Japan. A Framework for Action 2005–15 was formulated (later referred to as Hyogo Framework for Action [HFA]), which was adapted by 169 countries [27] for reducing vulnerability and for DRR assessment [70]. apprised annual average losses from natural phenomenon's, namely, earthquakes, tsunamis, floods, and cyclones amounts to about hundreds of billions of dollars, necessitating US$6 billion of investment annually in disaster risk management. UNDP advocates risk-informed approach to development and have disbursed over $1 billion in the year 2018 to strengthen community resilience to shocks and crisis. Studies reveal that disasters and human developments are intimately related to each other. Population explosion beyond the capacity of urban authorities, which fails to cater to the need of basic

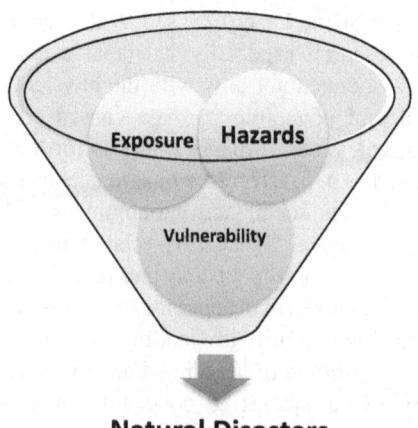

Natural Disasters

FIGURE 6.1 Elements of natural disaster.

infrastructure, leads to the accumulation of population to the risk-prone areas. Generally poor community becomes socially excluded and are pushed to settle in the marginal areas with least access to resources, early warnings, and preventions, thus increasing their vulnerability. Urbanization can lead to modifications in hazard patterns as well as redefines new types of risks.

In December 2019, a new type of pandemic was identified firstly in Wuhan City, China, which spread quickly across the world [24] (WHO, Jan 2020). The novel coronavirus (COVID-19) is a new type of infectious disease that triggers severe acute respiratory syndrome (SARS) [48]. COVID is the acronym for CO—corona, VI-virus, D-disease. COVID-19 is highly contagious and spreads due to human-to-human contact [11]. The virus spreads through the droplets of saliva or discharge of the nose while coughing, sneezing, or talking [79]. This pandemic has hit almost every part of the world and has caused a detrimental impact on health, social, and the economic life of masses [77]. COVID-19 was declared as a global emergency on January 30, 2020 [57]. Globally there are 15,012,731 confirmed cases while 619,150 people have lost their lives, and in South-East Asia 1,571,317 confirmed cases are reported till July 2020 [73].

2. Hazards, vulnerability, and resilience

Hazard is probability of occurrence of some threatening phenomenon or condition with a potential to cause damage to people, injuries or other health impacts, even loss of lives, damage to their properties, disruption of socio-economic conditions, and environmental degradation in a given area for specific time period. Assessment of the hazard can be made depending upon the level of severity and its impact in the particular area for a specific duration. A natural hazard in our context is the occurrence of geophysical events (hurricanes, earthquakes, forest fires, droughts, and so forth) individually or in combination in different areas (namely, coastlines, hillsides, earthquake faults, and so forth) in different times (specific months, particular time of the day. and so forth). When these consequences pose significant impact on community as well as their habitat then such events becomes disaster. With time, various conceptualizations of the term natural hazards have evolved [51], namely, attributed hazards as a result of some minute perturbations leading to the abrupt change in the landscape [8], expressed it as the presence of harmful elements in the physical environment.

Ref. [7] referred hazard as frequency of occurrence of the phenomenon of returning period whereas vulnerability as the degree of damage done by the hazard depending on its severity. Vulnerability is the condition or circumstances of community which makes it susceptible to the hazard. After examining different definitions of natural disaster it is being clear to identify different vulnerabilities to cope with the impact of hazards on the environment and its interaction with people. In this context several definitions for vulnerability are stated viz [42]. defined the vulnerability as the extent to which a system is sensitive to and is unable to cope with adverse climate changes [1]. Vulnerability as a measure of shock or stress a social-ecological system is exposed to and its level of susceptibility as well as capacity to adapt [37] is formulated as

$$\text{Vulnerability} = \frac{\text{sensitivity to stress}}{\text{state relative to threshold}} \times \text{probability of exposure to stress}$$

Frequency as well as severity of natural hazards have even more increased and have given a deeper impact on the world in terms of human and their economic loss. Greater resilience is the need of an hour to cope up with abrupt environmental changes [10]. has interpreted resilience as the ability to self-organize, learn, and adapt. Attaining resilience timely after the impact of disaster is one of the prime objectives of United Nations International Strategy for Disaster Reduction (UNISDR) [68] defined resilience as the ability of as exposed community to resist, absorb, adjust, and recover timely and effectively [56]. states resilience as the capacity of community to absorb and recover from exposure of the hazardous event which measures the rate of recovery.

History reveals that no matter the severity or intensity of disaster/crisis, irrespective of the phenomenon like floods, earthquakes, fires, hurricanes, irradiations, bombing, and so forth, earth has always reestablished like a phoenix. Big cities like Hiroshima and Nagasaki, Baghdad, Moscow, Budapest, Tokyo, and so forth are the examples, while from the 12th to 19th century world have witnessed the complete abandon of 42 cities including Centralia, Pennsylvania; Varosha, Cyprus; Prypiat, Ukraine; and so forth after destruction [72].We cannot escape from the facts that even the most competent and rich government can never guarantee that such disruptive events will not occur. But all the policymakers can frame the effective resilience strategy to mitigate crisis [74]. Resilience strategy focuses on response and recovery operations than on preparedness and mitigation functions [28]. Effective governmental policies for disaster resilience can only be framed after understanding social, cultural, economic, and environmental factors.

3. Community-based disaster management (CBDM) approach

Local community is the first responder to the disaster. So it becomes very important to provide adequate education and awareness regarding preparedness and mitigation techniques at community level especially to the people residing in the vulnerable areas. Community in the current study refers to group of people subjected to common risks or threat [19,40,53,61] and with the capacity to recognize disruptive events. L.J. Carr, a disaster sociologist in

1932 asserted that these communities should be directly involved in planning and preparing for response to disaster, mitigating its effects, thus reducing its risk through community-based disaster management (CBDM) strategy [52]. Community plays a vital role in reducing casualties; disaster loses and increases reduction capacity. CBDM is employed as an effective tool to build the capacity of society for disaster reduction. CBDM is the constructive approach for extrapolating and mitigating future disasters [60]. In [54], it shows that this community approach can help in assessing levels of disaster risk in advance and also enhances the social capacity of preparedness, response, and recovery. In general, CBDM is the approach in which local communities are supported to analyze their vulnerabilities to both natural and human induced hazards and providing resources as well as developing strategies to mitigate the identified disasters. CBDM empowers communities to frame their own strategies with the proactive approach rather than completely relying on the governmental aids. Disaster reduction capacity includes different elements, namely, risk management capacity, early warnings, prevention, disaster information, emergency rescue, CBD reduction at urban and rural levels, science and technology reduction and social mobilization [41]. In the program forum of International Decade for Natural Disaster Reduction (IDNDR) held in Geneva, Switzerland, from July 5—9, 1999, greater attention was paid to community-based action in the disaster reduction. The Geneva Mandate on Disaster Reduction, held in United Nations (1999) on disaster reduction addressed various concerns of risk management related to education, social and economic vulnerability; land use, environment protection etc. It was realized that local communities have greater understanding regarding their habitat and environment and bears more knowledge regarding ways to mitigate the vulnerabilities. Creating leadership among the community can enhance independence as well as self-reliance among them. Geneva Mandate in United Nations(1999) emphasized to construct "disaster—resistance communities" as a disaster reduction goal by expanding risk reduction networks at different levels [69].

Sustained government support, technical assistance and education material is required to apply CBDM approach effectively. But due to insufficient computing techniques of risk assessment as well as lack of legislation, funding etc. various challenges are faced at the time of follow up actions [55]. Community participation is vital for implementing sustainable disaster reduction program at local level which is possible only when local people own the program and shows continuous engagement in such activities. Involvement of community is important in predisaster mitigation as well as in postdisaster response and recovery process.

United Nations office for Disaster Risk Reduction (UNDRR) asserts that the effect of disaster especially on loss of lives can be reduced by making community equipped and prepared.

4. Outline of Total Disaster Risk Management (TDRM)

Asian Disaster Reduction Center is working to establish community resilience among its member countries and establishes the network among them through various programs including exchange of ideas and personnel. For disaster risk reduction at global level, ADRC is cooperating among various UN as well as international bodies, namely, United Nations Office for Disaster Risk Reduction (UNDRR), Asian Disaster Response Unit (ADRU), and so forth. With the prime focus of attaining sustainable development, ADRC in collaboration with United Nations Office for the Coordination of Humanitarian Affairs Kobe (UN-OCHA/Kobe) have framed Total Disaster Risk Management (TDRM) approach as an effective strategy to substantially reduce the adverse effects of natural disaster, particularly in Asia. ADRC asserts that natural disasters are considered to be biggest obstacles to the sustainable development particularly in Asia. Due to geographical as well as geological features, Asia confronts various disasters annuallyFig. 6.2.

The concept of TDRM focuses on following elements [16]:

(1) development of multistakeholder partnership and citizen participation,
(2) advancement in efficient exchange of information,
(3) making assessment and evaluation of risks,
(4) implementation of risk reduction tools and techniques, and
(5) drafting legal framework of policies, their implementation, and maintenance for disaster reduction.

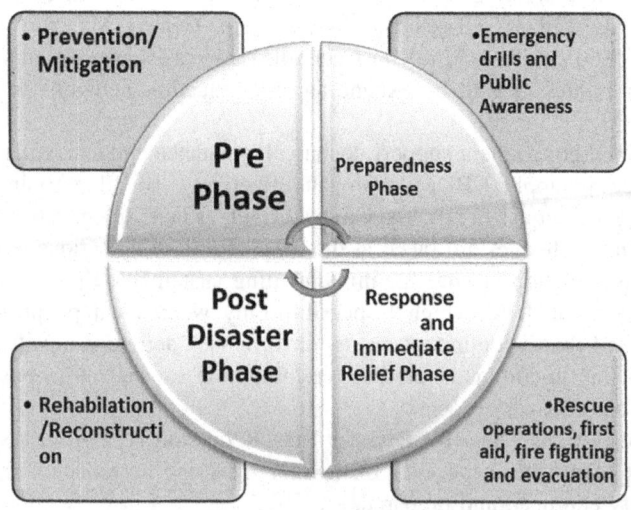

FIGURE 6.2   Disaster reduction cycle.

5. Disaster reduction cycle

Disaster reduction cycle illustrates the plan of government and nongovernment organizations to prevent or reduce the impact of disaster, react during and immediately after the event occurred and finally to take steps for recovery from the disaster. Appropriate and necessary actions taken during the cycle can result into effective early warnings, greater preparedness, reduced impact or even prevention of disaster in the next iterative cycle [44]. The four key activities taken up for disaster reduction are as follows as shown in Fig. 6.2:

(a) **Before the disaster occurs (predisaster phase)**: Activities in predisaster phase aims at preventing or at least minimizing negative effects of future potential hazards. This phase includes building codes and zoning, flood proofing of homes, construction of dams against floods, buying insurance etc.

(b) **Preparedness phase**: Disaster preparedness is possible by vulnerability analysis based on previous experiences and application of strategies likely to reduce the possibilities that hazard will become disaster. It includes the preparation of disaster management plans at community level. Such preparations may include stocking of food items and water in advance, preparation of community maps, which can reflect vulnerable areas and even evacuation routes, and identification of emergency response team, task forces, and volunteers. Activities like mock drills and public awareness are conducted in this phase.

(c) **Response and immediate relief phase**: This phase includes all the initiatives taken during or immediately following the event. It is ensured that all the needs of the victims are met and sufferings can be minimized. It is the most important stage where all the plans are finally brought to the action. All the efforts are made to save lives and preventing further property losses. It is also known as *disaster relief phase*. All activities are focused on understanding need of the community, making critical rapid assessments, providing food and nonfood items, first aid and shelter. In this critical phase information is often confusing so search and rescue activities are very strenuous.

(d) **Postdisaster Phase**: This is the time to bring community back to its normal life or predisaster stage. This is the recovery/rehabilitation/ reconstruction phase. Recovery activities can start when disaster is stabilized. It focuses on meeting with the basic needs of people until more permanent and sustainable alternatives can be found. Recovery measures include reconstruction of damaged properties, trauma counseling, temporary housing, and economic impact studies. Documentation of lessons so learned is done.

6. Case studies

6.1 Community-based disaster management approach in Bangladesh

Bangladesh is a low lying deltaic country formed by three major rivers, i.e., Padma (Ganges), Brahmputra, and Meghna with the long coastal line. The country has low topography with two-thirds of its part lies less than 5 m above sea level. Bangladesh contains 310 rivers and tributaries -[84]. Owing to the geography and climatic pattern, Bangladesh is one of the most disaster prone countries [31]. apprises that in every four to 5 years, country is 65% flooded. Location of the nation is also seismically active, thus keeping the country prone to earthquake as well as tsunami [31]. Studies reveal that approximately 10 million Bangladeshis experience one or more natural hazards per year [26]. Common natural hazards in Bangladesh are tornadoes, floods, earthquakes, drought, tsunamis, arsenic contamination, salinity intrusion and landslides [20]. The country ranks ninth in terms of Global Climate Risk Index 2019 among 10 most affected countries globally for meteorological impact on economy and human fatalities [23]. Bangladesh is located in a tectonically active region. Due to seismic faults some of the major cities of Bangladesh including Dhaka, Chittagong, and Sylhetare always under the threat of massive destruction [45]. Earthquake disaster risk index of Stanford University has declared Dhaka as the most vulnerable cities to earthquakes.

Fig. 6.3 presents an overview of the total number of people killed by various natural disasters from the year 2000–2020. Data has been collected from the source [25]; which is the acronym for data that the Center for Research on the Epidemiology on Disasters (CRED) has been collecting from countries since 1987. It is evident from Fig. 6.3 that Bangladesh has lost its highest lives in catastrophic cyclones.

Over the last two decades, disaster management approach of Government of Bangladesh has shifted from reactive and relief to disaster reduction approach with special emphasis on community-based management. This proactive governmental approach focuses on hazard identification, mitigation, and community preparedness [20]. This paradigm shift approach has certainly shown progress in disaster management, which is evident from the decline in number of deaths as shown in Table 6.1.

In the year 2007, a catastrophic cyclone struck southwest coastal zone of Bangladesh which took 4234 lives, affected over 8.9 million people and caused US$ 2.3 billion of total damage. In 2009, cyclone Aila along with its associated storm surge killed 190 people affecting 3.9 million people and causing US$ 270 million of total damage. In the view of a new model, the Government of Bangladesh (GoB) drafted National Plan for Disaster Management (NPDM) (2010–2015) and then its successor (2016–20) for disaster risk reduction and emergency management in which the government has given special emphasis

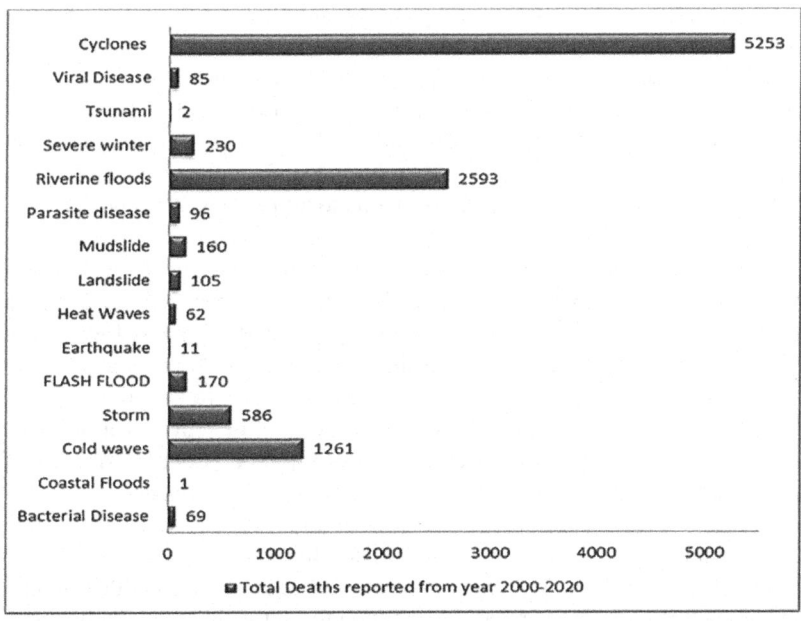

FIGURE 6.3   Disaster mortality losses reported from the year 2000−20 [25].

TABLE 6.1 Data of cyclones struck Bangladesh from 2007 to 20.

Year	Event name	Total deaths	Total affected
2007	Sidr	4,234	8,978,541
2008	Cyclone Reshmi	15	200
2009	Cyclone Aila	190	3,935,341
2009	Cyclone Bijli	7	19,209
2013	Tropical cyclone Mahasen	17	1,498,644
2015	Cyclone Komen	45	2,600,000
2016	Cyclone Roanu	28	1,203,555
2017	Cyclone Mora	7	3,300,012
2019	Cyclone Fani	39	10,045
2019	Tropical cyclone Bulbul	40	251,506
2020	Cyclone Amphan	26	1,100,000

Courtsey: Guha-Sapir D. EM-DAT: the emergency events database. Brussels, Belgium: Universitécatholique de Louvain (UCL) - CRED; n.d. www.emdat.be. [Assessed 1 July 2020].

to community participation from planning to implementation of disaster management policies at a local level. As a result, the death toll in successive cyclone Mahasen in 2013 was 17, during cyclone Komen in 2015 was 45 and most recently cyclone Amphan struck Bangladesh and took 26 lives and caused US$ 5.8 million of total damage [25]. The decline in the death toll and total damage from year 2007−2020 is evident from the above data which thereby indicates the success level of the national policies as adapted.

In NPDM (2010−2015) [45]; a concept of community involvement is introduced and various activities and roles are assigned to the community, namely, in the coastal zones of the country Cyclone Preparedness Program (CPP) is introduced. Under this program Bangladesh Meteorology Department (BMD) gives early warnings regarding cyclones and its associated storm surges to CPP volunteers, those further passes on the information to local people via microphones, sirens, using loudspeakers of religious institutions (Mosque, Temple etc.), or by beating drums [32]. CPP is run by Ministry of Disaster Management and Relief of Bangladesh which operates over 350 unions and 40 upazilas in 13 coastal districts [4]. CPP volunteers are further divided into teams namely: warning, rescue, first aid, response and shelter. People of coastal areas have planted palms and tall trees around their homes and along the roadside as preparedness action for mitigating risk of tidal surge. Local government also seeks community participation in community risk assessment (CRA) and Risk Reduction Action Plan (RRAP).

[45] has given priority to community representatives for decision making on disaster risk reduction through legal framework and have agreed to provide financial assistance to local authorities for developing coordination with communities, civil societies, and migrants for the management of disaster at the local level.

6.2 Empowering community for disaster risk reduction in Nepal

Nepal is a landlocked country also known as *Himalayan Mountain* between China and India. It is 83% of the total area is covered by mountains and hills while rest 17% of area is flat along the Indian border. Nepal has a wealth of nature and biodiversity but is equally prone to adverse effects of natural as well as human induced hazards, namely, floods, landslides, earthquakes, and so forth. Nepal is characterized by irregular topography, extreme weather conditions, and complicated geological structures with an active tectonic location. The country ranks fourth in terms of Global Climate Risk Index 2019 with 10.50 CRI scores for meteorological impact on the economy and human fatalities [15]. UNDRR (2019) cited Nepal at 11th rank in terms of global risk for earthquake occurrence and its impact [67]. Nepal is among the underdeveloped countries with GDP per capita US$ 744.7 with annual GDP growth of 7.5% [65]. Fig. 6.4 presents graphically the disaster mortality losses as reported in the year 2000−2020.

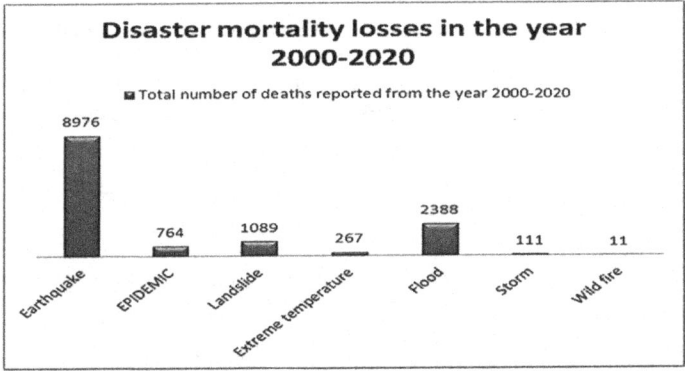

FIGURE 6.4   Disaster mortality losses reported in the year 2000—2020 [25].

Alarming mortality and severe economic loss annually due to natural disaster intrigued interests of government and disaster preparedness activities were taken into consideration. National action plan was enacted in 1996 in which emphasis of role of community in disaster preparedness process at a local level was addressed. Three Year Interim Plan (2007—10) recognized the need to foster the collaboration among government, NGO's and private sectors for providing relief and rescue to the affected people in due time. Government of Nepal (GoN) approved the National Strategy for Disaster Risk Management in 2009 in which collaboration with various stake holders as well as local community was given a special emphasis. Following the directions of Ministry of Home Affairs (MOHA), about 60 District Disaster Relief Committees prepared their District Disaster Preparedness Plan (DDPP). Red Cross societies and other security agencies were also involved in this plan [3]. Table 6.2 summarizes the Socioeconomic impact of disaster on Nepal from 2000 to 2019.

Temporal and spatial variability in climatic patterns has enhanced the variations in the river flow and have led to the increased frequency as well as intensity of floods. Midwestern regions of Nepal are regarded as flood prone areas, which suffer an immense loss of lives, severe damage to properties and production losses annually [18,38,49]. Southern Terai belt, inner Terai and the valley witness regular flooding in the months of June and September due to monsoonal precipitation. Flood event in the year 2008 forced 142 households to evacuate from Holiya Village Development Committee while in 2010, flash flood in Banke district left hundreds of people missing. Devkota & Cockfield, 2014 reported that about 11 people were killed and 2000 houses were struck by floods in Dang region [17]. In 2017%, 80% of the Terai region and its surrounding districts suffered flash floods leading to US$ 584.7 million total damage [46]. In April 2006, United Nations Development Program (UNDP) launched Community-Based Disaster Management Project (CBDMP) to

TABLE 6.2 Socioeconomic impact of disaster on Nepal from 2000 to 2019.

Disaster	Total no. of deaths	No. Injured	Total no. of affected	Total damages (x1,000 US$)
Drought	—	503,000	—	—
Earthquake	8,976	20,449	5,810,099	5,174,000
Epidemic	764	—	65,080	—
Extreme temperature	267	200	25,200	123
Landslide	1089	141	374,896	15,000
Flood	2388	958	4,114,069	883,729
Storm	111	787	15,029	—
Wildfire	11	—	—	—

Courtesy: Guha-Sapir D. EM-DAT: the emergency events database. Brussels, Belgium: Universitécatholique de Louvain (UCL) - CRED; n.d. www.emdat.be. [Assessed 1 July 2020].

support communities especially from 42 wards of Syangja, Tanahu, Chitwan, Makawanpur, Sarlahi and Sindhuli districts. These, as stated districts reported, are vulnerable to floods, landslides and river cut annually. Fig. 6.5 shows the year wise mortality loss in Nepal due to floods as reported in the year 2000–2020.

Understanding the gravity of menace of a flood, CBDMP extended financial and technical support to the center as well as local government [64]. CBDMP empowers communities to reduce geophysical vulnerabilities by the constructing dykes, spurs, and embankments in their wards. Local people engaged themselves to make bamboo spurs to check the flow of water and stone spurs along the banks of rivers for disaster risk reduction.

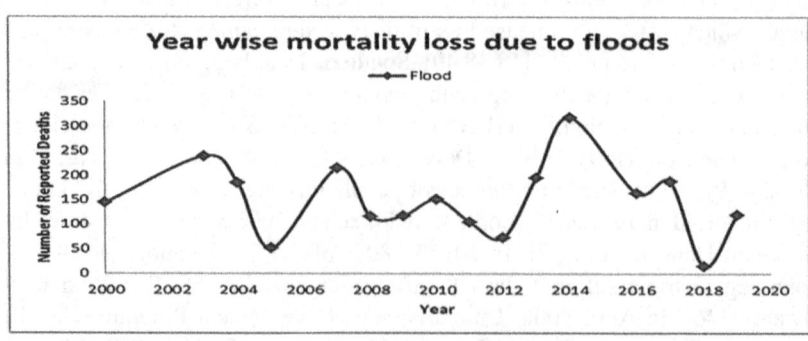

FIGURE 6.5 Year wise mortality loss due to floods reported in the year 2000–20 [25].

Landslide is another natural threat for Nepal resulting in major causalities and damage. Aggravated infrastructural development like construction of buildings in the vulnerable areas, undue increase in construction of roads, expansion of transport infrastructure causes soil erosion, loss of vegetation which in turn leads to destabilization of valley's slope toes [39]. There can be many reasons for landslide like water induced, earthquake, monsoon etc. Ramche Landslide, Rasuwa, activated in 1983 and reactivated on August 14, 2003, Jure Landslide, Sindhupalchowk, 2014, Taplejung Landslides 2015, Flood and landslide in Bhotekoshi 2016 are some of the examples of landslide disaster.

Nepal is situated on the region of a tectonic collision of Indian and Tibetian plates which results in the formation of various seismic faults and fractures. Among them, there are three main fault systems i.e., Main Central Thrust (MCT), the Main Boundary Thrust (MBT) and the Himalayan Frontal Faults (HFF) which makes Nepal vulnerable to earthquake disaster. Recently Gorkha earthquake in 2015 with eight Richter of magnitude struck the country which took 8831 lives, affecting 5.6 million of people and caused US$ 5.17 billion of total damage [25]. As a consequence, thousands of earthquake induced landslides occurred resulting in loss of mountain topsoil and affecting land productivity. As agriculture is the main source of food security and employment for Nepalis, thus natural calamities affect human settlements, agricultural land, employment etc.

For re-building Nepal after Gorkha Earthquake in 2015, several nongovernmental organizations (NGOs) and community-based organizations (CBOs) came forward and extended their help for reconstruction activities. Global Slum Dwellers International network affiliated NGO's namely The National Federation of Squatter Communities and the National Federation of Women's Savings Collectives are among such organizations those have worked with GoN for providing immediate relief as well as long term recovery [3]. The NGO Federation of Nepal (NFN) runs 6233 member NGO's in about 77 districts with the objective of Sustainable Development Goals (SDGs) bridging the gap between governmental policies and local groups.

An NGO known as "Disaster Preparedness Network Nepal (DPNet Nepal) was established in 1996 which is still active to work closely with the government for risk management of disasters such as earthquake, floods, drought, fire, epidemic etc. those occurs frequently in a rural and urban community. Ministry of Water Resources (GoN) established the Department of Water Induced Disaster Management (DWIDM) on Feb 7, 2000 which takes care of water induced disasters like landslide, flood etc. Water Induced Disaster Prevention Technical Center (DPTC) presently known as DWIDM was established under an agreement between the GoN and the Government of Japan on October 7, 1991 [71]. For disaster risk reduction, DWIDM has been organizing training/workshops/seminars for educating local community about the disaster and its preparedness. DWIDM prepares inventory map highlighting hazardous zones of entire country so that community can easily locate hazard and evacuation places.

In 1994, National Society for Earthquake Technology in Nepal (NSET) was established realizing the need to initiate community-based earthquake risk reduction program. The Kathmandu Valley Risk Mitigation Program is among various projects under NSET which involves community leaders, local masons, traders and other stake holders in vulnerability assessments process for providing aseismic school buildings [81]. In April 2006, a disaster risk reduction strategy was promoted under ActionAid Nepal program in which schools were considered as the major resource for disaster mitigation. It was realized that buildings of schools can be used as the temporary shelter at the time of the event and beside students can be used as resource for creating mass awareness as disaster preparedness [27]. Disaster Management Committees (DMC) prepares a community-based disaster preparedness plan in which an analysis of vulnerabilities, risk management as well as community participation is done. Special arrangements for first aid skills, mock drills are made for community preparedness.

6.3 Reporting on community-based disaster management in Indonesia

In terms of area, Indonesia is the 16th largest archipelago state consisting of 17,508 islands scattered on both sides of the equator. Indonesia is susceptible to various natural calamities mainly because of its location between Asian and Australian continents and within the Indian and Pacific oceans. Due to heavy rainfalls, the county is prone to soil erosion, which further leads to landslides. Hydrometeorological activities result in various disasters like floods, droughts, and an outbreak of communicable diseases etc. The country is most susceptible to frequent earthquakes owing to its location on the edges of Pacific, Eurasian and Australian tectonic plates (EC, ADPC, UNESCAP; 2018) [87]. In addition to this, Indonesia has world's 75% dormant and active volcanoes making the country vulnerable to fire ash and tsunami. Fig. 6.6 shows the pie chart of number of reported natural disasters, disaster mortality losses, directly reported cases in Indonesia from 2000 to 2019.

The 2004 earthquake which occurred near the Aceh and Nias islands which subsequently triggered a tsunami in the Indian Ocean is considered the deadliest event in history. Tsunami reached about 15 countries bordering Indian Ocean but Indonesia was worst affected by it. This event killed over 230,000 Indonesians, affected above 2.5 million people and caused US$ 11.4 billion of total damage. One-third of fatalities were reported in the capital of Aceh province alone [76,78]. Considering the magnitude of reconstruction required post-2004 Tsunami Government of Indonesia created the Agency for the Reconstruction and Rehabilitation of Aceh and Nias BRR (Badan

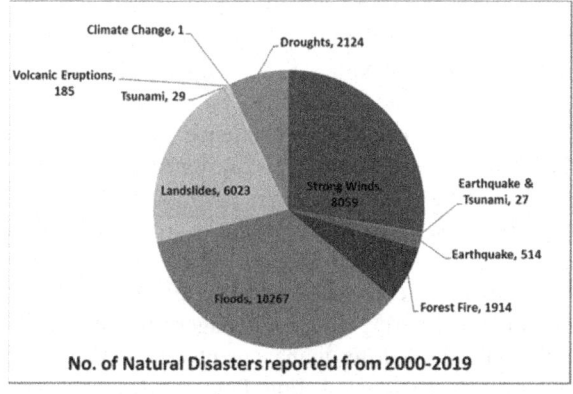

(i)

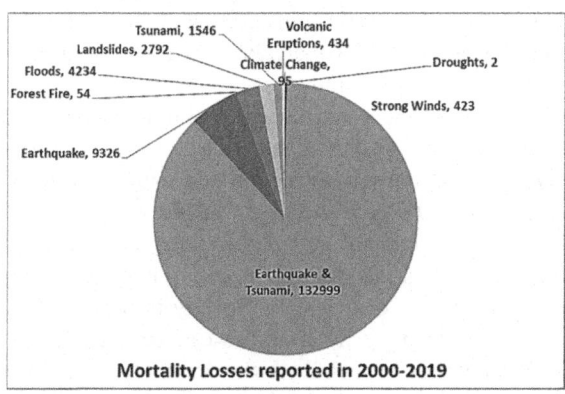

(ii)

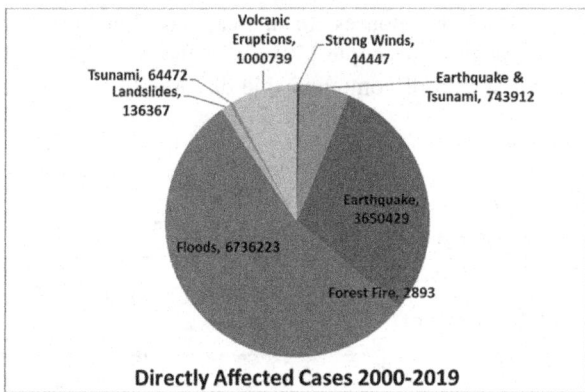

(iii)

FIGURE 6.6 Shows (1) number of reported natural disasters(2) disaster mortality losses (3) from directly reported cases from 2000 to 2019. *Courtesy: Badan Nasional Penanggulangan Bencana (BNPB). Data informasi bencana Indonesia; n.d. http://bnpb.cloud/dibi/. [Assessed 7 July 2020].*

Rekonstruksidan Rehabilitasi Aceh dan Nias) in 2005 to manage recon-struction.Earthquake in Yogyakarta and Central Java in 2006 was another similar destructive event that killed 5782 people and injured about 36,299 people [13]. These events acted as the catalyst for Indonesian government to rethink about existing strategies for resilience development at all levels. Impact of these devastating events was the enactment of the Indonesian Disaster Management Law (DM Law) in 2007. This DM Law governs the disaster management system from preparedness programmes to response and recovery. Later in 2008, National Disaster Management Authority [5] was established to improve the coordination among government agencies, NGOs and other stakeholders. To bring disaster management into the public domain, the role of community in disaster management was realized. Government of Indonesia launched its National Action Plan for Disaster Risk Reduction (NAP-DRR) for 2006–09 in which special emphasis was given to Community-based Disaster Risk Management (CBDRM). The need for building community resilience was perceived in Indonesia during the HFA in 2005 [22]. In the ministry of Indonesia, departments like Home Affairs Department, Social Department and Energy and mineral resource department is working closely with the CBDRM approach as shown in Fig. 6.7.

Community efforts in the reconstruction and rehabilitation process in the Aceh province proved to be very successful in history.

Increased human settlements as well as other anthropogenic activities, namely, deforestation, illegal land capturing of swampy areas, wetlands, etc., have deteriorated the river's conveyance capacity in the country. As a result, Indonesia is worst affected by floods those can be categorized as riverine, flash, tidal and urban floods, especially during monsoons. Flash flooding is one of the frequent events particularly in the region of Seulawah Mountains in Aceh Province. This water inundation carries slit, sediments and other debris which can easily ruin crops, houses etc. and can have long term direct as well as other indirect impacts on people [50]. After the onset of floods there are greater chances of spread of communicable diseases like fecal oral diseases,

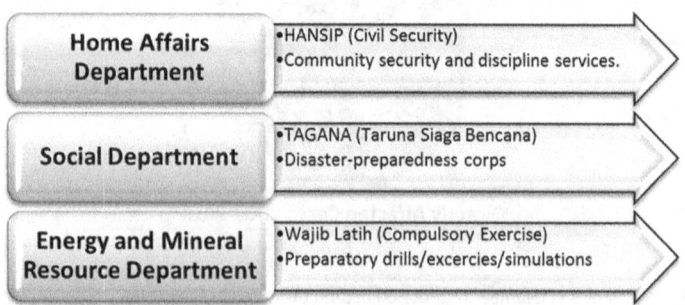

FIGURE 6.7   CBDRM approach of Ministry of Indonesia.

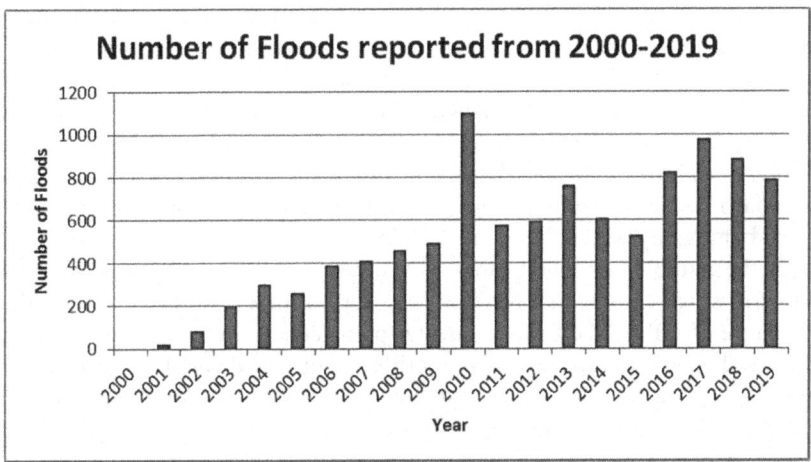

FIGURE 6.8   Number of floods reported from the year 2000–2019. *Courtesy: Badan Nasional Penanggulangan Bencana (BNPB). Data informasi bencana Indonesia; n.d. http://bnpb.cloud/dibi/. [Assessed 7 July 2020].*

vector-borne diseases [2]. The year 2010 accounts for a maximum number of occurrence of floods i.e., 1101 events in which about 500 people died, 353,523 people got directly affected whereas 625,202 people got indirectly affected [5]. Badan Nasional Penanggulangan Bencana (BNPB) reports that Bali has witnessed the maximum number of mortality losses i.e., 15,042 followed by provinces Jawa Tengah, JawaTimur with losses 1007, 996 respectively due to floods from the year 1815–2019 [5]. Fig. 6.8 presents the year wise frequency of floods in Indonesia from 2000–2019.

Mt. Merapi volcano in the central Java and Yogyakarta province is one of the most active volcanoes and consists of dangerous composites [34]. The year 2010 Mt. Merapi volcanic destruction is considered largest-ever eruption killing 322 and affecting 137,140 people [25] with the estimated damages of about US$360 million as per World Bank office in 2012 [86]. Fly ashes produced due volcanic eruption leads to major losses of crops. Volcanic dust, hot ashes and poisonous gases destroyed vegetation, livestock and all the essentials of local communities.

For reconstruction and rehabilitation post-Mt. Merapi year 2010 eruptions, Government of Indonesia again realized the importance of community participation through Rekompak program. Rekompak program was already proved to be very successful and effective during the post-2006 Aceh Tsunami and Yogyakarta earthquake reconstruction. Rekompak is the acronym for Rehabilitasidan Rekonstruksi Pemukiman Berbasis Masyarakat (Community-based Housing Rehabilitation and Reconstruction) which gives community an equal partnership to work for reconstruction and rehabilitation with local government [89]. In the Rekompak program, local community is empowered to reconstruct and rehabilitate their own homes as well as public infrastructure under the assistance of task forces which comprises of technical experts, construction supervisors and finance specialists. To minimize the level of corruption funds are directly channeled to communities. Funding was raised by Java Reconstruction Funds (JRF) which was established post-2006 Aceh Tsunami and Yogyakarta earthquake in July 2006. The success of Rekompak approach lies in fostering the sense of ownership and decision making among local community which fills them with the strength to control even situations after the future misfortune. Another key aspect of the Rekompak program is the focus on community preparedness for disaster resilience. Under the guidance of village facilitators, disaster preparedness committees are formed which regularly conduct evacuation drills, training and simulations in which the whole village participates. Communities are trained to identify the hazards and to learn mitigation and preparedness techniques using simulations (World Bank, 2012).

Indonesia is among the "middle-income countries" which is frequently eroded by disasters impelling its government decline to request for humanitarian aid from various international agencies. Different national-international organizations, namely, UNICEF, AusAid, Japan International Cooperation Agency (JICA), International Red Cresent, International Organization of Migration (IOM), European Commission, Asian Disaster Preparedness Center (ADPC), and so forth has extended their support from time to time for disaster mitigation as well as resilience. This scenario has opened the door for many domestic faith-based organisations to play an active role in disaster management. Muhammadiyah, claims to be the largest as well as oldest domestic NGO, which is running several hospitals, clinics, universities, schools etc. nationwide. Muhammadiyah, a social welfare organization was founded in 1912 in Yogyakarta, Central Java [9]. Ironically, people consider bad calamity (sayyi'ah) as the anger of God toward sinner and local communities used to perform offering ceremonies to satisfy the spirits. Here faith-based organizations play an important role in providing psycho-social recovery [75].

Past experiences of 2004 Aceh-Niahs earthquake and tsunami, and the earthquake in Yogyakarta and Central Java in 2006 lead Muhammadiyah'sto establish Muhammadiyah Disaster Management Center (MDMC) with the focus to achieve disaster mitigations and recovery through Sekolah Siaga Bencana (Disaster-Prepared School) and Rumah Sakit Siaga Bencana (Disaster-Prepared Hospital). For disaster response and mitigation, MDMC is providing disaster preparedness training at school level Child Disaster Awareness for School and Communities (CDASC), through Hospital and Community Preparedness for Disaster Management (HCPDM) it is providing training to hospital staff, Volcano Community-Hospital Ring (VaCHRi), thus strengthening the community toward disaster [6]. Faith-Based communities thus play an important role in disaster management.

6.4 India's community-based disaster risk reduction plan

India is prone to a wide range of disasters, in particular earthquake, landslides, floods, cyclones, tsunamis, drought, extreme heatwaves and wildfires. Owing to its unique geo-climatic as well as socioeconomic conditions, India is among the 10 worst disaster prone countries. In 2018, India ranked fifth in the Global Climate Risk Index chart with 18.17 CRI scores [15]. India can be divided into five regions i.e., Himalayan region, the alluvial plains, Indian deserts, the hilly part of the peninsula, and the coastal zone. Each region is vulnerable to one or more disasters, namely, the Himalayan region frequently witnesses earthquakes and landslides because Indian tectonic goes below Eurasian plates in this region. Plane regions are susceptible to floods whereas due to change in oceanic pressure, coastal areas are more prone to cyclones, storms and tectonic movement under the ocean floor causes tsunamis. Himalayan region is a source of various rivers which makes Uttar Pradesh and Bihar victims of floods every year. Western parts of the country including Rajasthan, Gujarat and few areas of Maharashtra are drought prone. Indian Meteorological Department in 2019 declared that year 2018 was the sixth driest year with only 56% of long term average monsoon [82] while the year 2019 stood seventh warmest year since 1901. Another anomaly was seen in the behavior of the Arabian Sea in 2019, where out of eight cyclones, five intense storms were originated over the Arabian Sea -[88].

India has a7516 km long coastline which is surrounded by Arabian Sea on its left and Bay of Bengal to its right. Studies show that India is exposed to about 10% of the world's cyclones where majority of them originate from Bay of Bengal followed by Arabian Sea in a ratio of 4:1 [21,58]. States along coastline i.e., Gujarat, Maharashtra, Goa, Karnataka, Kerala, Tamil Nadu, Puducherry, Andhra Pradesh, Orissa and West Bengal along with Andaman, Nikobar and Lakshwadeep Islands are frequently affected by cyclones [21]. reports that Tamil Nadu -Andhra Pradesh, Orissa- West Bengal coasts witness's highest and intense storms during October—December (NE monsoon). Eastern India is more vulnerable to tropical cyclones as compared tothe western part [35]. The destructive winds accompanying cyclones cause greater damage. Table 6.3 summarizes the data of cyclones struck India from 2007—2020.

Drought is a complex ecological challenge that causes a serious threat to social, economic as well as environmental conditions. The Indo-Gangetic Plain was among the regions that experienced severe drought in 2015 with a rainfall deficit of 25.8% [43].From 2014 to 15 India faced a 12% decline and from 2015 to 16,a 14% deficit in rainfall was observed [66]. [25] report apprises that in 2015 about 330 million of people got affected and India faced huge damage of US$ 3000 million.As per the report presented by Ministry of Agriculture and Farmer's Welfare, Government of India, 266 districts

TABLE 6.3 Data of cyclones struck India from 2007 to 2020.

Year	Event	States
2007	Sidr	West Bengal, Orissa
2009	Cyclone Aila	West Bengal
2009	Cyclone Phyan	Gujarat, Madhya Pradesh, Maharashtra provinces
2010	Cyclone Laila	Andhra Pradesh, Tamil Nadu
2010	Cyclone Jal	Andhra Pradesh
2011	Cyclone Thane	Tamil Nadu, Puducherry
2012	Cyclone Nilam	Andhra Pradesh, Tamil Nadu
2013	Cyclone Phailin	Orissa, Andhra Pradesh, Jharkhand, Bihar, West Bengal, Chhattisgarh
2013	Helen	Andhra Pradesh
2013	Tropical cyclone Mahasen	Andhra Pradesh
2014	Cyclone Hudhud	Andhra Pradesh, Orissa, Chhattisgarh
2016	Cyclone Vardah	Tamil Nadu; Andhra Pradesh
2017	Cyclone Mora	Manipur, Mizoram
2017	Cyclone Ockhi	Kerala, Tamil Nadu, Andhra Pradesh, Lakshadweep islands
2018	Tropical storm Titli	Andhra Pradesh and Odisha
2018	Cyclone Gaja	Tamil Nadu
2018	Tropical storm Phethai	Andhra Pradesh, Odisha
2019	Cyclone Fani	Odisha
2019	Tropical cyclone Bulbul	South Assam, Meghalaya, Tripura, Mizoram, West Begal, Odisha state
2020	Cyclone Amphan	West Bengal, Odisha state
2020	Cyclone Nisarga	Maharashtra

Courtesy: Guha-Sapir D. EM-DAT: the emergency events database. Brussels, Belgium: Universitécatholique de Louvain (UCL) - CRED; n.d. www.emdat.be. [Assessed 1 July 2020].

across 11 states were officially declared drought in 2015—16. Andhra Pradesh, Karnataka, Maharashtra and Uttar Pradesh were among the worst affected states [66]. Fig. 6.9 presents the graphical view of losses due to drought in India from the year 2000-2018.

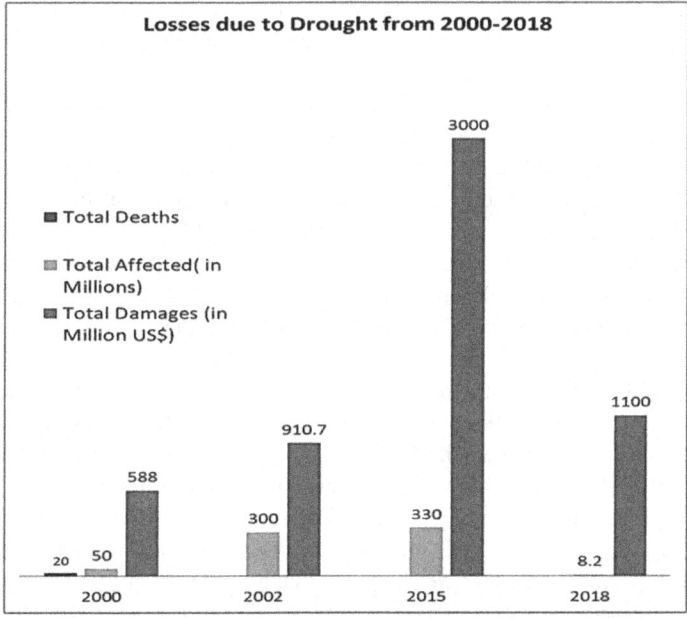

FIGURE 6.9 Losses due to drought reported from 2000 to 2018. *Courtesy: Guha-Sapir D. EM-DAT: the emergency events database. Brussels, Belgium: Universitécatholique de Louvain (UCL) - CRED; n.d. www.emdat.be. [Assessed 1 July 2020].*

Flood is another havoc which the country experiences every year. According to Ref. [21] almost one – eighth part of the country is flood prone owing to its geographical reasons. Monsoons, silted rivers, erodible mountains especially Himalayan range are the factors of floods. Floods bring string of miseries, namely, outbreak of diseases, scarcity of drinking water, agriculture losses etc. along with it. As per the assessment made by the Ministry of Environment and Forest, Government of India [85], by 2030 Himalayan regions of India may witness an increment in temperature up to 2.6°C and increase in intensity by 2%–12% with respect to 1970, consequently rise in glacier melts will increase the frequency of floods in India. In the mountain erodible areas floods can also trigger landslides. While in Western Ghats and coastal zones, a rise of about 1.7–1.8°C in temperature with respect to 1970, may be observed.

Northeast India, the northern portion of Bihar, Uttarakhand, Himachal Pradesh, Jammu, Kashmir, Gujarat, and Andaman and Nicobar Islands are considered to be the most active seismic zone [21]. The Bhuj earthquake of 2001 in the Kutchh district of Gujarat is the most devastating event in the history where death toll of 20,005 was reported whereas about 6.3 million people were affected and 1.79 people were homeless. Gujarat 2001 earthquake caused US$ 2623 million of total damage [25].

Realizing the need for community as well as other stakeholder participation in disaster risk reduction,the Government of India enacted National Policy on Disaster Management in 2009 emphasizes on disaster prevention, preparedness and mitigation. The need for capacity building among local authorities including Panchayati Raj/Gram Sabha, Municipalities is urgently imperative. The 73^{ed} and 74th constitutional amendments recognize Panchayati Raj and Municipalities as institutions of *self-government*.

Government of India sensitized that capacity building of vulnerable communities along with professionals and personals should be prioritized for developing sustainable livelihoods. Indian Government is disseminating adequate knowledge as well as providing training to the community to tackle any misfortune event.

Under [36], 169 districts across selected most vulnerable 17 states of India, preparedness and mitigation plans at the state, district, block, village and ward were chalked out. At the village level, Village Disaster Management Committee (VDMC) is formed which is responsible for framing and planning disaster preparedness programs across the region. VDMC consists of local NGOs, local representatives, members of youth groups, namely, National Service Scheme (NSS) and Nehru Yuva Kendra Sangathan (NYKS), women groups, local government representatives. At the village level, seasonal calendars are prepared to indicate the best times for conducting mock drills, exercises for disaster preparedness. Similarly, Gram Panchayat Disaster Management Committee (GPDMC) is responsible for the administrative unit intermediate between block and village [36].

State wise various initiatives are taken for disaster mitigation and preparedness. Uttar Pradesh's government used local media as a tool to educate people about the disaster. Various *NukadNataks* (street shows), puppet shows are organized from time to time for creating awareness about various challenges as well as cope up techniques for disaster. The local government of Maharashtra has used auto rickshaws as the carrier for spreading disaster awareness. To spread awareness among children, under SarvSikshaAbhiyan (2000—01) a compulsory subject on Disaster Management was introduced for school students. During the year 2002—03 local government of Jaipur introduced 7 days teacher training program UDAY II in which general awareness about disaster was imparted [29].

Over many years National Cadet Corps (NCC), National Service Scheme (NSS), Nehru Yuva Kendra Sangathan (NYKS), Bharat Scouts are volunteering and extending help for disaster management. National Disaster Management Authority in 2016 initiated ApdaMitra Scheme to train the community for disaster response, thus giving first responders to the disaster a central stage. Volunteers under this scheme stood front line during various misfortune events, namely, Kolhapur flood (2019), Cyclone Fani, Odisha (2019) Sitamadhi flood, and Bihar (2019) [90].

In nutshell, initiatives by GOI of organizing awareness campaigns, mason training, gender equity in disaster management, forming youth groups, namely, NSS, NCC, NYKS, and Bharat Scouts etc. are among the highlights of its disaster management programs. Under Sarv Siksha Abhiyan introduction of Disaster Management as the compulsory subject in the curriculum is another initiative to educate the masses.

6.5 Japan's disaster risk reduction plan

Japan is the island country with four major islands surrounded by more than 4000 small islands in the western Pacific Ocean. Japan Islands are located on the convergent boundaries of Pacific, North American, Eurasian, and the Philippine Sea plates making it tectonically active region. An earthquake under the ocean floor can further lead to tsunamis. In addition to this, Japan has about 10% of active volcanoes of the world those can erupt any time [83]. Despite the small area, Japan has a variety of climates ranging from subarctic to subtropical. Japan faces strong winds from typhoons every year, which can further trigger both floods and landslides.

Japan faces serious natural disasters every year owing to its geophysical conditions. The great Kanto earthquake (1923) is considered to be the worst disaster in the history of Japan that struck Kanto plains and devastated cities of Tokyo and Yokohama which caused mortality losses of about 100, 000 [80]. It further triggered tsunami waves as well as a fire break out, which made this disaster an even more destructive and Japan suffered from the total damage loss of about US$ 600 million [25]. Great Hanshin-Awaji Earthquake in 1995 was another deadly event which hit the city of Kobe, Hyogo prefecture in which about 6437 people were killed 43,792 injuries were recorded and due to fire spread about 104,906 houses were completely burned [91]. The Great East Japan Earthquake in 2011 is the largest recorded earthquake with a magnitude of 9.0 Richter's hit at the coast of Sanriku. Additionally, vibrations near the boundary of Pacific Plate and plates beneath Tohoku area caused under seafloor movements which further triggered a tsunami. In this massive disaster, more than 15,000 people lost their lives while 2681 people got missing as of April 10, 2013. Total damage of about 16.9 trillion yen was faced by Japan [30].

Typhoons add into the miseries of Japanese every summer. Winds associated with typhoons are so much powerful that can blow trees, buildings along with them and can trigger floods as well as landslides. Typhoon Makurazaki in 1945 hit Kyushu, Kanto, and killed 3746 whereas more than 1.3 million people got affected and about US$ 400,000 of total damage was reported. Typhoon Vera (1959) was one of the disastrous events which affected the almost entire nation, more than 5000 mortality losses were reported while about 1.5 million were affected and total damages of about US$600000 was suffered by the country [25]. Table 6.4 lists the storms struck Japan from the year 2000−2019.

TABLE 6.4 Data of storms stuck Japan from 2000 to 19.

Year	Event name
2000	Kirogi, Saomai
2001	Pabuk, Danas
2002	Halong, Chata'an, Rammasun, Sinlaku
2003	Etau, Maemi, Meari (Quinta)
2004	Dianmu (Helen), Meari (Quinta), Tokage (Siony), Chaba, Ma-on (Rolly), Songda (Nina), Megi (Lawin), Aere (Marce)
2005	Nabi (Jolina), Mawar
2006	Shanshan
2007	Man-Yi, Wipha/Goring, Fitow
2009	Etau, Melor
2011	Roke, Talas, Ma-on
2012	Bolaven, Jelawat, Sanba
2013	Wipha, Man-Yi
2013	Toraji, Fitow
2014	Neoguri, Phanfone, Nakri, Halong, Vongfong
2015	Nangka, Chan-Home, Etau, Noul (Dodong), Goni (Ineng), Dujuan
2016	Mindulle, Lionrock, Malakas, Chaba
2017	Lan/Paolo, Talim, Noru
2018	Jebi, Trami, Kong-Rey
2019	Hagibis, Lingling, Tapah, Faxai

Courtsey: Guha-Sapir D. EM-DAT: the emergency events database. Brussels, Belgium: Universitécatholique de Louvain (UCL) - CRED; n.d. www.emdat.be. [Assessed 1 July 2020].

The disaster Management plan of Japan includes National, Prefectural and Municipal level as shown in Fig. 6.10.

After the Great Hanshin-Awaji Earthquake in 1995, establishing community-based disaster prevention program was a major objective of Government of Japan. Under the umbrella of Disaster Safe Welfare Communities "BOKOMI" emergency mock drill programmes, first aid training are provided to local vulnerable community. Various NGO's like plus Arts have come forward and have extended their help for community awareness about disasters.

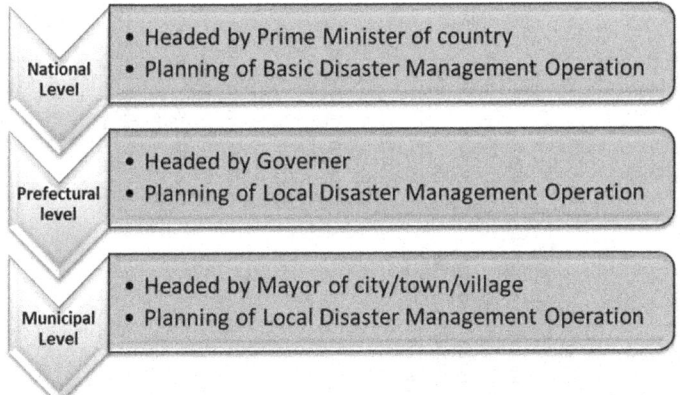

FIGURE 6.10 Scheme of disaster management operation plan.

The community disaster risk management (CDRM) plan includes the involvement of the vulnerable community in identification of disaster risks as well as hazard areas. To share experiences about the Community-Based Disaster Management Plans, the cabinet office held a [12] in Osaka City on March 16, 2019. In this forum, a network of local government officials those are intended to work on CDRM and were named as Chikubo'z. In addition to the involvement of various NGOs, volunteers etc., a wide community associated with research is also working on planning and mitigation of disasters. This network comprises about 57 academic societies till March 2019 [91].

7. Risk mitigation analysis

Our research work reveals that as referred countries are working on the principle of *Prevention is better than cure* and also seeks community participation in community risk valuation and preparedness. All Asian countries have developed a consensus to organize training/workshops/seminars for educating local community about disaster and its preparedness. Special arrangements for first aid skills, mock drills for community preparedness have also been made from time to time. A comparative analysis of disaster mitigation and preparedness strategies of countries under study is given in Table 6.5.

In addition to funds generated by the government of respective countries various other agencies, namely, United Nations Development Program (UNDP), United Nations Educational, Scientific and Cultural Organization (UNESCO), United Nations Children's Fund (UNICEF), European Civil Protection and Humanitarian Aid Operations (ECHO), World Health Organization (WHO), and also extends financial support to as referred countries. Our current study reveals that preparedness and mitigation strategies of

TABLE 6.5 Comparative analysis of disaster mitigation and preparedness strategies.

	Bangladesh	Nepal	Indonesia	India	Japan
Global climate risk index (2017)	9	4	50	14	36
Programmes initiated	Cyclone Preparedness Program (CPP), National Plan for Disaster Management [45], Bangladesh Climate Change Strategic Action Plan (BCCSAP), National Plan for Disaster Management 2010–15.	National Strategy for Disaster Risk Management (2009), District Disaster Preparedness Plan (DDPP), Community-Based Disaster Management Project (CBDMP), Disaster Preparedness Network Nepal (DPNet Nepal), Department of Water Induced Disaster Management (DWIDM), National Society for Earthquake Technology-Nepal (NSET), ActionAid Nepal Program	Agency for the Reconstruction and Rehabilitation of Aceh and Nias BRR, National Action Plan for Disaster Risk Reduction (NAP-DRR), Rekompak program	National Policy on Disaster Management 2009, [36]; National Disaster Management Plan (NDMP), 2019, Indigenous Technical Knowledge (ITK), India Disaster Resource Network (IDRN), India Disaster Knowledge Network (IDKN)	Science and Technology Research Partnership for Sustainable Development (SATREPS), Enterprise Resilience Rated Loan Program, Community Disaster Management Plan forum, Three Year Emergency Response Plan for Disaster Prevention, Disaster Mitigation, and Building National Resilience (Dec 2018)

Disaster management tools				
Disaster Management Act (2012), Sendai Framework for Disaster Risk Reduction (SFDRR) 2016 −30	Natural Calamity (Relief) Act, 1982, Sendai Framework for Disaster Risk Reduction (SFDRR) 2016−30, Disaster Risk Reduction and Management Act, 2017, Natural Calamity (Relief) Act, 1982, Local Self Governance Act, 1999, National Action Plan for Disaster Risk Management 1996, National Strategy for Disaster Risk Management (NSDRM), 2009, National Disaster Risk Reduction Strategic Action Plan, 2018 −2030 (NDRRSAP)	Indonesian Disaster Management Law (DM Law) in 2007	ApdaMitra Scheme, The Disaster Management Act, 2005, PradhanMantriGraminAwaasYojana 2015	[12]; Amended Disaster Relief Act (2018),Sendai Framework for Disaster Risk Reduction (SFDRR) 2016 −30, Road Traffic Act, the Flood Control Act, and the Port and Harbor Act, Making Local Areas Resistant to Tsunami Act No. 123, Promotion of Tsunami Countermeasures Act No. 77, Making Local Areas Resistant to Tsunami, Law No. 123 of 2011, Art. 10.

Continued

TABLE 6.5 Comparative analysis of disaster mitigation and preparedness strategies.—cont'd

	Bangladesh	Nepal	Indonesia	India	Japan
Authorities working	National Disaster Management Council (NDMC), Inter-Ministerial Disaster Management Coordination Committee (IMDMCC); National Disaster Management Advisory Committee (NDMAC); National Platform for Disaster Risk Reduction (NPDRR); Earthquake Preparedness and Awareness Committee (EPAC); and Focal Point Operation Coordination Group of Disaster Management (FPOCG).	National Risks Reduction and Management Authority (NRRMA)	National Disaster Management Authority [5]	National Disaster Management Authority 2005, Cabinet Committee on Management of Natural Calamities, Panchayati Raj/Gram Sabhaand Municipalities, Gram Panchayat Disaster Management Committee (GPDMC), District Disaster Management Authority	Disaster Safe Welfare Communities "BOKOMI", Chikubo'z, a Network of Local Government Officials Working on Community Disaster Risk Management Plans, Japan Meteorological Agency (JMA), Fire and Disaster Management Agency, Establishment of the Megaflood Management Committees

	Bangladesh	Nepal	Indonesia	India	Japan
Community participation	Community Risk Assessment (CRA) and Risk Reduction Action Plan (RRAP). NGO Coordination Committee on Disaster Management (NGOCC), Bangladesh Red Crescent Society (BRCS)	Nepal Red Cross, NGO Federation of Nepal (NFN)	Indonesia NGO, faith-based or religious Organizations, Civil Society Organizations (CSOs)	National Service Scheme (NSS) and Nehru Yuva Kendra Sangathan (NYKS), Woman groups, Bharat Scouts, Red cross society	Japanese Red Cross Society, Empowerment of Female Fire Corps Volunteers, NGO's, NPO's
Funding agencies	National/District Disaster Management Fund, Grameen Bank, Proshika, Bangladesh Rural Advancement Committee (BRAC), Association for Social Advancement (ASA)	Princep Disaster Relief Fund, CDP Nepal Earthquake Recovery Fund, Prime Minister Relief Fund, Natural Calamity Relief Fund	Java Reconstruction Funds (JRF), Contingent Fund (Dana Darurat) Government Regulation 44/2012	Chief Minister's Disaster Relief Fund, MGNREGS, PradhanMantriAwasYojana, (PMAY), National Disaster Response Fund (NDRF), National Calamity Contingency Fund, Calamity Relief Fund	Disaster relief fund, Art. 24: Inclusion of funds for the redemption of principal and interest related to small disaster bonds in the standard budget request

Continued

TABLE 6.5 Comparative analysis of disaster mitigation and preparedness strategies.—cont'd

	Bangladesh	Nepal	Indonesia	India	Japan
Disaster response and mitigation	Organizing mock drills for disaster preparedness, Early warnings, Construction and maintenance of cyclone and flood shelters, supply of drinking water and food, first aid.	School Safety Program (SSP), reconstruction of earthquake-resistant schools	SekolahSiagaBencana (Disaster-Prepared School) and RumahSakitSiagaBencana (Disaster-Prepared Hospital), Child Disaster Awareness for School and Communities (CDASC), Hospital and Community Preparedness for Disaster Management (HCPDM, 2009)	People participation through gram sabhas, Organizing mass awareness programmes like nukadnatak, puppet show, disaster education through SarvSikshaAbhyan (2000–01), early warnings, risk assessment at local level, organizing training programmes and mock drills	Japan has implemented an extensive program of building tsunami walls of up to 4.5 m (13.5 ft) high in front of populated coastal areas, tsunami preparedness day- every year on November 5, rescue program workshop, Nuclear Emergency Core Hospitals, community tsunami preparedness measures, School Tsunami Preparedness

government of respective countries stress on community participation in disaster management. Government of Bangladesh has drafted [45] in which priority is given to community representatives for decision making on disaster risk reduction through legal framework and have agreed to provide financial assistance to local authorities for developing coordination with communities, civil societies and migrants for the management of disaster at the local level. Local government also seeks community participation in CRA and RRAP. The prime focus of ActionAid- Nepal program is Disaster Management and Policy Advocacy, Community Empowerment for Disaster Management, Community Led Reconstruction Program. Government of Indonesia have also given key importance to community participation through Rekompak program. Rekompak is the acronym for Rehabilitasidan Rekonstruksi Pemukiman Berbasis Masyarakat (Community-based Housing Rehabilitation and Reconstruction) which gives community an equal partnership to work for reconstruction and rehabilitation with local government. Even the faith-based communities like Muhammadiyah's are playing an active role in disaster management in Indonesia. In 2016 National Disaster Management Authority of India initiated ApdaMitra Scheme to train community for disaster response. Over many years NCC, NSS, NYKS, Bharat Scouts are volunteering and extending help for disaster management. In Japan under the umbrella of Disaster Safe Welfare Communities "BOKOMI" emergency mock drill programmes, first aid training are provided to a local vulnerable community.

Thus, disaster mitigation strategies are adopted by the government of the respective countries. Mutual collaboration for disaster management is also extended and financial support from UN bodies, Asian Development Bank (ADB), World Bank, and so forth is also fetched.

8. Conclusion

Alternative measures for disaster management are undertaken by the governments of Asian countries giving key importance to community involvement in vulnerability assessment, disaster preparedness, mitigation, and resilience. The vulnerable community is first responder to the disaster thus the need of providing them a central stage in disaster management is realized worldwide. Effective community participation, as well as involvement of NGOs, volunteers and other stakeholders, are important elements of DRR. Community participation is vital for implementing a sustainable disaster reduction program at a local level which is possible only when local people own the program and show continuous engagement in DRR activities. The involvement of the community is important in predisaster mitigation as well as in the postdisaster response and recovery process. In this chapter, a brief study of CBDM strategies adopted by different Asian countries, namely, Bangladesh, Nepal, Indonesia, India, and Japan have been presented. As studied countries have come to a consensus that community participation is vital for implementing sustainable disaster reduction program at local level which is possible only when local people own the program and shows continuous engagement in such activities. The involvement of community is important in predisaster mitigation as well as in postdisaster response and recovery process. Thus, CBDM is the dominant approach accepted worldwide.

Appendix A. Supplementary data

Supplementary data to this article can be found online at https://doi.org/10. 1016/B978-0-12-824038-0.00003-1.

References

[1] Adger WN. Vulnerability. Global Environ Chang 2006;16:268−81.

[2] Ahern M, Kovats RS, Wilkinson P, Few R, Matthies F. Global health impacts of floods: epidemiologic evidence. Epidemiol Rev 2005;27:36−46.

[3] ALNAP. Nepal earthquake response: lessons for operational agencies. London: ALNAP/ ODI; 2015.

[4] Azad MAK, Uddin MS, Zaman S, Ashraf MA. Community-based disaster management and its salient features: a policy approach to people-centred risk reduction in Bangladesh. Asia Pac J Rural Dev 2020:1−26. https://doi.org/10.1177/1018529119898036.

[5] Badan Nasional Penanggulangan Bencana (BNPB). Data informasi bencana Indonesia; n.d. http://bnpb.cloud/dibi/. [Assessed 7 July 2020].

[6] Baidhawy Z. The role of faith-based organization in coping with disaster management and mitigation: Muhammadiyah's experience. Journal Indones Islam 2015;09(Number 02):167−94.

[7] Blaikie P, Cannon T, Davis I, Wisner B. At risk: natural hazards, peoples vulnerability, and disasters. London: Routledge; 1994.

[8] Burton I, Kates RW. The perception of natural hazards in resource management. Nat Resour J 1964;3:412−41.

[9] Bush R. Muhammadiyah and disaster response: innovation and change in humanitarian assistance. In: Brassard C, et al., editors. Natural disaster management in the Asia-Pacific: policy and governance, disaster risk reduction. Japan: Springer; 2015. p. 33−49. https:// doi.org/10.1007/978-4-431-55157-7_3 [Chapter 3].

[10] Carpenter S, Walker B, Anderies JM, Abel N. From metaphor to measurement: resilience of what to what? Ecosystems 2001;4(8):765−81.

[11] Cascella M, Rajnik M, Cuomo A, Dulebohn SC, Di Napoli R. Features, evaluation and treatment coronavirus (COVID-19). StatPearls Publishing LLC; 2020. https://www.ncbi. nlm.nih.gov/books/NBK554776/.

[12] Community-Based Disaster Risk Reduction. National disaster management guidelines, national disaster management authority (NDMA). India: Ministry of Home Affairs, Government of India; 2019.

[13] Community-based disaster risk management: experiences from Indonesia. International Organization for Migration (IOM) Indonesia; May 2011.

[14] Coping with Major Emergencies. WHO strategies and approaches to Humanitarian action. WHO document WHO/EHA/95.1; 1995. https://apps.who.int/iris/bitstream/handle/10665/ 61335/WHO_EHA_95.1.pdf?sequence=1&isAllowed=y.

[15]   Eckstein D, Hutfils M-L, Winges M. GLOBAL CLIMATE RISK INDEX 2019 who suffers most from extreme weather events? Weather-related loss events in 2017 and 1998 to 2017. Germanwatche.V.; 2018. p. 1−36. https://www.germanwatch.org/en/cri.

[16]   de Guzman M. Recap of the programme on the 7th of August 2002. In: Regional workshop on total disaster risk management; 2002. p. 1−13. https://www.adrc.asia/publications/TDRM/19.pdf.

[17]   Devkota RP, Cockfield G. Perceived community-based flood adaptation strategies under climate change in Nepal. Int J Glob Warming 2014;6(1):113−24.

[18]   Devkota RP, Bahracharya B, Maraseni TN, Cockfield G, Upadhyay BP. The perception of Nepal's Tharu community in regard to climate change and its impacts on their livelihoods'. Int J Environ Stud 2011;68(6):937−46.

[19]   Dheria A, et al. Evaluating implications of flood vulnerability factors with respect to income levels for building long-term disaster resilience of low-income communities. Int J Disaster Risk Reduct 2020;48:101608. Elsevier.

[20]   Disaster Management Bureau. National disaster management policy. Ministry of food and disaster management. Government of the People's Republic of Bangladesh; 2008.

[21]   Disaster Management of India, GOI-UNDP disaster risk reduction programme (2009−2012); n.d.

[22]   Djalante R, Thomalla F, Sinapoy MS, Carnegie M. Building resilience to natural hazards in Indonesia: progress and challenges in implementing the Hyogo Framework for Action. Nat Hazards 2012;62:779−803.

[23]   Eckstein D, Hutfils ML, Winges M. Global climate risk index 2019 who suffers most from extreme weather events? Weather-related loss events in 2017 and 1998 to 2017. Germanwatche.V.; 2018. p. 1−36. https://www.germanwatch.org/en/cri.

[24]   European Centre for Disease Prevention and Control. Outbreak of acute respiratory syndrome associated with a novel coronavirus, China: first local transmission in the EU/EEA − third update. Stockholm: ECDC; 2020.

[25]   Guha-Sapir D. EM-DAT: the emergency events database. Brussels, Belgium: Universitécatholique de Louvain (UCL) - CRED; n.d. www.emdat.be. [Assessed 1 July 2020].

[26]   Ernst MJ, Islam MF, Gerard JG, Taher M. Bangladesh comprehensive disaster management programme: mid-term review. 2007.

[27]   Gautam D. Good practices and lessons learned disaster risk reduction through schools-kathmandu. National Disaster Risk-reduction Centre Nepal (NDRC Nepal); 2010, ISBN 978-99946-800-9-2.

[28]   Geis DE. By design: the disaster resilient and quality of life community. Nat Hazards Rev 2000;1(3):151−60.

[29]   Good Practices in Community-Based Disaster Risk Management, GOI-UNDP Disaster risk Management Programme, National Disaster Management Division, Ministry of Home Affairs, Government of India (2002−2009). https://tnsdma.tn.gov.in/app/webroot/img/document/library/42-Good-Practices.pdf.

[30]   Private sector strengths applied: good practices in disaster risk reduction from Japan. UNISDR; 2013.

[31]   Government of the People's Republic of Bangladesh. Bangladesh climate change strategy and action plan 2009. Dhaka: Ministry of Environment and Forests; Government of the People's Republic of Bangladesh; 2009.

[32]   Hossain MA. Community participation in disaster management: role of social work to enhance participation. J Anthropol 2013;9(1):159−71.

[33] IPCC Working Group II. Impacts, adaptation and vulnerability, contribution of IPCC WGII to the IPCC fourth assessment report. Cambridge: Cambridge University Press; 2007.

[34] Thouret JC, Lavigne F, Kelfoun K, Bronto S. Toward a revised hazard assessment at Merapi volcano, Central Java. J Volcanol Geoth Res 2000;100.

[35] Kumar KKS, Tholkappian S. Relative vulnerability of Indian coastal districts to sea-level rise and climate extremes. Int Rev Environ Strat 2005;6(1):3−22.

[36] Local Level Risk Management: Indian Experience, an Initiative Under GOI-UNDP Disaster Risk Management Programme, National Disaster Management Division, Ministry of Home Affairs, Government of India (2002−2007).

[37] Luers AL, Lobell DB, Sklar LS, Addams CL, Matson PA. A method for quantifying vulnerability, applied to the agricultural system of the Yaqui Valley, Mexico. Glob Environ Chang 2003;13:255−67.

[38] Marahatta S, Dongol BS, Gurung GB. Temporal and spatial variability of climate change over Nepal (1976−2005). Kathmandu, Nepal: Practical Action Nepal; 2009.

[39] McAdoo B, Quak M, Gnyawali KR, Adhikari BR. Roads and landslides in Nepal: how development affects environmental risk. Nat Hazards Earth Syst Sci 2018;18(12):3203−10.

[40] McAslan A. Community resilience. Understanding the concept and its application. Torrens Resilience Institute; 2011. www.torrensresilience.org/.

[41] McBean G, Rodgers C. Climate hazards and disasters: the need for capacity building. Wiley Interdiscip Rev Clim Change 2010;1(6):871−84.

[42] McCarthy JJ, Canziani OF, Leary NA, Dokken DJ, White KS, editors. Climate change 2001: impacts, adaptation and vulnerability. Cambridge: Cambridge University Press; 2001.

[43] Mishra V, Aadhar S, Asoka A, Pai S, Kumar R. On the frequency of the 2015 monsoon season drought in the Indo-Gangetic Plain. Geophys Res Lett 2016;43:12102−12. https://doi.org/10.1002/2016GL071407.

[44] National Disaster Management Authority. Guidelines for community-based disaster management. 2014. Retrieved from: http://ww.ndma/gov.in/pdf/draftnationalpolicyguideline sonCBDM. [Accessed 11 June 2020].

[45] National Plan for Disaster Management (2016-2020). Building resilience for sustainable human development. Government of the People's Republic of Bangladesh Ministry of Disaster Management and Relief; 2007. p. 1−77.

[46] National Planning Commission. Nepal flood 2017: post flood recovery needs assessment. Kathmandu: Government of Nepal; 2017.

[47] Natural Disasters Organization Australian emergency manuals series (manual 3). Emergency Management Australia (EMA); 1996. https://reliefweb.int/sites/reliefweb.int/files/resources/D66AEE7768867E49C1256C3A002CD1EF-ema-manual-1996.pdf.

[48] Nicola M, Alsafi Z, Sohrabi C, Kerwan A, Al-Jabir A, Iosifidis C, Agha M, Agha R. The socioeconomic implications of the coronavirus pandemic (COVID-19): a review. Int J Surg 2020;78:185−93. Elsevier.

[49] Sapkota S, Paudel MN, Thakur NS, Nepali MB, Neupane R. Effect of climate change on rice production: a case of six VDCs in Jumla District, Nepal. J Sci Technol 2011;11:57−62.

[50] Sarker AA, Rashid AKMM. Landslide and flashflood in Bangladesh disaster risk approaches in Bangladesh, disaster risk reduction. Japan: Springer; 2013.

[51] Scheidegger. Hazards: singularities in geomorphic systems. Geomorphology 1994;10:19−25.

[52] Seddiky A, et al. International principles of disaster risk reduction informing NGOs strategies for community-based DRR mainstreaming: the Bangladesh context". Int J Disaster Risk Reduct 2020;48:101580.

[53] Selby D, Kagawa F. Disaster risk reduction in school curricula: case studies from thirty countries. Geneva, Switzerland: United Nations Children Fund UNICEF 5/7 avenue de la paix, 1211; 2012.

[54] Shi P, Liu J, Yao Q, Tang D, Yang X. Integrated disaster risk management of China. 2007. http://www.oecd.org/dataoecd/52/14/38120232.pdf.

[55] Sims JH, Baumann DD. Education programs and human response to natural hazards. Environ Behav 1983;15(2):165−89. https://doi.org/10.1177/097133360701900201.

[56] Smith K. Environmental hazards: assessing risk and reducing disaster. London: Routledge; 1992.

[57] Sohrabi C, Alsafi Z, O'Neill N, Khan M, Kerwan A, Al-Jabir A, Iosifidis C, Agha R. World Health Organization declares global emergency: a review of the 2019 novel coronavirus (COVID-19). Int J Surg 2020;76:71−6. Elsevier.

[58] Thattai DV, Sathyanathan R, Dinesh R, Harshit LK. Natural disaster management in India with focus on floods and cyclones. ICCIEE 2017, IOP Conf Series: Earth & Environ Sci 2017;80:012054.

[59] Training Package, WHO/EHA Panafrican Emergency Training Centre. Addis Ababa Updated March 2002 by EHA; n.d. https://apps.who.int/disasters/repo/7656.pdf.

[60] Twigg J. Characteristics of a disaster-resilient community: a guidance note. London: DFID Disaster Risk Reduction Interagency Coordination Group. DFID; 2007.

[61] UN development agenda. UN system task team on the POST-2015; n.d. https://www.un.org/en/development/desa/policy/untaskteam_undf/thinkpieces/3_disaster_risk_resilience.pdf.

[62] UNCHS. Settlement planning for disasters, Nairobi. 1981.

[63] UNDP- a global report - reducing disaster risk a challenge for development. New york, USA: United Nations Development Programme Bureau for Crisis Prevention and Recovery; 2004. www.undp.org/bcpr.

[64] UNDP Nepal community-based disaster management practices, 2006−2008.

[65] UNDRR. Disaster risk reduction in Nepal: status report 2019. Bangkok, Thailand: United Nations Office for Disaster Risk Reduction (UNDRR), Regional Office for Asia and the Pacific; 2019.

[66] UNICEF. When coping crumbles- Droughts in India 2015−2016. 2016.

[67] UNICEF- for every child, Nepal; n.d. https://www.unicef.org/nepal/emergency. [Assessed 23 June 2020].

[68] UNISDR (United Nations International Strategy for Disaster Reduction). Hyogo framework for action 2005−2015: building the resilience of nations and communities to disasters: extract from the final report of the world conference on disaster reduction (a/Conf.206/6. Geneva: United Nations; 2007. http://www.preventionweb.et/files/1037_hyogoframework foractionenglish.pdf.

[69] United Nations. The Geneva mandate on disaster reduction. In: IDNDR international programme forum. 5−9 July 1999, Geneva; 1999.

[70] United nations development programme- annual report. 2018. https://www.undp.org/content/undp/en/home/librarypage/corporate/annual-report-2018.html.

[71] UN-SPIDER, Nepal Department of Water Induced Disaster Management; n.d. http://www.un-spider.org/institutions-guides/nepal-department-water-induced-disaster-management. [Assessed on 23 June, 2020].

[72] Vale L, Campanella T. The resilient city: how modern cities recover from disasters. New York: Oxford University Press; 2004.

[73] WHO Coronavirus Disease. (COVID-19) dash board as accessed on 2020/7/23, 7:02pm CEST; n.d. https://covid19.who.int/.

[74] Wildavsky AB. Searching for safety. Berkeley, CA: University of California Press; 1988.

[75] Wisner B. Untapped potential of the world's religious communities for disaster reduction in an age of accelerated climate change; an epilogue & prologue, religion, vol. 40; 2010. p. 128−31 (2).

[76] World Bank. Lessons from the reconstruction of post-tsunami Aceh: build back better through ensuring women are at the center of reconstruction of land and property. Washington D.C.: World Bank; 2011.

[77] World Health Organization as Accessed on 24 July 2020. https://www.who.int/health-topics/coronavirus#tab=tab_1.

[78] World Bank. Novel coronavirus (2019-nCoV) situation report − 1. 2011.

[79] Xu R, Cui B, Duan X, Zhang P, Zhou X, Yuan Q. Saliva: potential diagnostic value and transmission of 2019-nCoV. Int J Oral Sci 2020;12. https://www.nature.com/articles/s41368-020-0080-z.

[80] Yohta K. Geographical study of the disaster in Japan. Review article of the special issue on geography in Japan after the 1980s (Part II). Geogr Rev Jpn B 2014;86(2):132−7.

[81] Yodmani S. Disaster risk management and vulnerability reduction: protecting the poor. In: Social protection workshop 6: protecting communities—social funds and disaster management under the Asia and Pacific forum on poverty: reforming policies and institutions for poverty reduction held at the Asian Development Bank, Manila; 5−9 February 2001.

[82] Annual climate summary 2018. Ministry Of Earth Sciences, India Meteorological Department, Government of India, Climate Monitoring & Analysis Group; May 1, 2019. http://www.indiaenvironmentportal.org.in/files/file/Annual%20Climate%20Summary%202018.pdf.

[83] Community participation in disaster preparedness planning: a comparative study of Nepal and Japan - final report. Asian Disaster Reduction Center prepared by MaiyaKadel; 2011.

[84] Government of the People's Republic of Bangladesh. Comprehensive disaster management programme (phase II) inception workshop working paper. Disaster Management and Relief Division, Ministry of Food and Disaster Management; 2010.

[85] INCCA. Indian network for climate change assessment, climate change and India- A 4×4 assessment, A sectoral and regional analysis for 2030. Ministry of Environment and Forest, Government of India; November 2010.

[86] MDF-JRF Working Paper Series: Lessons Learned from Post-Disaster Reconstruction in Indonesia. Adapting community driven approaches for post-disaster recovery: experiences from Indonesia. Jakarta, Indonesia: MDF - JRF Secretariat The World Bank Office; December 2012. p. 1−91.

[87] Monitoring and reporting progress on community-based disaster risk management in Indonesia partnerships for disaster reduction-South East Asia phase 4. European Commission, ADPC, UNESCAP; April 2008.

[88] Statement on climate of India during 2019, India meteorological department. Climate Research and Services (CRS); January 6, 2020. http://www.indiaenvironmentportal.org.in/files/file/Statement%20on%20Climate%20of%20India%20during%202019.pdf.

[89]   The World Bank, GFDRR. Building Indonesia's resilience to disaster: experiences from mainstreaming disaster risk reduction in Indonesia program. 2016. http://documents1. worldbank.org/curated/en/318951507036249300/pdf/106245-REVISED-PUBLIC-Building-Indonesia-s-Resilience-to-Disaster.pdf.

[90]   White GF. In: Knees AV, Smith SC, editors. Optimal flood damage management: retrospect and prospect. Water research. Baltimore: John Hopkins Press; 1966.

[91]   White paper disaster management in Japan. 2019. http://www.bousai.go.jp/kaigirep/ hakusho/pdf/R1_hakusho_english.pdf.

Chapter 7

ZnO nanoparticles: a facile synthesized agent for removing dye from aqueous solution in an ecofriendly way

Priyanka Debnath, Arghadip Mondal, Naba Kumar Mondal
Environmental Chemistry Laboratory, Department of Environmental Science, The University of Burdwan, Bardhaman, West Bengal, India

1. Introduction

Nanotechnology is a looming field of technology and science that mainly deals with development and modification various nanomaterials. The objects, in 1−100 nm size range can be defined as nanoparticles [1]. When a bulk material becomes nanoparticle many conventional properties changes as the nanoparticle has bigger surface to volume ratio. This distinct character makes the nanoparticle more reactive in nature than other molecule. Diverse application of different metal and nonmetal nanoparticles are ranging from medical treatments to industry production. Moreover, the extensive use of nanoparticles into materials used every day, for example, cosmetics or clothes, makes it a highlighted area for research [2]. Research shows that nontoxicity, high stability, and high photocatalytic activity makes titanium dioxide (TiO_2) one of the most utilized and studied metal oxide nanoparticle [3]. Whereas, zinc oxide (ZnO) nanoparticle is another extensively studied adsorbent as well as catalyst for water pollution, as it is having a large excitation binding energy (60 meV), high photosensitive activity, and cost effectivity [4−6].

Organic pollutants are alarming threats, causing water pollution and harm to both humans and environment. Industrial sectors, which are mainly associated with significant organic wastewater issues, are mainly the dye or textile industry [7]. There are many industrial processes including various textile industries such as pulp and paper, cosmetics, pharmaceuticals, and food processing industries, which use different kinds of organic dyes

Cognitive Data Models for Sustainable Environment. https://doi.org/10.1016/B978-0-12-824038-0.00007-9

163

vigorously [8]. Untreated raw materials, various intermediates of dye along with undegradable dye residues are the major harmful chemicals in the wastewater released from industries, which may cause serious health issue if ingested [9]. Conventional methods such as coagulation-flocculation, chemical precipitation, ion exchange, and so forth have been widely used to treat wastewater, but all these methods have their own drawbacks [10–12]. Hence, the evolution of an authentic, efficient, and cost-friendly method or modification of the existing ones for either deletion or effective reductions of contaminants in the wastewater is the need of the hour. Both photocatalytic degradation and adsorption has already been proven to be an effective wastewater treatment technique, since it requires no external heat, comparatively low in cost and also chemically stable, and most importantly, they do not produce any toxic agents [13]. ZnO is the most frequently used catalysts for photocatalytic degradation as well as a good adsorbent for water treatment, which offers good photochemical and adsorption stability [14–16]. Hence, this chapter presents a review focused on the synthesis as well as the photocatalytic activity and adsorption capacity of zinc oxide nanoparticles for removal of dyes from wastewater.

2. Synthesis of ZnO nanoparticles

Synthesis of ZnO nanoparticles in large quantities by both physical and chemical procedures in a limited time frame are available. Among them, hydrothermal, sol-gel, chemical precipitation, electrochemical, and photochemical reduction are some preferable methods [17]. However, both physical and chemical synthesis processes have some drawbacks as it not only requires high amount of energy and chemicals as precursor, and also produces some byproducts, which are harmful for the environment [18]. Moreover, the involvement of harmful reducing and stabilizing agents are brought concern to nonbiodegradability associated with chemical and physical types of synthesis consequently that may lead to hamper large-scale applications. Therefore, the ecofriendly methods for producing nanoparticles gained huge attention. Basically, green synthesis of nanomaterials through proper regulations directly helps to uplift their environmental friendliness. Various plants and their parts as extracts and cellular extracts of different microorganisms like bacteria and fungi are already reported, which showed that ZnO nanoparticles can be synthesized following the green synthesis methods where the biological substances act as reducing, capping as well as a stabilizing agent, thus reducing the pollution load in the environment. Moreover, the use of various plants and microbes are safer as it produces less toxic byproducts to release into the environment (Fig. 7.1).

A brief description of biological synthesis of zinc oxide nanoparticle using plant parts and microbes are given below.

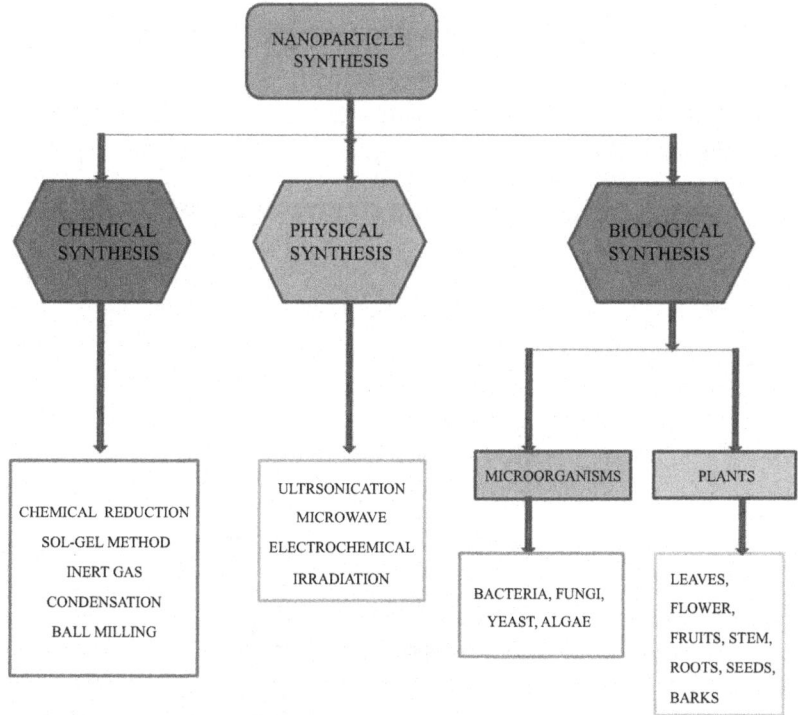

FIGURE 7.1 Green method for the synthesis of nanoparticles.

2.1 Synthesis of ZnO nanoparticle using plant parts

Plants contain the biomolecules which plays the main role for the production of nanoparticles by reducing the metal salt to its nano form. Nanoparticles which are frequently used in different industries, mainly synthesized by following the traditional methods which in turn increases the pollution load in the environment. The usage of green synthesis of nanoparticles is highly required to reduce the pollution load caused by other available conventional methods. Nanoparticle synthesis using plant extracts is quite simple and easy process rather than other green methods for large scale production [19]. For production of nanoparticles using plant extract, the extract is simply mixed with a solution of the metal salt at a certain ratio in a particular temperature [20]. Different biomolecules like phenols, carbohydrates, proteins etc. and other coenzymes which are present in the plants is reported to have potential to reduce metal salt into nanoparticles [19]. Various plants such as *Tabernaemontana divaricata*, *Calotropis gigantea* (L.), *P. trifoliate* (L.), and *Aloe vera* have been used to synthesis ZnO nanoparticles that are listed in the following table (Table 7.1, Fig. 7.2).

TABLE 7.1 Biosynthesis of ZnO nanoparticle using different plant source.

SL. No.	Source	Part	Shape/ Morphology	Size	References
1.	Raphanus sativus	Root	Hexagonal wurtzite	15–25 nm	[21]
2.	Tecoma castanifolia	Leaf	Spherical	70–75 nm	[22]
3.	Tabernaemontana divaricata	Leaf	Hexagonal wurtzite	20–50 nm	[23]
4.	Anacardium occidentale	Leaf	Hexagonal	33 nm	[24]
5.	Myristica fragrans	Leaf	Rod like shape	100 nm	[25]
6.	Azadirachta indica	Leaf	Spherical	30–35 nm	[26]
7.	Calotropis gigantean (L.)	Latex	Spherical to hexagonal	25–48 nm	[27]
8.	Olea europaea	Leaf	Spherical	41 nm	[28]
9.	Nephelium lappaceum L.	Peel	Spherical	25–40 nm	[29]
10.	Trifolium pratense	Flower	Flower like shape	100–190 nm	[30]
11.	Corriandrum sativum	Leaf	Spherical	9–19 nm	[31]
12.	Vitex trifolia L.	Leaf	Spherical	30 nm	[32]
13.	Ruta graveolens	Stem	Wurtzite	28 nm	[33]
14.	Citrus paradise	Peels	Spherical	12–72 nm	[34]
15.	Citrus aurantifolia	Fruits	Roughly spherical	9–10 nm	[35]
16.	Hybanthus enneaspermus (L.)	Stem	Spherical	50 nm	[36]
17.	P. trifoliate	Fruit	Spherical	8.48–32.51 nm	[37]
18.	Calotropis gigantea	Leaf	Spherical	30–35 nm	[38]
19.	Aloe vera	Leaf	Spherical	25–40 nm	[39]

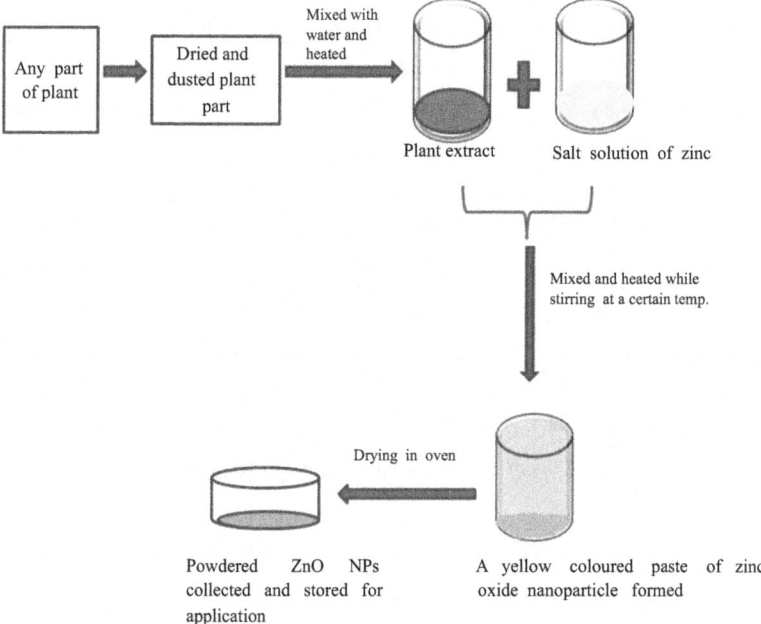

FIGURE 7.2 Schematic of biosynthesis of zinc oxide nanoparticle by plant extract.

2.2 ZnO nanoparticles synthesis using microbes

Biosynthesis of ZnO nanoparticles by using microorganisms is another effective process for the formation of nanoparticle. Different variety of bacterial as well as fungal species can be utilized for the preparation of various metal or metal oxide nanoparticles. The intracellular enzyme present in the microbe plays the role of reducing agent for the synthesis of nanoparticles [40]. It is already reported that the generation of the metal or metal oxide nanoparticles is probably due to the presence of the reductase enzyme inside of the fungal cell [41]. Many microorganisms that have the ability to synthesize ZnO nanoparticles are described in Table 7.2.

3. Different characterization techniques of ZnO nanoparticles

The characterization of nanoparticles deals with both the material and chemical properties of nanoparticles. The techniques used to characterize nanomaterials include electron microscopy, scanning electron microscopes (SEM), transmission electron microscopy (TEM), atomic force microscopy, X-ray diffraction, X-ray scattering, X-ray fluorescence spectrometry, and so forth [51]. Moreover, UV/Vis spectroscopy, an important tool for identifying

TABLE 7.2 Synthesis of ZnO nanoparticles using microbes.

SL No.	Source	Shape/ Morphology	Size	References
1.	*Aeromonas hydrophila*	Spherical, oval	57.72 nm	[42]
2.	*Aspergillus fumigates*	Spherical and hexagonal	1–6 nm	[43]
3.	*Lactobacillus plantarum*	Roughly spherical	7–19 nm	[44]
4.	*Rhodococcus pyridinivorans*	Roughly spherical	100–120 nm	[44]
5.	*Serratia ureilytica*	Spherical	50–61 nm	[45]
6.	*Sargassum muticum*	Hexagonal wurtzite	30–57 nm	[46]
7.	*S. aureus*	Acicular	10–50 nm	[47]
8.	*Candida albicans*	Quasi-spherical	20–30 nm	[48]

and characterizing optical properties of nanoparticles, which gives preliminary idea about the size, concentration, and agglomeration state of the nanoparticle [52]. Spectroscopy is actually based on a simple method which quantifies the amount of light gets absorbed and scattered by a sample. Whereas, the nanostructure morphology confirmation reveals by the analysis of SEM and TEM [53]. TEM images the electron beam transmission through a sample and it is used to measure the differences in particle and grain size, size distribution and also the shape of nanoparticles. Similarly, SEM also can be employed for revealing the details about nanoparticle shape and surface. The ZnONPs have been further investigated by XRD and FTIR along with UV-vis, SEM and TEM for characterization. FTIR interfaces to investigate the surface adsorption of functional groups on nanoparticles [54]. FTIR works by producing an infrared absorption spectrum and identifying the chemical bonds present in a molecule which makes it another valuable tool for chemical identification. While, XRD gives idea about crystalline structure, and size of the crystal. Some characterization images of biosynthesized ZnO nanoparticles are given below.

The UV-vis image showing a distinct peak at a particular wavelength (~ 368 nm) which confirms ZnO nanoparticles production (Fig. 7.3A). Spherical shape of the biosynthesized nanoparticle is confirmed by Fig. 7.3B. Whereas, the sharp peaks in Fig. 7.3C affirms the crystalline nature of the synthesized nanoparticles. Fig. 7.3D showing a variety of functional groups which exists on the nanoparticle surface.

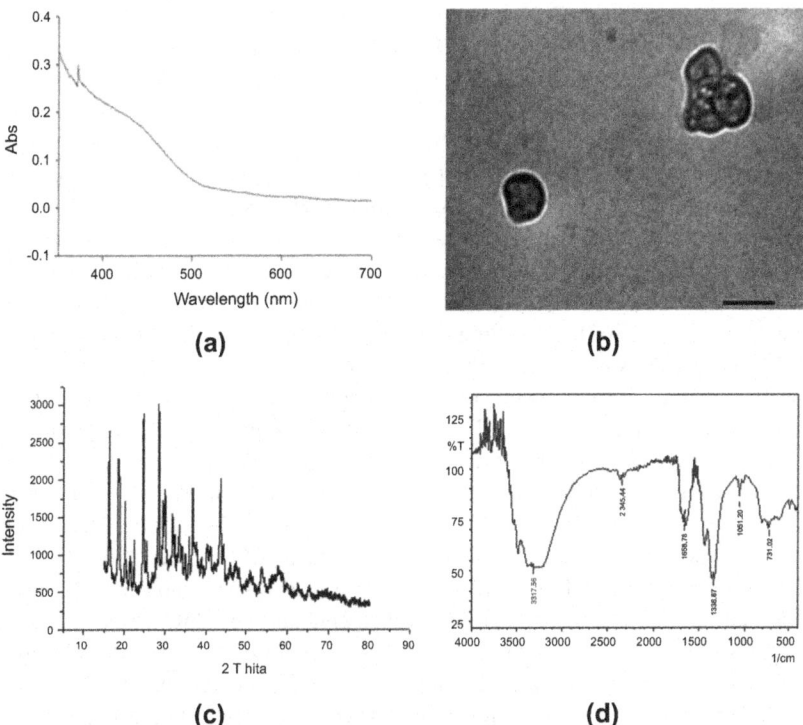

FIGURE 7.3 Showing (A) UV-vis, (B) TEM, (C) XRD, (D) FTIR images of zinc oxide nanoparticle synthesized by *Hibiscus rosa-sinensis* leaf extract [55].

4. Effect of ZnO nanoparticles on dye solution

4.1 Photocatalytic activity of ZnO NPs for dye degradation in wastewater

Different methods for dye degradation have been used in recent times for wastewater treatment, among which the photocatalysis by ZnO nanoparticles is found quite promising for water purification process. Recently, the photocatalytic activity of ZnO nanoparticles was investigated after being immobilized on synthetic carbon in the elimination of acid blue 113 dye from wastewater and establishing a correlation between concentration of photocatalyst with contact time with the process efficiency in removal of acid blue 113 dye [56]. Similarly, the photocatalytic activity of ZnONPs, synthesized using *C. ramiflora* was examined against a pollutant dye, Rhodamine B, and results showed that degradation efficiency of almost 98% was attained within 200 min in sunlight irradiation [57]. Moreover, another study showed highest percentage of decolorization for ZnO when the photocatalytic activity for degradation of azo dyes was compared to TiO_2 [58]. The study also reported about the sensitivity of ZnO to the pH; stronger acidic condition facilitates the photocatalytic activity but lower pH lowers the potential [59]. Furthermore, a ZnO-clay nanocomposite has been synthesized to increase the stability against the photocorrosion, which showed higher photocatalytic activity under sunlight exposure than UV radiation [60]. Similarly, the Pd-doped ZnO photocatalysts prepared by three different process which are microwave irradiation, photo reduction and borohydride reduction showed a significant their photocatalytic activity differs in each cases for an azo dye, congo red dye degradation respectively [61]. Study also reported chemically synthesized ZnONP showed good photocatalyst activity on methylene blue (MB) and RO dye in aqueous solutions [62]. Accordingly, Undoped ZnO and transition metal doped ZnO nanoparticles synthesized by thermal decomposition have been evaluated for their photocatalytic behavior using methyl orange as probe dye [63]. Another study reported the pH (2−10) exhibits impact on the degradation capacity of ZnO, on methyl orange concentration and catalyst loading [12]. The maximum efficiency was observed for the system at pH value of 8. Similar behavior was reported for the ZnO as photocatalyst for decolorization of azo dyes [64,65]. Although Isai and Srivastava [66] showed that 2% Fe−ZnO higher photocatalytic activity than ZnO as results showed maximum degradation 86% and 92% of MB dye using ZnO and Fe−ZnO, respectively. A recent study reported the photodegradation capability of ZnO NPs of some cationic dyes, using 125 and 100 mg of ZnO NP can remove 100% of the methylene blue solution and rhodamine B within 60 and 50 min respectively. Whereas biosynthesized flower-shaped ZnO NPs using the leaf extract of *Calliandra haematocephala* were utilized for methylene blue dye degradation [67]. Result suggested degradation capacity can reach up to 88% after only 270 min of reaction. Similarly, a discrete flower-shaped ZnONPs by leaf *Peltophorum pterocarpum* leaf extract was synthesized and showed that to degrade 95% of methylene blue dye after 120 min under sunlight irradiation [68]. Moreover, *Trianthema portulacastrum* extract was used for synthesis of ZnONPs by other researchers and the catalytic activity ZnONPs were evaluated by degrading Synozol Navy Blue-KBF textile dye under solar irradiation that causes 91% degradation in 159 min [52]. Accordingly, the efficiency of ZnONPs as a photocatalyst for removal of MO dye showed 90% degradation, which can be obtained after 120 min [42]. The following table showing the comparative study of adsorption capacity of ZnO nanoparticles (Table 7.3).

4.2 ZnO NPs as an adsorbent for effective removal of dyes

Organic dyes contain highly toxic and harmful chemicals which should be removed from wastewater discharged from different industries before it go out in the aquatic environment. Almost all organic dyes possess complex molecular structure which makes them nondegradable through conventional wastewater treatment processes. Adsorption processes, on the other hand, has a wide

TABLE 7.3 Comparative study of photocatalytic degradation of dyes using ZnO NPs.

Name of the dye	Photocatalytic degradation (%)	References
Congo red	93.7	[42]
Methyl orange	~100	[43]
Rhodamine B	98	[57]
Methylene blue	86	[66]
Reactive Yellow X-RG	90.6	[44]
Reactive black-5	91.35	[45]
Reactive blue 81	100	[46]
Methyl orange	97.7	[51]
Methyl violet	~100	[47]
Direct blue 129	95	[48]
Eosin yellow	90	[49]

adaptability in recent days for simple use and efficiently removing dyes. Different metal or metal oxide nanomaterials have reported to being used as adsorbent for environmental remediation through wastewater treatment. Investigation showed biosynthesized ZnO NP showed maximum adsorption capacity of 2963, 3307, and 1554 mg/g for adsorption of malachite green, acid fuchsin and congo red respectively [77]. Furthermore, Cr-doped ZnO was synthesized which can remarkably remove methyl orange from water [78]. Report said a high adsorption capacity of 310.56 mg/g has been achieved for this adsorption process in a short time. Moreover, the impact of Ni doped ZnO NPs for both anionic and cationic dyes, i.e., fast green and Victoria blue dyes was studied by Saharan et al. [80]. Additionally, Mittal et al. [81] reported the gum arabic grafted polyacrylamide (GA-cl-PAM) hydrogel mediated ZnONP and concluded that 0.4 g/L adsorbent has the ability to adsorb approximately 99% malachite green at neutral ph. The performance of ZnO/polyaniline nanocomposite (ZnO-PANI-NC) is also evaluated for quick adsorption of methyl orange from aqueous solution and 240.84 mg/g of maximum adsorption capacity was recorded [82]. Whereas, Kataria and Garg [83] synthesized ZnONPs by hydrothermal method and the maximum adsorption capacity were obtained for congo red and brilliant green as 71.4 and 238 mg/g respectively. Not a long ago, the same researchers produced flower-shaped zinc oxide nanoparticle which can achieve up to 163 mg/g of Victoria blue B dye at optimum conditions [84]. Whereas, composite nanofibers comprised of polyacrylonitrile (PAN), zinc oxide, and hinokitiol (HT) exhibited adsorption capacity of ZnO-HT-PAN-H as 267.37 and 245.76 mg/g for RB 19 and RR 195, respectively [85]. Moreover, a composite of chitosan–zinc oxide of 33 nm size showed adsorption capacity of 11 mg/g for removal of malachite green [86]. Similarly, Hassan et al. [87] also prepared chitosan/silica/ZnO nanocomposite to eliminate MB from wastewater. Result showed removal capacity of MB reached up to 293.3 mg/g. Furthermore, the maximum dye-adsorption capacity of ZnONP toward Basic Red 12, Acid Orange seven was investigated and the result was found as 15.64, 6.78 and 6.38 mg/g, respectively [87]. Simultaneously, a ZnO hybrid beads was formulated which was reported to have high dye-adsorption performance for C.I. 41 basic blue dye [88]. The maximum dye sorption was 16.5 mg/g using 100 ppm dye concentration. Similarly, batch adsorption experiments with different parameters were conducted for removal of RB8 from aqueous solutions using ZnONPs. Khoshhesab et al. [89] reported the maximum adsorption capacity was noted as 27.6 mg/g. Whereas, Mahmoodi and Najafi [90] tested the sorption capability of both amine-functionalized ZnONP (AFZON) and normal ZnONP on Acid Blue 25, Direct Red 23 and Direct Red 31. Result revealed both of these adsorbents showed good dye-adsorption capacity that is 20, 12 and 15 mg/g for ZnONP and 1250, 1000 and 1429 mg/g for AFZON respectively. Previously, the possible potential of nickel doped zinc oxide nanoparticles for removing some specific dyes, i.e., methyl orange and tartrazine has tested and at pH 4 and 6 maximum dye-adsorption capacity was obtained respectively for TA and MO [91]. Some biosynthesized ZnO NPs and their adsorption efficiencies are listed in the table below (Table 7.4).

TABLE 7.4 Comparative study of zinc oxide nanoparticles used for various dye removal.

Name of the dye	Maximum adsorption capacity (mg/g)	References
Malachite green Acid fuchsin Congo red	2963 mg/g 3307 mg/g 1554 mg/g	[77]
Methyl orange	310.56 mg/g	[78]
Methyl orange	240.84 mg/g	[82]
Congo red Brilliant green	71.4 mg/g 238 mg/g	[83]
Congo red	90.8 mg/g	[79]
Orange G Rhodamine B	153.8 mg/g 128.2 mg/g	[91]

TABLE 7.4 Comparative study of zinc oxide nanoparticles used for various dye removal.—cont'd

Name of the dye	Maximum adsorption capacity (mg/g)	References
Methylene blue	83.9 mg/g	[92]
Crystal violet	81.6 mg/g	
Reactive yellow	22.73 mg/g	[93]
BBR-250	32.82 mg/g	[94]
Congo red	95.5 mg/g	[55]

5. Conclusion

The current review aimed to discuss the generation, characterization, and application of ZnO nanoparticles. ZnONP can be produced by both biological and chemical methods, whereas the biological method is more helpful due to its ecofriendly nature. Various plant parts and microbes were utilized for the production of ZnO nanoparticles. ZnO is characterized by various analyzing techniques to know the exact size, shape, structure, elemental composition, and functional groups, which reacts on the adsorbent's surface. The bio-synthesized ZnO nanoparticles also have huge applications in removing organic dyes from aqueous solutions. After testing multiple combinations between adsorbents and dye molecules through batch study, it was observed that adsorption capacity depends on many other factors such as pH, reaction rate, initial concentration, adsorbent dose, and so forth, which are important in order to obtain an effective material for industrial applications. Studies reveal their photostable, biologically, and chemically inert nature makes ZnO NPs a suitable photocatalyst and an effective adsorbent for the removal of wastewater-containing dyes. The photocatalytic activity for the degradation and demineralization of dyes as well as other toxic organic pollutants under UV, visible, or solar light radiation also thoroughly precipitable. Evidently, ZnO has proved to be a powerful tool for water purification, via photocatalysis and adsorption, as well as other environmental applications.

6. Future perspectives

The primary shortcoming of these dye-adsorption or degradation studies is that their use is still limited in the laboratory stage, mostly without large-scale industrial studies. There are also no clear ideas or concerns about the byproducts that were released into the aquatic environment. Moreover, very little attempts for thorough economic and market analyses makes it more difficult for daily use. The main future target should be to advance this dye-adsorption process to the large scale using real wastewater. But the implementation of this adsorption proves that in a large industrial scale it would be a challenging effort to demonstrate and also would require a sufficient economical and technical aid. Accordingly, the evaluation and regeneration of all the novel adsorbents as well as the study of their end of life is needed to be highlighted. However, various methods have been approached for the synthesis of ZnONPs and using the synthesized ZnONPs as adsorbents as well as photocatalyst for removing toxic dye molecules from wastewater. Other than the aforementioned techniques, many other wastewater treatment methods are being adopted for dye removal depending upon the composition of wastewater. Hybrid treatments, other than chemical, biological, and physical treatment is gaining popularity now. Sensor-based devices such as optical sensors, wireless sensors, ZigBee, and microcontrollers are used for monitoring the water quality. But only some research has been implemented using these artificial intelligence options based on these techniques.

Appendix A. Supplementary data

Supplementary data to this article can be found online at https://doi.org/10.1016/B978-0-12-824038-0.00007-9.

Acknowledgments

Authors are extending their sincere gratitude to all the faculty members and staff of Department of Environmental Science, The University of Burdwan, for their moral support and valuable suggestions for preparing this manuscript. This study was funded by Swami Vivekananda Merit cum Means Fellowship, Govt. of West Bengal.

References

[1] Horikoshi S, Serpone N. Microwaves in nanoparticle synthesis. 1st ed. Wiley-VCH Verlag GmbH & Co. KGaA; 2013.

[2] Dubchak S, Ogar A, Mietelski JW, Turnau K. Influence of silver and tianium nanoparticles on arbuscularmycorhiza colonization and acumulation of radiocaesium in *Helianthus anus*. Spanish J Agric Res 2010;8(1):103−8.

[3] Khataee AR, Kasiri M. Photocatalytic degradation of organic dyes in the presence of nanostructured titanium dioxide: influence of the chemical structure of dyes. J Mol Catal Chem 2010;328:8−26.

[4] Adnan MAM, Julkapli NM, Abd Hamid SB. Review on ZnO hybrid photocatalyst: impact on photocatalytic activities of water pollutant degradation. Rev Inorg Chem 2016;36(2). https://doi.org/10.1515/revic-2015-0015a.

[5] Kumar SG, Rao KSRK. Zinc oxide based photocatalysis: tailoring surface-bulk structure and related interfacial charge carrier dynamics for better environmental applications. RSC Adv 2015;5:3306−51.

[6] Kołodziejczak-Radzimska A, Jesionowski T. Zinc oxide—from synthesis to application: a review. Materials 2014;7:2833−81.

[7]   Lellis B, Zani C, João F-P, Pamphile A, Polonio JC. Effects of textile dyes on health and the environment and bioremediation potential of living organisms. Biotechnol Res & Innov 2019;3(2):275−90.

[8]   Luo X, Deng F. Nanomaterials for the removal of pollutants and resource reutilization. A volume in Micro and Nano Technologies, ISBN 978-0-12-814837-2. https://doi.org/10.1016/C2017-0-01288-0.

[9]   Lam S-M, Sin J-C, Zuhairi AA, Mohamed AR. Degradation of wastewaters containing organic dyes photocatalysed by zinc oxide: a review. Desalination & Water Treat 2012;41(1−3):131−69.

[10]   Pardeshi SK, Patil AB. A simple route for photocatalytic degradation of phenol in aqueous zinc oxide suspension using solar energy. Sol Energy 2008;82:700−5.

[11]   Qamar M, Muneer M. A comparative photocatalytic activity of titanium dioxide and zinc oxide by investigating the degradation of vanillin. Desalination 2009;249:535−40.

[12]   Kansal SK, Singh MD. Studies on photodegradation of two commercial dyes in aqueous phase using different photocatalysts. J Hazard Mater 2007;141:581−90.

[13]   Koe WS, Lee JW, Chong WC. An overview of photocatalytic degradation: photocatalysts, mechanisms, and development of photocatalytic membrane. Environ Sci Pollut Control Ser 2020;27:2522−65.

[14]   Samadi M, Zirak M, Naseri A, Khorashadizade E, Moshfegh AZ. Recent progress on doped ZnO nanostructures for visible-light photocatalysis. Thin Solid Films 2016;605:2−19.

[15]   Benhebal H, Chaib M, Leonard A, Lambert SD, Crine M. Photodegradation of phenol and benzoic acid by sol−gel-synthesized alkali metal-doped ZnO Materials. Sci Semicond Process 2012;15:264−9.

[16]   Di Mauro A, Farrugia C, Abela S, Refalo P, Grech M, Falqui L, Nicotra G, Sfuncia G, Mio A, Buccheri M-A, Rappazzo G, Brundo M, Scalis E, Pecoraro R, Iaria C, Privitera V, Impellizzeri G. Ag/ZnO/PMMA nanocomposites for an efficient water reuse. ACS Appl Bio Mater 2020. https://doi.org/10.1021/acsabm.0c00409.

[17]   Mustapha S, Ndamitso MM, Abdulkareem AS. Application of TiO$_2$ and ZnO nanoparticles immobilized on clay in wastewater treatment: a review. Appl Water Sci 2020;10:49.

[18]   Kharissova OV, Kharisov BI, Oliva GCM, Méndez YP, López I. Greener synthesis of chemical compounds and materials. Royal Society Open Science; 2019. p. 6191378.

[19]   Singh J, Dutta T, Kim K. Green synthesis of metals and their oxide nanoparticles: applications for environmental remediation. J Nanobiotechnol 2018;16:84.

[20]   Khandel P, Yadaw RK, Soni DK. Biogenesis of metal nanoparticles and their pharmacological applications: present status and application prospects. J Nanostruct Chem 2018;8:217−54.

[21]   Liu D, Liu L, Yao L, Peng X, Li Y, Jiang T, Kuang H. Synthesis of ZnO nanoparticles using radish root extract for effective wound dressing agents for diabetic foot ulcers in nursing care. J Drug Deliv Sci Technol 2020;55:101364.

[22]   Sharmila G, Thirumarimurugan M, Muthukumaran C. Green synthesis of ZnO nanoparticles using Tecoma castanifolia leaf extract: characterization and evaluation of its antioxidant, bactericidal and anticancer activities. Microchem J 2018. https://doi.org/10.1016/j.microc.2018.11.022.

[23]   Raja A, Ashokkumar S, Marthandam RP, Jayachandiran J, Khatiwadae CP, Kaviyarasuf K, Ramang RG, Swaminathanh, M M. Eco-friendly preparation of zinc oxide nanoparticles using Tabernaemontana divaricata and its photocatalytic and antimicrobial activity. J Photochem Photobiol B Biol 2018;181:53−8.

[24]   Zhao C, Zhang X, Zheng Y. Biosynthesis of polyphenols functionalized ZnO nanoparticles: characterization and their effect on human pancreatic cancer cell line. J Photochem Photobiol B Biol 2018;183:142—6.

[25]   Ashokan AP, Paulpandi M, Dinesh D, Murugan K, Vadivalagan C, Benelli G. Toxicity on dengue mosquito vectors through myristica fragrans-synthesized zinc oxide nanorods, and their cytotoxic effects on liver cancer cells (HepG2). J Clust Sci 2016;28(1):205—26.

[26]   Ahmed S, Annu, Chaudhry SA, Ikram S. A review on biogenic synthesis of ZnO nanoparticles using plant extracts and microbes: a prospect towards green chemistry. J Photochem Photobiol B Biol 2017;166:272—84.

[27]   Panda K, Golari D, Venugopal A, Achary V, Phaomei G, Parinandi N, Panda B. Green synthesized zinc oxide (ZnO) nanoparticles induce oxidative stress and DNA damage in Lathyrus sativus L. root bioassay system. Antioxidants 2017;6(2):35.

[28]   Hashemi S, Asrar Z, Pourseyedi S, Nadernejad N. Green synthesis of ZnO nanoparticles by Olive (Olea europaea). IET Nanobiotechnol 2016;10(6):400—4.

[29]   Karnan T, Selvakumar SAS. Biosynthesis of ZnO nanoparticles using rambutan(Nephelium lappaceum L.) peel extract and their photocatalytic activity on methylorange dye. J Mol Struct 2016;1125:358—65.

[30]   Dobrucka R, Długaszewska J. Biosynthesis and antibacterial activity of ZnO nanoparticles using Trifolium pratense flower extract. Saudi J Bio Sci 2016;23(4):517—23.

[31]   Hassan SSM, El-Azab WIM, Ali HR, Mansour MSM. Green synthesis and characterization of ZnO nanoparticles for photocatalytic degradation of anthracene. Adv Nat Sci — Nanosci 2015;6(4). 045012.

[32]   Elumalai K, Velmurugan S, Ravi S, Kathiravan V, Adaikala Raj G. Bio-approach: plant mediated synthesis of ZnO nanoparticles and their catalytic reduction of methylene blue and antimicrobial activity. Adv Powder Technol 2015;26(6):1639—51.

[33]   Lingaraju K, Raja Naika H, Manjunath K. Biogenic synthesis of zinc oxide nanoparticles using Ruta graveolens (L.) and their antibacterial and antioxidant activities. Appl Nanosci 2015;6:703—10.

[34]   Kumar B, Smita K, Cumbal L, Debut A. Green approach for fabrication and applications of zinc oxide nanoparticles. Bioinorg Chem Appl 2014 2014:1—7.

[35]   Ramesh P, Rajendran A, Subramanian A. Synthesis of zinc oxide nanoparticle from fruit of Citrus aurantifolia by chemical and green method. Asian J Pharm Clin Res 2014;2(4):189—95.

[36]   Shekhawat MS, Ravindran CP, Manokari MA. Biomimetic approach towards synthesis of zinc oxide nanoparticles using Hybanthus enneaspermus (L.) F. Muell. . Trop Plant Res 2014;1(2):55—9.

[37]   Nagajyothi PC, Minh An TN, Sreekanth TVM, Lee J, Joo Lee D, Lee KD. Green route biosynthesis: characterization and catalytic activity of ZnO nanoparticles. Mater Lett 2013;108:160—3.

[38]   Vidya C, Hirematha S, Chandraprabha MN, Antonyraja MAL, Gopala IV, Jaina A, Bansal K. Green synthesis of ZnO nanoparticles by Calotropis Gigantea. Int J Curr Eng Technol 2013:118—20.

[39]   Gunalan S, Sivaraj R, Rajendran V. Green synthesized ZnO nanoparticles against bacterial and fungal pathogens. Prog Nat Sci: Mater 2012 2012;22(6):693—700.

[40]   Sandhu RS, Aharwal RP, Kumar S. Green synthesis: a novel approach for nanoparticles synthesis. Int J Pharmaceut Sci Res 2019;10(8):3550—62.

[41] Narayanan KB, Sakthivel N. Synthesis and characterization of nano-gold composite using Cylindrocladiumfloridanum and its heterogeneous catalysis in the degradation of 4-nitrophenol. J Hazard Mater 2011;189:519−25.

[42] Goudarzi M, Mousavi-Kamazani M, Salavati-Niasari M. Zinc oxide nanoparticles: solvent-free synthesis, characterization and application as heterogeneous nanocatalyst for photodegradation of dye from aqueous phase. J Mater Sci Mater Electron 2017;28:8423−8.

[43] Chidambaram J, Rahuman A, Kirthia AV, Marimuthu S, Santhoshkumar T, Bagavana A, Gaurav K, Karthik L, Rao KVB. Novel microbial route to synthesize ZnO nanoparticles using Aeromonas hydrophila and their activity against pathogenic bacteria and fungi. Spectrochim Acta A Mol Biomol Spectrosc 2012;90:78−84.

[44] Tarafdar JC, Raliya R. Rapid, low-Cost, and ecofriendly approach for iron nanoparticle synthesis using Aspergillus oryzae TFR9. J Nanopart 2013;1−4.

[45] Selvarajan E, Mohanasrinivasan V. Biosynthesis and characterization of ZnO nanoparticles using Lactobacillus plantarum VITES07. Mater Lett 2013;112:180−2.

[46] Kundu D, Hazraa C, Chatterjee A, Chaudhari A, Mishra S. Extracellular biosynthesis of zinc oxide nanoparticles using Rhodococcus pyridinivorans NT2: multifunctional textile finishing, biosafety evaluation and in vitro drug delivery in colon carcinoma. J Photochem Photobiol B 2014;140:194−204.

[47] Dhandapani P, Siddarth AS, Kamalasekaran S, Maruthamuthu S, Rajagopal G. Bio-approach: ureolytic bacteria mediated synthesis of ZnO nanocrystals on cotton fabric and evaluation of their antibacterial properties. Carbohydr Polym 2014;103:448−55.

[48] Azizi S, Ahmad MB, Namvar F, Mohamad R. Green biosynthesis and characterization of zinc oxide nanoparticles using brown marine macroalga Sargassum muticum aqueous extract. Mater (Basel) 2014 2014;116:275−7.

[49] Rauf MA, Owais M, Rajpoot R, Ahmada F, Khana N, Zubair S. Biomimetically synthesized ZnO nanoparticles attain potent antibacterial activity against less susceptible S. aureus skin infection in experimental animals. RSC Adv 2017;7. 36361-36S.

[50] Shamsuzzaman, Mashrai A, Khanam H, Aljawfi RN. Biological synthesis of ZnO nano-particles using C. albicans and studying their catalytic performance in the synthesis of steroidal pyrazolines. Arab J Chem 2017;10(2):S1530−6.

[51] Mourdikoudis S, Pallares RM, Nguyen TK. Characterization techniques for nanoparticles: comparison and complementarity upon studying nanoparticle properties. Nanoscale 2018;10:12871−934.

[52] Khan I, Saeed K, Khan I. Nanoparticles: properties, applications and toxicities. Arab J Chem 2019;12(7):908−31.

[53] Emil E, Alkan G, Gurmen S, Rudolf R, Jenko D. Tuning the morphology of ZnO nano-structures with the ultrasonic spray pyrolysis process. Metals 2018;8:569.

[54] Petit T, Puskar L. FTIR spectroscopy of nanodiamonds: methods and interpretation. Diam Relat Mater 2018;89. https://doi.org/10.1016/j.diamond.2018.08.005.

[55] Debnath P, Mondal NK. Effective removal of Congo red dye from aqueous solution using biosynthesized zinc oxide nanoparticles. Environ Nanotechnol Monit & Manag 2020:100320.

[56] Dargahi A, Samarghandi MR, Vaziri Y, Ahmadidoust G, Ghahramani E, Shekarchi AK. Kinetic study of the photocatalytic degradation of the acid blue 113 dye in aqueous solutions using zinc oxide nanoparticles immobilized on synthetic activated carbon. J Adv Environ Health Res 2019;7(2):75−85.

[57] Varadavenkatesan T, Lyubchik E, Pai S, Pugazhen A, Vinayagam R, Selvaraj R. Photo-catalytic degradation of Rhodamine B by zinc oxide nanoparticles synthesized using the leaf extract of Cyanometraramiflora. J Photochem Photobiol B Biol 2019;199:111621.

[58] Sakthivel S, Neppolian B, Shankar MV, Arabindoo B, Palanichamy M, Murugesan V. Solar photocatalytic degradation of azo dye: comparison of photocatalytic efficiency of ZnO and TiO_2. Sol Energy Mater Sol Cell 2003;77:65−82.

[59] Daneshvar N, Salari D, Khataee AR. Photocatalytic degradation of azo dye acid red 14 in water on ZnO as an alternative catalyst to TiO_2. J Photochem Photobiol Chem 2004;162:317−22.

[60] Bel Hadjltaief H, Ben Ameur S, Da Costa P, Ben Zina M, Elena Galvez M. Photo-catalyticdecolorization of cationic and anionic dyes over ZnO nanoparticle immobilized on natural Tunisian clay. Appl Clay Sci 2018;152:148−57.

[61] Guy N, Cakar S, Ozacar M. Comparison of palladium/zinc oxide photocatalysts prepared by different palladium doping methods for Congo red degradation. J Colloid Interface Sci 2016;466:128−37.

[62] Intarasuwan K, Amornpitoksuk P, Suwanboon S, Graidist P. Photocatalytic dye degradation by ZnO nanoparticles prepared from $X_2C_2O_4$ (X = H, Na and NH_4) and the cytotoxicity of the treated dye solutions. Separ Purif Technol 2017;177:304−12.

[63] Kaur J, Singhal S. Facile synthesis of ZnO and transition metal doped ZnO nanoparticles for the photocatalytic degradation of Methyl Orange. Ceram Int 2014;40(5):7417−24.

[64] Lizama C, Freer J, Baeza J, Mansilla HD. Optimized photodegradation of reactive blue 19 on TiO_2 and ZnO suspensions. Catal Today 2002;76:235−9.

[65] Akyol A, Yatma HC, Bayramoglu M. Photocatalytic decolorization of Remazol Red RR in aqueous ZnO suspensions. Appl Catal B Environ 2004;54:19−24.

[66] Isail KA, Shrivastava VS. Photocatalytic degradation of methylene blue using ZnO and 2% Fe-ZnO semiconductor nanomaterials synthesized by sol-gel method: a comparative study. J Water & Environ Nanotechnol 2019;4(3):251−62.

[67] Vinayagam R, Selvaraja R, Arivalagan P, Varadavenkatesan T. Synthesis, characterization and photocatalytic dye degradation capability of Calliandrahaematocephala-mediated zinc oxide nanoflowers. J Photochem Photobiol B Biol 2020;203:111760.

[68] Pai S, Sridevi H, Varadavenkatesan T, Vinayagam R, Selvaraj R. Photocatalytic zinc oxide nanoparticles synthesis using Peltophorumpterocarpum leaf extract and their characteriza-tion. Optik 2019;185:248−55.

[69] Zhang J, Deng S, Liu S, Chen J, Han B, Wang Y, Wang Y. Preparation and photocatalytic activity of Nd doped ZnO nanoparticles. Mater Technol 2014;29:262−8.

[70] Kuriakose S, Choudhary V, Satpati B, Mohapatra S. Facile synthesis of Ag−ZnO hybrid nanospindles for highly efficient photocatalytic degradation of methyl orange, PCCP. Phys Chem Chem Phys 2014;16:17560−8.

[71] Li L, Liu L, Li Z, Hu D, Gao C, Xiong J, Li W. The synthesis of CB [8]/ZnO composites materials with enhanced photocatalytic activities. Heliyon 2019;5:e01714.

[72] Gholami M, Shirzad-Siboni M, Yang J-K. Application of Ni-doped ZnO rods for the degradation of an azo dye from aqueous solutions. Korean J Chem Eng 2016;33:812−22.

[73] Nodehi A, Atashi H, Mansouri M. Improved photocatalytic degradation of reactive blue 81 using NiOdoped ZnO−ZrO2 nanoparticles. J Dispers Sci Technol 2019;40:766−76.

[74] Jeyasubramanian K, Hikku GS, Sharma RK. Photo-catalytic degradation of methyl violet dye using zinc oxide nano particles prepared by a novel precipitation method and its anti-bacterial activities. J Water Process Eng 2015;8:35−44.

[75] Fardood S, Ramazani A, Moradi S. Green synthesis of zinc oxide nanoparticles using arabic gum and photocatalytic degradation of direct blue 129 dye under visible light. J Mater Sci: Mater Electron 2017;28:13596–601.

[76] Rambabu K, Bharath G, Banat F, Show PL. Green synthesis of zinc oxide nanoparticles using Phoenix dactylifera waste as bioreductant for effective dye degradation and antibacterial performance in wastewater treatment. J Hazard Mater 2020:123560.

[77] Zhang F, Chen X, Wu F, Ji Y. High adsorption capability and selectivity of ZnO nanoparticles for dye removal. Colloid Surface Physicochem Eng Aspect 2016;509:474–83.

[78] Meng A, Xing J, Li Z, Li Q. Cr-doped ZnO nanoparticles: synthesis, characterization, adsorption property, and recyclability. ACS Appl Mater Interfaces 2015;7:27449–57.

[79] Chen X, Zhang F, Wang Q, Han X, Li X, Liu J, Lin H, Qu F. The synthesis of ZnO/SnO$_2$ porous nanofibers for dye adsorption and degradation. Dalton Trans 2015;44:3034–42.

[80] Saharan P, Chaudhary GR, Lata S, Mehta SK, Mor S. Ultra fast and effective treatment of dyes from water with the synergistic effect of Ni doped ZnO nanoparticles and ultrasonication. Ultrason Sonochem 2015;22:317 [and].

[81] Mittal H, Morajkar PP, Alili AA, Alhassan SM. In-situ synthesis of ZnO nanoparticles using gum Arabic based hydrogels as a self-template for effective malachite green dye adsorption. J Polym Environ 2020;28:1637–53.

[82] Deb A, Kanmania M, Debnath A, Bhowmik KL, Saha B. Ultrasonic assisted enhanced adsorption of methyl orange dye onto polyaniline impregnated zinc oxide nanoparticles: kinetic, isotherm and optimization of process parameters. Ultrason Sonochem 2019;54:290–301.

[83] Kataria N, Garg VK. Removal of Congo red and Brilliant green dyes from aqueous solution using flower shaped ZnO nanoparticles. J Environ. Chem. Eng 2017;5(6):5420–8.

[84] Kataria N, Garg VK, Jain M, Kadirvelu K. Preparation, characterization and potential use of flower shaped zinc oxide nanoparticles (ZON) for the adsorption of Victoria Blue B dye from aqueous solution. Adv Powder Technol 2016;27(4):1180–8.

[85] Phan D-N, Rebia RA, SaitoY KD, Khatri M, Tanaka T, Lee H, Kim I-S. Zinc oxide nanoparticles attached to polyacrylonitrile nanofibers with hinokitiol as gluing agent for synergistic antibacterial activities and effective dye removal. J Ind Eng Chem 2020;85:258–68.

[86] Muinde VM, Onyari JM, Wamalwa B, Wabomba JN. Adsorption of malachite green dye from aqueous solutions using mesoporous chitosan–zinc oxide composite material. Environ Chem & Ecotoxicol 2020;2:115–25.

[87] Khosla E, Kaur S, Dave PN. Ionic dye adsorption by zinc oxide nanoparticles. Chem Ecol 2015;31:173–85.

[88] Hassan HS, Elkady MF, El-Shazly AH, Bamufleh HS. Formulation of synthesized zinc oxide nanopowder into hybrid beads for dye separation. J Nanomater 2014:14. https://doi.org/10.1155/2014/967492. 967492.

[89] Khoshhesab ZM, Gonbadi K, Behbehani GR. Removal of reactive black 8 dye from aqueous solutions using zinc oxide nanoparticles: investigation of adsorption parameters. Desalination & Water Treat 2015;5(6):1558–65.

[90] Mahmoodi NM, Najafi F. Preparation of surface modified zinc oxide nanoparticle with high capacity dye removal ability. Mater Res Bull 2012;47(7):1800–9.

[91] Klett C, Barry A, Balti I, Lelli P, Schoenstein F, Jouini N. Nickel doped zinc oxide as a potential sorbent for decolorization of specific dyes, methylorange and tartrazine by adsorption process. J Environ Chem Eng 2014;2(2):914–26.

[92] Saini J, Garg VK. Removal of Orange G and Rhodamine B dyes from aqueous system using hydrothermally synthesized zinc oxide loaded activated carbon (ZnO-AC). J Environ Chem Eng 2017;5:884–92.

[93] Dil EA, Ghaedi M, Asfaram A. The performance of nanorods material as adsorbent for removal of azo dyes and heavy metal ions: application of ultrasound wave, optimization and modeling. Ultrason Sonochem 2017;34:792–802.

[94] Banu KN, Santhi T. Development of tri-metal oxide nano composite adsorbents for the removal of reactive yellow-15 from aqueous solution. Int j sci nat 2013;4:381–9.

[95] Chaudhary S, Kaur Y, Umar A, Chaudhary GR. Ionic liquid and surfactant functionalized ZnO nanoadsorbent for recyclable proficient adsorption of toxic dyes from waste water. J Mol Liq 2016;224:1294–304.

Chapter 8

Optimization of rural indoor kitchen structure and minimizing the pollution load: a sustainable environmental modeling approach

Deep Chakraborty, Naba Kumar Mondal
Environmental Chemistry Laboratory, Department of Environmental Science, The University of Burdwan, Bardhaman, West Bengal, India

1. Introduction

Emissions from traditional cookstoves while burning of crude biomass cooking fuel used in rural households around the developing countries results in pollution in indoor environment. Household air pollution comprises a range of air pollutants like particulate matter, gaseous, aerosols, black carbon, volatile organic carbons, polycyclic aromatic hydrocarbon, heavy metals, and so forth. It is clear from the previous studies that they could find concentrations of these pollutants are much higher in magnitude than their natural presence in ambient air [1−3]. The majority of developing countries' population uses biomass-generated fuel energy for their everyday use, i.e., cooking, space heating, and so forth [4]. Countries such as India, Sri Lanka, Nepal, and Pakistan used biomass as fuel (72%, 88%, and 67%) for regular household cooking, according to the WHO in 2006 [4]. Nearly 3 billion people depend on crude biomass cooking fuel for their daily cooking purposes and the figures are projected to rise by 2030 [5,6]. Emissions produced by the combustion of this biomass fuel have a major effect on the local environment and public health [7,8], particularly in remote areas in India where conventional stoves would be utilized as daily cooking and room heating purposes. The resultant exposure from indoor pollution has been well established to occur around 4% of the worldwide burden of health [9]. However, the relationship between both the features of the kitchen room and the concentration of indoor pollution is

Cognitive Data Models for Sustainable Environment. https://doi.org/10.1016/B978-0-12-824038-0.00011-0

not very well defined. Some researchers reviewed stove efficiency indicators for indoor air pollution [10], although several experiments have found that better heating systems will mitigate exposures by lowering pollution and exhaust the emissions [11–13]. But, there is still a lacking of a straight link of the connection among kitchen room characteristics and indoor air quality improvement.

Modeling methods for forecasting pollutant loads between pollution sources and environmental factors, including climate experiments used in numerous scientific papers [14–16]. Previous researchers [17] proposed ways to enhance the indoor air environment in modern houses after evaluating the key factors affecting the indoor environment of the air conditioning system, though some researchers [18] have simulated and improved the mechanical properties of composites applying the response surface methodology (RSM) simulation techniques. However, in rural settings, these types of modeling approaches are absent where urban sectors having many potentials. Modeling techniques have many positive advantages, including (1) predicting likely impacts of household air pollutants before actually undertaking expensive and time-consuming scientific experiments; (2) assessing the significance and influences of vital kitchen room patterns, such as airflow and environmental conditions; and (3) provide a mechanism to establish guidelines or standards for kitchen room design. There really is a growing trend toward setting criteria regarding stove efficiency as being part of a global efforts to promote healthier ovens for wood stoves [19]. There is nowadays a green building concept documented in some research studies [20] to minimize the health-hazardous pollutants, especially in urban areas and in developed countries. But, in rural settings, no such concept is available to reduce the pollutants by improving their household structural pattern.

Keeping this in mind the aforementioned gap in this research observed a first approach to address toward the needs of a simple RSM which may predict indoor kitchen structural pattern to minimize the pollution load during cooking with unprocessed biomass fuel. This modeling approach can help villagers to build a rural kitchen with suitable ventilation patterns while they can also use traditional stoves and fuel which may provide them a sustainable indoor environment and health.

2. Material and methods

2.1 Area of study and research design

The current investigation was performed neighboring rural region of Burdwan, West Bengal, a district of East India, which chose 33 households based on the following criteria: (1) the position must be 10 km from the highway to reduce the impact of automobile emissions; (2) no air contaminants sources such as power plants, cement plants, iron, and steel plants, rice mills inside 10 km; and

FIGURE 8.1   Picture of a typical rural kitchen and traditional cookstove.

(3) only agricultural fuel is used for household purposes by the villagers. Thirteen households were omitted from this analysis due to nonsignificant results.

2.2 Study period

Data collection was mainly performed at night, while most of the residents prepared their daily food after coming home from work. During the day observations were also made for those that seem to prepare their food twice a day. So, each household has been visited six times a month and the average was taken to put into the RSM model. A typical picture of the rural kitchen and cook stove is depicted in Fig. 8.1A and B.

2.3 Measurement of indoor air quality

The measurement of three toxic gas i.e., CO_2, CO, and O_3 were monitored by small handheld gaseous indicators (gaZguard Tx make). GaZguard Tx CO, CO_2, and O_3 meter having ranges 0—200 ppm, 0—5000 ppm, 0—5 ppm respectively. All instruments have used the diffusion/aspiration sampling technique. Each instrument having a resolution of 0.1 ppm except O_3 having a resolution of 0.01 ppm. Both CO and O_3 have used electrochemical sensors while CO_2 used the InfraRed sensor. All instruments cum gaseous indicators were kept by maintain a distance of 1.5 m from the cookstove and about 1.0 m from the floor [21,22]. Every device (GaZguard Tx model) was configured and standardized as per ISO 9001—2008 before being used. HTC-1 model device were used to measure RT and RH while women executed the cooking process.

2.4 Kitchen and living room ventilation pattern

For the dimension of kitchen volumes, the traditional measuring tape was used. The airflow pattern was estimated taking into account the total window numbers in the both kitchen and living room. For the parallel positions of the windows and doors in both the kitchen and living room, cross-ventilation was observed. The cooking hours were reported by women who were practiced cooking in survey questionnaires [21].

2.5 About Response Surface Methodology (RSM) and Central Composite Design (CCD)

RSM is the technique to evaluate the correlation between different factors and the different factors influencing response and to examine the impact of these method parameters on coupling responses. It is a method for defining and representing the association among mean responses and input variables that gives the two-dimensional (2D) or three-dimensional (3D) graphical outputs.

CCD is a mathematical approach which uses the multivariate nonlinear model. Moreover, CCD has been commonly used by suitable experiments to refine process variables to evaluate output model equations and its relation to operational conditions [23,24]. It is also helpful in researching the relationships of the different parameters that control the mechanism. In this present analysis, the CCD was implemented to define the optimum process factors for optimizing the rural kitchen structural pattern. This very same CCD has been used to balance a second-order model, which needs just a few modeling simulations [25]. The CCD model comprised of a factorial runs of $2n$ with $2n$ axial runs and n_c center runs (six repeats, 0, 0, 0, ..., 0). While the total of factors n rises the number of rounds for a complete repeat of the design also rises, which is given in Eq. (8.1).

$$N = 2^n + 2n + n_c \tag{8.1}$$

The methods of optimization includes three key phases: (1) carrying out the statistically calculated experimentations, (2) assessing coefficients in specific models, and (3) forecasting the response and examination the acceptability of the model [26,27]. Here an empirical model has been framed to compare the relationships between rural household configurations and pollutants as provided in Eq. (8.2) to analyze the outcome of factor interactions.

$$Y = \beta_0 + \sum_{i=1}^{k} \beta_i x_i + \sum_{i=1}^{k} \sum_{j=1}^{k} \beta_{ij} x_i x_j + \sum_{i=1}^{k} \beta_{ii} x_{ii}^2 + \varepsilon \tag{8.2}$$

Y represented as the depended factor; β_0 represent the intercept; coefficients are β_i, β_{ij}, and β_{ii} for the linear outcome, and dual interactions; independent factors are x_i, x_j; and ε represent as an error factor.

TABLE 8.1 Variables with ranges and levels for the measurement of various indoor pollutants at kitchen.

Name (factor)	Units	Low	High
Window number (X_1)	Number	1	2
Volume of kitchen (X_2)	m³	2.80	7.50
Cooking hour (X_3)	Hour	1	5

2.6 Experimental model design

Three independent variables including window number, kitchen volume, and cooking hour were selected and taken into consideration to optimize the pollutant concentration arising from biomass stove. Independent variables with their ranges and levels are tabulated in Table 8.1 which was used during the analysis. According to the scheme alluded to in Table 8.1, a total of 20 trials were run in triplicates. For regression model and graphical interpretation of the data collected, Design Expert 7.0.0 software was used [28]. From the model equations and the 3D graphical output, the actual values of the specific variables can be defined. The discrepancy was justified by the multiple determination coefficients of dependent variables. The optimum value was estimated by R^2 and the model equation and the relationship between the operational parameters within the definite range was subsequently clarified.

2.7 Ethical permission

The ethical approval of the study (IEC/BU (2016/1)) was allowed by the Board of the Ethical Committee of the University of Burdwan.

3. Results and discussion

Upon assessing the inclusion and exclusion factors, the CCD has been used to optimize the concentration of contaminants as a function of multiple parameters, i.e., window number, kitchen size, and cooking duration. The ANOVA output of the models consist of Fisher's F test with its related probability $P > F$, the correlation analysis R^2 and the lack of fitness, which calculated the effectiveness of the regression models [29]. A high F value and a low P value were used for a meaningful corresponding coefficient [30,31].

3.1 Nonlinear regression models

The interrelationships between kitchen room variables and the contaminants (Y) of the validated output model observed in Eqs. (8.3)–(8.7). The variables were similar to the coefficients and the signs $(+, -)$ of the model specific. The unique weight of the parameters in the formula would be shown by the co-efficients. Although the indications $(+)$ and $(-)$ implies the positive and negative influence on the dependent response (Y). The coefficients represented the importance of the variables in the model, which implies the significant role of the parameters in the degree of emission level.

The best model for any set of data depends upon the F value, in which the F value is being used to evaluate the importance of applying new model terms to those that are in the equation. The small P value of "Prob > F" (<0.0001, less than 0.05) suggests that the inclusion of second-order features strengthened the model.

$$Y = +38.84 + 1.03x_1 - 0.65x_2 - 4.27x_3 + 5.06x_1x_2 - 0.374x_1x_3 \\ - 6.53x_2x_3 - 1.36x_2^2 + 7.11x_3^2 \tag{8.3}$$

$$Y = +511.88 + 90.57x_1 - 432.44x_2 - 8.05x_3 + 37.44x_1x_2 + 190.16x_1x_3 \\ - 400.69x_2x_3 + 78.45x_2^2 + 149.28x_3^2$$

$$\tag{8.4}$$

$$Y = +0.062 + 2.415E - 003x_1 - 1.937E - 003x_2 - 2.505E - 003x_3 \\ - 3.338E - 003x_1x_2 + 4.339E - 003x_1x_3 - 5.262E - 003x_2x_3 \\ + 4.13x_2^2 - 0.98x_3^2$$

$$\tag{8.5}$$

$$Y = +30.16 - 0.557x_1 + 0.61x_2 - 0.16x_3 - 2.28x_1x_2 + 0.89x_1x_3 \\ - 1.26x_2x_3 + 1.58x_2^2 - 0.25x_3^2 \tag{8.6}$$

$$Y = +81.45 - 1.55x_1 + 1.21x_2 - 1.03x_3 - 6.01x_1x_2 + 2.99x_1x_3 \\ - 4.43x_2x_3 + 4.13x_2^2 - 0.98x_3^2 \tag{8.7}$$

Where, the definition of x_1, x_2, and x_3 are shown in Table 8.1. The statistical equation between one of the responses (e.g. carbon monoxide) and the process variables considered is derived from the coefficients arising from the Design Expert 7.0 program output.

As indicated by the coefficients in the case of Eq. (8.3) for CO, the significance of the relationships is $x_1x_2 > x_1x_3 > x_2x_3$. The synergistic effect of the model is to increase the efficiency of the indoor kitchen environment, whereas the antagonistic effects indicate an opposite sense. As showed, the interactions x_1x_2 having a positive effect on the other hand x_1x_3 and x_2x_3 were showed the antagonistic effects that implied window number and volume of the kitchen has more positive effects on the concentration of carbon monoxide in the kitchen environment. That means if the ventilation were poor in the kitchen room the concentration of CO can rise to an alarming level. The same kind of theory was also speculated by some previous researchers, they also found an improper ventilated rural house can cause an increase in indoor pollutant concentration [32,33].

The output model of CO_2 (Eq. 8.4) and O_3 (Eq. 8.5) has shown similar types of results. Both model outputs revealed that window number and cooking time were two important factors that have a positive effect on the pollutants followed by "window number-kitchen room volume" and "kitchen room volume-cooking hour." Moreover, temperature (Eq. 8.6) and relative humidity model (Eq. 8.7) showed a similar pattern where the importance of interactions followed a pattern like $x_1x_3 > x_2x_3 > x_1x_2$. Here, cooking frequency showed an important part in the case of kitchen room temperature and relative humidity, because during the survey it was observed that all the traditional stoves were unvented, and it may disperse the heat all over the kitchen room and it will in terns affect the indoor environment.

The three-dimensional contour plots represent the associations at all points with the same response to generate constant lines. In the model, the interactions between three variables (window number, kitchen room volume, and cooking hour) form a curvature model, and the contour diagram having nonparallel straight lines were presented. Interactions plots of the output were given in Figs. 8.2–8.6.

3.2 Descriptive statistics and ANOVA analysis of the response variables

From Table 8.2 it is evident that mean pollution concentrations in the respective household were very high CO (44.8 ± 0.092 ppm), CO_2 (680 ± 136.96 ppm), and O_3 (0.06 ± 0.0019 ppm). Similar types of high concentration were also reported previously in the same district by Ref. [21]. Another study from the same province [22] also highlighted that a correlation present between the impact of biomass smoke on indoor air pollution and high blood pressure. Indoor air pollution also causes a high risk of carcinogenicity among the child belongs to biomass using households [1,2,34]. So there is a necessity for a remediation approach whether in reality or modeling.

In statistics, the coefficient of determination R^2 is the proportion of uncertainty that the mathematical model accounts for in a given data. It gives a test of how accurately the model can forecast future results. The R^2 values of 0.99, 0.76, 0.73, 0.99, and 0.99 are obtained from the model for carbon monoxide, carbon dioxide, ozone, temperature, and relative humidity, respectively. R^2 determined the fitness of the models and the F test assessed their statistical significance. According to Ref. [30]; the high value of R^2 represents the stronger model. In all the models the lack of fit also suggests nonsignificant values. The ANOVA was introduced to ensure a successful regression model and the results showed in Table 8.3. The "F value" of models was 8152.37, 8.43, 3.68, 194.63, and 114.61 implied that all models of CO,

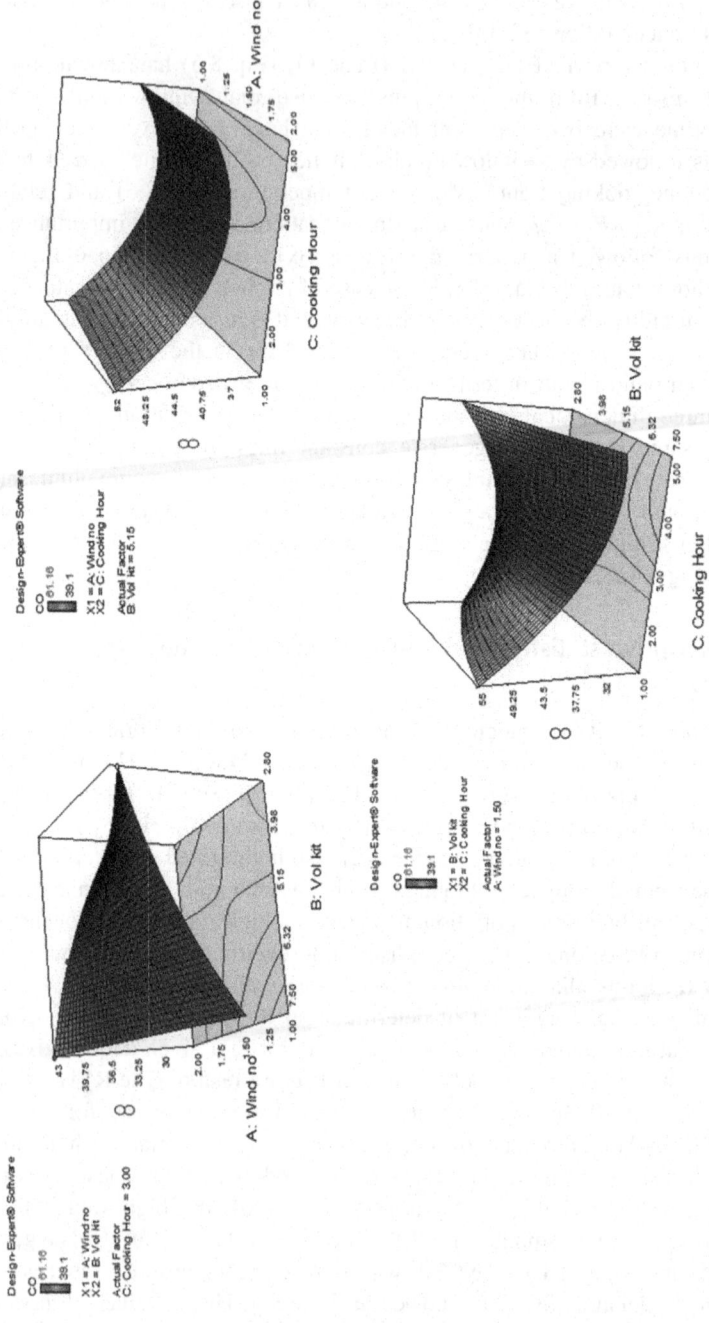

FIGURE 8.2 Response surface model output and contour interaction plots of different kitchen factors and the concentration of Carbon monoxide (CO).

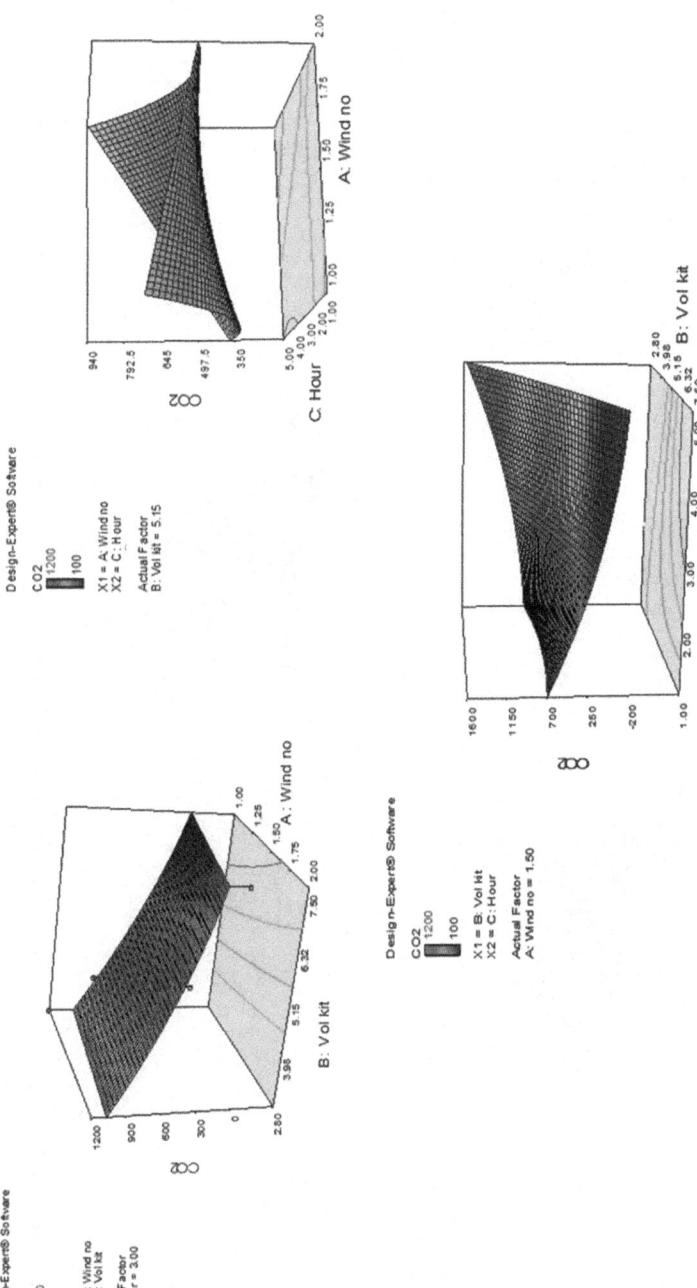

FIGURE 8.3 Response surface model output and contour interaction plots of different kitchen factors and the concentration of Carbon dioxide (CO_2).

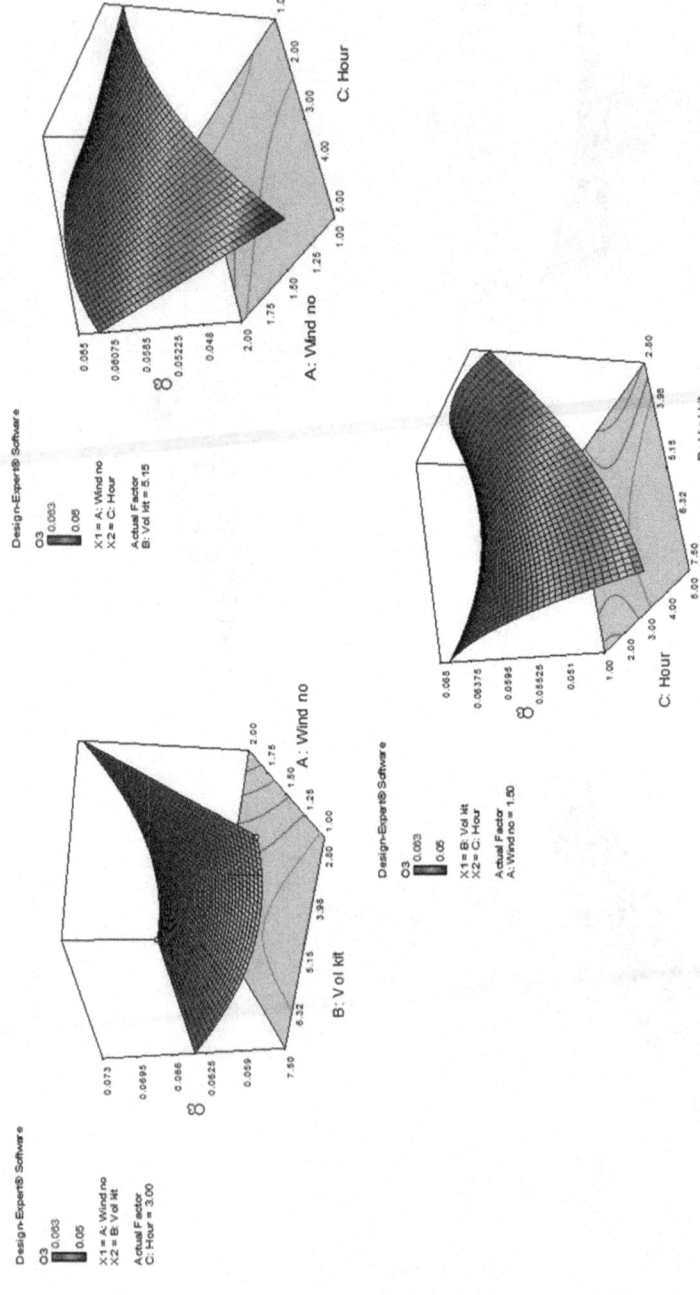

FIGURE 8.4 Response surface model output and contour interaction plots of different kitchen factors and the concentration of Ozone (O_3).

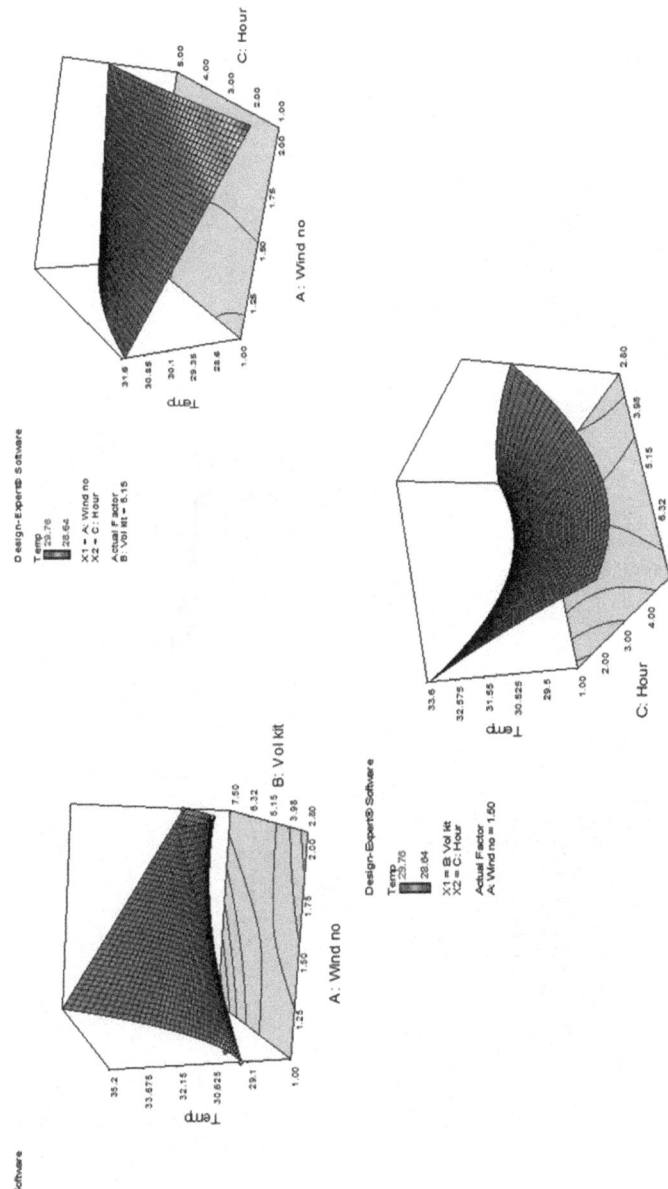

FIGURE 8.5 Response surface model output and contour interaction plots of different kitchen factors and the Temperature.

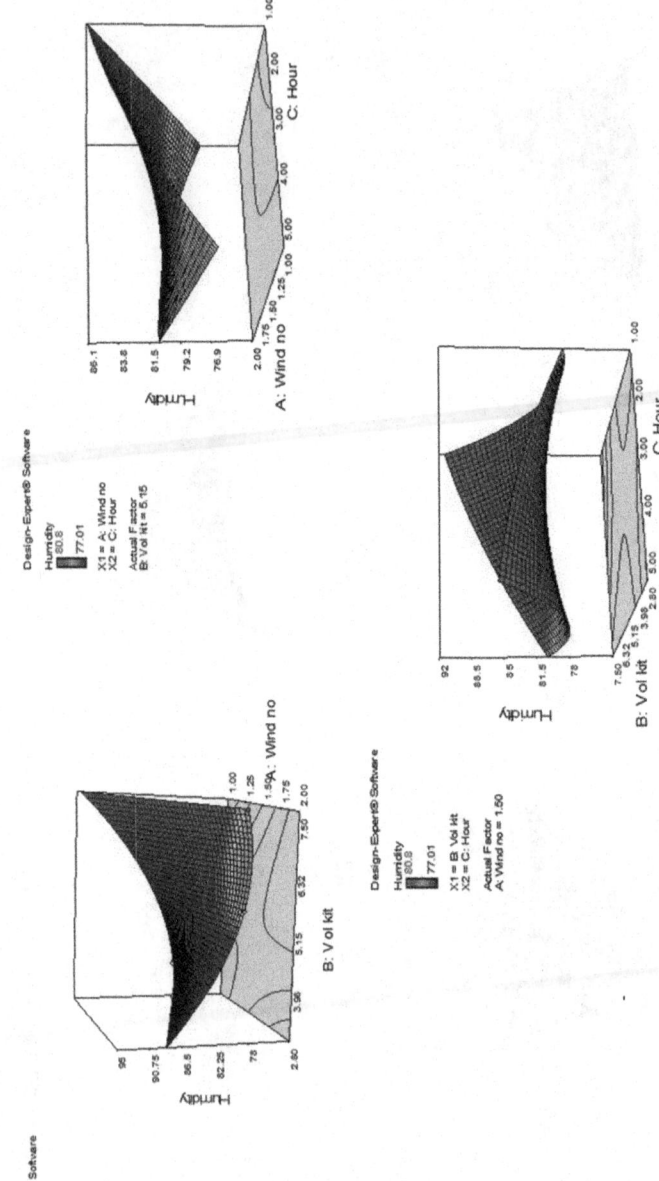

FIGURE 8.6 Response surface model output and contour interaction plots of different kitchen factors and the Relative Humidity.

TABLE 8.2 Descriptive statistics of the model applied.

	CO (ppm)	CO_2 (ppm)	O_3 (ppm)	Temperature (°C)	RH (%)
Mean	44.8	680	0.06	29.36	79.2
Std. Dev.	0.092	136.98	0.0019	0.033	0.15
C.V. %	0.21	20.14	3.232,958	0.11	0.18
R-Squared	0.9998	0.8598	0.728	0.993	0.9881
Adj R-Squared	0.9997	0.7578	0.5302	0.9879	0.9795
Adeq precision	357.149	9.732	8.449	54.411	38.127

CO_2, O_3, room temperature, and relative humidity were significant (Table 8.3). It also emerged that there is merely a 0.01% probability that model F-values of this magnitude will arise due to noise in the case of CO. From Table 8.3 it can be observed that CO, room temperature, and relative humidity models were well regressed with the experimental values, where it showed the R^2 values 0.9998, 0.8598, 0.728, 0.993, and 0.9881 for CO, CO_2, O_3, temperature, and relative humidity, respectively. Ref. [35] in their study also formulated the models to predict the conditions of comfort. They have used variables namely temperature, CO_2 level, and relative humidity in a specified range as input conditions.

3.3 Optimization of the desirable condition by using RSM

In the optimization study, the different response methods were conducted to find out the optimized goals of any four combination between window number, the volume of the kitchen, cooking hour, and concentration of pollutants. For each operational parameter, required targets were chosen and responses were obtained from the "optimization" menu in Design Expert applications. The potential aims of this menu included: optimization, reduction, target, range, zero (only for answers), and "fix to a definite value" (variables only). For each included variable, a maximum and a minimum level have been defined [36]. From the menu, "weights" can be applied to each target to change its basic desirability feature to type. Finally, the targets are merged into an aggregate feature of desirability. Desirability is a purposeful attribute varying from zero to one at the target. The target quest was randomly triggered at a start point by the program and takes the sharpest incline to the top. Desirability tests were run for every response. From desirability, it was also tested that if the kitchen room has two windows, the volume of kitchen 7.50 m^3 and if the cooking

TABLE 8.3 Analysis of variance of effects of kitchen factors on pollutants: CO, CO_2, O_3, room temperature, and relative humidity.

Pollutant types	Variation source	Sum of square	Degree of freedom	Mean square	F Value	P
CO	Model	554.08	8	69.26	8152.37	<0.0001
	Residual	0.093	11	8.496E-003		
	Lack of fit	0.093	10	9.325E-003	46.63	0.1135
	Pure error	2.000E-004	1	2.000E-004		
	Total	554.18	19			
CO_2	Model	1.266E+006	8	1.582E+005	8.43	<0.0010
	Residual	2.064E+005	11	18,763.35		
	Lack of fit	1.614E+005	10	16,139.69	0.36	0.8741
	Pure error	45,000.00	1	45,000.00		
	Total	1.472E+006	19			
O_3	Model	1.115E-004	8	1.394E-005	3.68	0.0246
	Residual	4.167E-005	11	3.788e-006		
	Lack of fit	4.117E-005	10	4.117E-006	8.23	0.2653
	Pure error	5.000E-007	1	5.000E-007		
	Total	1.532E-004	19			

	Source	Sum of squares	df	Mean square	F-value	p-value
Temperature	Model	1.67	8	0.21	194.63	<0.0001
	Residual	0.012	11	1.073E-003		
	Lack of fit	0.012	10	1.175E-003	23.51	0.1593
	Pure error	5.000E-005	1	5.000E-005		
	Total	1.68	19			
Relative humidity	Model	19.41	8	2.43	114.61	<0.0001
	Residual	0.23	11	0.021		
	Lack of fit	0.11	10	0.011	0.086	0.9933
	Pure error	0.13	1	0.13		
	Total	19.65	19			

activities continue greater than four hours, then the concentration of CO will be minimized. From the CO model output, it gave the desirability of 1.00. The desirability of other responses was observed as per the following 0.967, 0.860, 0.967, and 1.00 in the case of CO_2, O_3, room temperature, and relative humidity, respectively. It was mentioned that more the desirability close to one the results of the output models will be strong. The optimum conditions of the above desirability were given in Fig. 8.7A–E.

3.4 Predicted versus actual plots of indoor variables deriving through RSM

The response surface modeling approach also provided the predicted values upon experimental values. It is important because predicted values will give an idea of how much variability are present within experimental values. This can be found in Fig. 8.8A–E that the test values were well suited to the forecast values. This indicates that the results of the study were considerable. The model can be altered and improved in many fields, with increasing data and/or modifying the assumptions. The efficiency of the output is affected, as with any model, by the input quality. Moreover, there is no published data on these indoor pollutant parameters using the RSM method in the developing regions, and here is the first study from India. Here, we are reasonably confident in our assumptions, but these parameters can differ wildly, and therefore the adaptation of the model to a particular area involves region-specific inputs. The model version assumes that the supply of all indoor air pollutants comes from a single source of stoves with a consistent emission rate for the sake of simplicity and that all emissions from that supply enter the room. Many sources of indoor air pollution typically exist, and several stoves are used for various purposes in many households. Outdoor air pollution inputs were not blanketed, because those impact the concentrations of indoor air contaminants and could be responsible for the model as a discrete or beneficial distribution. Contributions from outside sources typically have limited relative effects on concentrations of indoor air quality for conventional stove consumers, but for houses with cleaner cooking technologies, a greater relative source of indoor air pollutants could be present.

4. Model improvements

The basic response surface model presented here can be used to forecast kitchen concentrations of air pollutants by window number, kitchen volume and cooking time data for biomass stove/fuel combinations, and details on standard cooking needs and cooking characteristics. This capability can be a valuable method for a tentative, cost-effective estimation of the possible impacts of indoor air emissions, as well as for relating health-based recommendations on air quality to the layout of indoor rural buildings. There are,

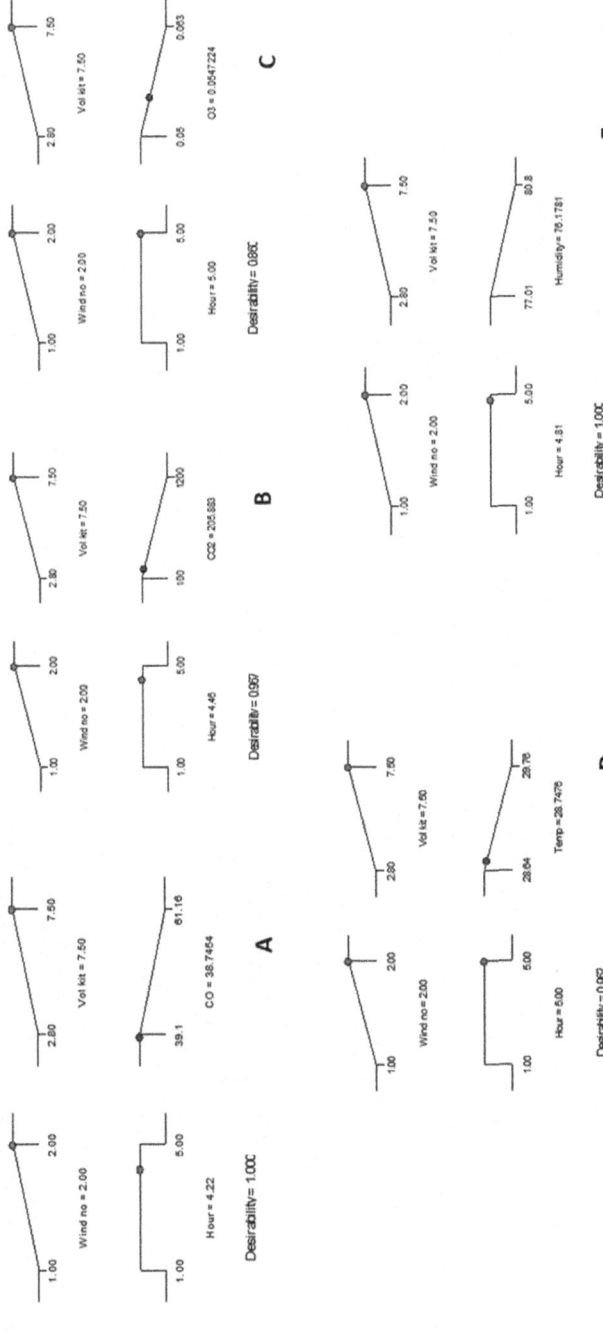

FIGURE 8.7 A–E Desirability plots of carbon monoxide, carbon dioxide, ozone, temperature, and relative humidity obtained from the model output.

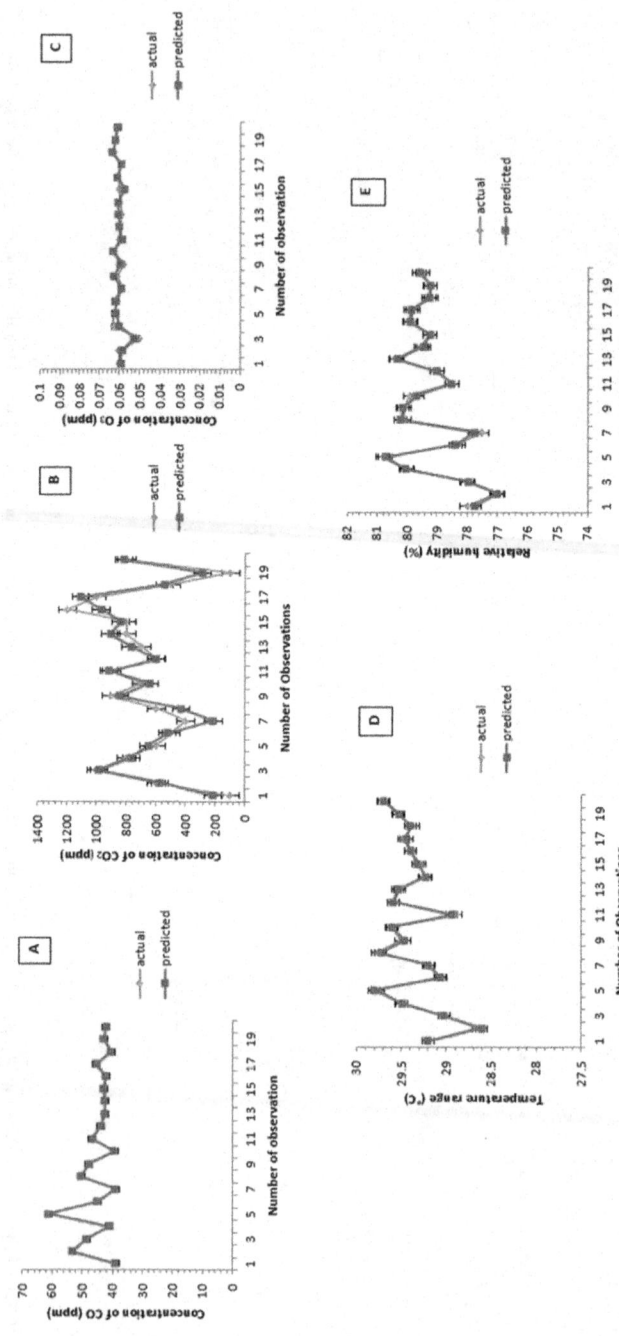

FIGURE 8.8 A–E Predicted vs Actual plot of carbon monoxide, carbon dioxide, ozone, temperature and relative humidity obtained from the model output.

however, things that could be improved to optimize the effectiveness of the proposed model are:

1. Better estimation of all phases of contaminants such as gases, particulates, and vapors will help to resolve the homogeneity of indoor air contamination levels in indoor kitchens required to improve the precision of the model.
2. With more interesting data such as per day cooking capacity, the accuracy of the model could be improved. In this research, described percentage of pollutants were not monitored here that exhausted outward before even being combined throughout the kitchen.
3. This study is unique and this model is being first used for the section of indoor air pollution remediation to find out sustainable health in the rural part of India. More validity with this model is needed to grow up the concept of green building in the rural community.

5. Conclusions

For determination of the reduction strategies of indoor air quality levels in the rural household's response surface model were applied. CO, CO_2, and O_3 concentrations were measured along with temperature and relative humidity. It was observed from the model output that the R^2 values of the models were 0.99, 0.76, 0.73, 0.99, and 0.99 for carbon monoxide, carbon dioxide, ozone, temperature, and RH, respectively. As per ANOVA analysis CO, room temperature, and RH models are well fitted. The optimum desirability of carbon monoxide, carbon dioxide, ozone, temperature, and relative humidity models were 1.00, 0.967, 0.860, 0.967, and 1.00, respectively. The simple response surface statistical equation formulated here might be used to predict kitchen room concentrations of air pollutants after providing the data of various information about the typical cooking frequency and kitchen features. The model would be also useful to determine to construct the kitchen room pattern in rural areas. This capacity building can be important for cost-effectiveness and also from the health point of view. With more comprehensive available data, the output of the model may be more useful like data on the amount of cooking energy/day, ventilation rate, emission rate, and putting data of other pollutants. These models may be useful for government bodies, researchers and, NGOs working on public health policymaking.

6. Limitation of the study

More study households could be selected. Volumes of the kitchen areas and ventilation patterns from various sources across India can be fed into this model for a wide range of the results. If a high number of experimental data can be run, the model may give more significant outputs. The study was conducted in the purely natural environment to obtain the crude values of the fate of the pollutants. Other pollutants like $PM_{2.5}$, PM_{10} were not included in this study.

Appendix A. Supplementary data

Supplementary data to this article can be found online at https://doi.org/10. 1016/B978-0-12-824038-0.00011-0.

List of abbreviations

ANOVA Analysis of Variation
CCD Central Composite Design
CO Carbon monoxide
CO₂ Carbon dioxide
O₃ Ozone
R² Regression Coefficient
RH Relative Humidity
RSM Response Surface Methodology

Acknowledgments

The author expressed their gratitude to the University Grants Commission (F. No. 42−434/ 2013(SR), dated 12th March, 2013 and DST-SERB F. No. [CRG/2019/004506, Dated 14.01.20120] for funding the research study.
Ethical considerations
Writers are conscious of and agree with best practices in writing ethics, especially concerning authorship, dual submission and misuse of figures, conflicting interests, and agreement with ethical research policies. Writers adhere to the standards for publishing since the research submitted is original and has not been written in any language elsewhere.

References

[1] Chakraborty D, Mondal NK. Reduction in household air pollution and associated health risk: a pilot study with an improved cookstove in rural households. Clean Technol. Environ. Policy 2021:1−17. https://doi.org/10.1007/s10098-021-02098-9.

[2] Chakraborty D, Mondal NK. Estimation of nitrogen dioxide (NO₂) due to burning of household biomass fuel and assessment of health risk among women in rural West Bengal. Curr. World Environ. 2021;16(SI1):1−8.

[3] Mondal NK, Chakraborty D. Vulnerability of rural health exposed by indoor pollution generated from biomass and fossil fuels. Mor. J. Chem. 2015;3(1):83−98.

[4] World Health Organisation. Fuels for life: household energy and health. Rehfuees E, WHO Library Cataloguing-in-Publication Data; 2006.

[5] International Energy Agency. World energy outlook. Paris: OECD/IEA; 2002.

[6] World Health Organisation. The world health report, reducing risks, promoting health life. Geneva, Switzerland: WHO; 2002.

[7]   Andreae MO, Merlet P. Emission of trace gases and aerosols from biomass burning. Global Biogeochem Cycles 2001;15(4):955−66.

[8]   Smith KR, et al. Indoor air pollution in developing countries and acute lower respiratory infection in children. Thorax 2000;55:518−32.

[9]   Bruce N, et al. Indoor air pollution in developing countries: a major environmental and public health challenge. Bull. World Health Organ. 2000;78(9):1078−92.

[10] Johnson M, et al. Modeling indoor air pollution from cookstove emissions in developing countries using a Monte Carlo single-box model. Atmos Environ 2011;45:3237−43.

[11] Armendáriz AC, et al. Reduction in particulate and carbon monoxide levels associated with the introduction of a Patsari improved cook stove in rural Mexico. Indoor Air 2008;18:93−105.

[12] Pennise D, et al. Indoor air quality impacts of an improved wood stove in Ghana and an ethanol stove in Ethiopia. Energy Sust Dev 2009;13:71−6.

[13] Saksena S, Thompson L, Smith K. The indoor air pollution and exposure database: household pollution levels in developing countries. 2003. http://ehs.sph.berkeley.edu/heh/hem/documents/iapi.pdf.

[14] Bond TC, et al. Quantifying immediate radiative forcing by black carbon and organic matter with the specific forcing pulse. Atmos Chem Phys Discuss 2011;10:15713−53.

[15] Hellweg S, et al. Integrating human indoor air pollutant exposure within lifecycle impact assessment. Environ Sci Technol 2009;43:1670−9.

[16] Nicas M. Quantitative surveying- application of mathematical modeling to estimate air contaminant exposure. In: Perkins JL, editor. Modern industrial hygiene. 2nd ed. Cincinnati: ACGIH; 2008.

[17] Dequan K, Rong W. Analysis on influencing factors of indoor air quality and measures of improvement on modern buildings. In: Proceedings of ICBBE, 16−18 May, Shanghai, China; 2008.

[18] Jayabal S, Natarajan U. Modeling and optimization of thrust force, torque and tool wear in drilling of coir fibre reinforced composites using response surface method. Int J Mach Mach Mater 2011;9:149−72.

[19] MacCarty N, Still D, Ogle D. Fuel use and emissions performance of fifty cooking stoves in the laboratory and related benchmarks of performance. Energy Sust Dev 2010;14:161−71.

[20] Samer M. Towards the implementation of the Green Building concept in agricultural buildings: a literature review. Agric Eng Int CIGR J 2013;15(2):25−46.

[21] Chakraborty D, Mondal NK, Datta JK. Indoor pollution from solid biomass fuel and rural health damage: a micro-environmental study in rural area of Burdwan, West Bengal. Int J Sust Built Environ 2014;3:262−71.

[22] Chakraborty D, Mondal NK. Hypertensive and toxicological health risk among women exposed to biomass smoke: a rural Indian scenario. Ecotoxicol Environ Saf 2018;161(15):706−14.

[23] Arulkumar M, Sathishkumar P, Palvannan T. Optimization of orange G dye adsorption by activated carbon of *Thespesia populnea* pods using response surface methodology. J Hazard Matter 2011;186:827−34.

[24] Kalavathy HM, et al. Modelling, analysis and optimization of activation parameter of H_3PO_4 activated rubber wood saw dust using response surface methodology (RSM). Colloids Surf B Biointerfaces 2009;70:35−45.

[25] Ahmad A, et al. Removal of Cu(II) and Pb(II) ions from aqueous solutions by adsorption on sawdust of Meranti wood. Desalination 2009;247(1−3):636−46.

[26] Kumar R, et al. Response surface methodology approach for optimization of biosorption process for removal of Cr (VI), Ni (II) and Zn (II) ions by immobilized bacterial biomass sp. *Bacillus brevis.* Chem Eng J 2009;146:401−7.

[27] Sahu JN, Acharya J, Meikap BC. Response surface modeling and optimization of chromium (VI) removal from aqueous solution using tamarind wood activated carbon in batch process. J Hazard Mater 2009;172:818−25.

[28] Stat-Ease Inc. Design-expert 7 for windows: software for design of experiments. Minneapolis, MN: DOE; 2009, http://www.statease.com.

[29] Roy P, Dey U, Chattoraj S, Mukhopadhyay D, Mondal NK. Modeling of the adsorptive removal of arsenic (III) using plant biomass: a bioremedial approach. Appl Water Sci 2015. https://doi.org/10.1007/s13201-015-0339-2.

[30] Chattoraj S, Sadhukhan B, Mondal NK. Predictability by Box-Behnken model for carbaryl adsorption by soils of Indian origin. J Environ Sci Health B 2013;48:626−36.

[31] Myers RH, Montgomery DC. Response surface methodology: process and product optimization using designed experiments. 2nd ed. USA: Wiley; 2002.

[32] Mondal NK, Saha SK, Datta JK, Banerjee A. Indoor air pollution: a household study in the village Faridpur and Ranchi colony, Durgapur, Burdwan District, West Bengal. World Environ Poll 2011;1(1):5−7.

[33] Mondal NK, Bhaumik R, Das CR, et al. Assessment of indoor pollutants generated from bio and synthetic fuels in selected villages of Burdwan, West Bengal. J Environ Biol 2013;34:963−6.

[34] Chakraborty D, Mondal NK. Assessment of health risk of children from traditional biomass burning in rural households. Expo Health 2017;10(1):15−26.

[35] Thirumal P, Amirthagadeswaran KS, Jayabal S. Optimization of IAQ characteristics of an air-conditioned car using GRA and RSM. J Mech Sci Technol 2014;28:1899−907.

[36] Chowdhury S, Chakraborty S, Saha PD. Response surface optimization of a dynamic dye adsorption process: a case study of crystal violet adsorption onto NaOH modified rice husk. Environ Sci Pollut Res 2013;20:1698−705.

Chapter 9

IoT-based health care data analytical paradigm using blockchain technology

T. Poongodi[1], R. Sujatha[2], M. Kiruthika[3], P. Suresh[4]

[1]*School of Computing Science and Engineering, Galgotias University, Greater Noida, Uttar Pradesh, India;* [2]*School of Information Technology & Engineering, Vellore Institute of Technology, Vellore, Tamil Nadu, India;* [3]*Department of Computer Science and Engineering, Jansons Institute of Technology, Coimbatore, Tamil Nadu, India;* [4]*School of Mechanical Engineering, Galgotias University, Greater Noida, Uttar Pradesh, India*

1. Introduction

The Internet of Things (IoT) reforms the conventional industries to smart industries with its intrinsic feature of a data-driven approach. The recent advancement in Information and Communication Technology (ICT) promotes the evolution of smart industries with the efficient data-driven strategy of decision making. In particular, IoT plays a significant role in connecting smart objects interlinks physical environment to cyberspace systems that consequently form cyber-physical systems (CPS). Moreover, IoT supports diversified industrial applications include logistics, the food industries, the pharmaceutical industries, the manufacturing industries, etc. [1]. The major objectives of IoT are to enhance production throughput, product quality, and operation efficiency by reducing downtime. IoT systems are characterized by, (a) decentralization, (b) heterogeneity, (c) network complexity, and (d) diversified devices and systems. Several challenges are also addressed such as resource constraints, poor interoperability, and vulnerabilities in the aspect of security and privacy. Hence, the arrival of blockchain technologies overcome the mentioned challenges of IoT systems. In a blockchain, the distributed ledger is maintained in the entire distributed system. The decentralized consensus mechanism in blockchain facilitates to create, share, and validate the transaction in a distributed system without any third-party intervention.

Cognitive Data Models for Sustainable Environment. https://doi.org/10.1016/B978-0-12-824038-0.00001-8

Copyright © 2022 Elsevier Inc. All rights reserved.

In the conventional transaction management processing, the centralized agency would validate the transaction; rather, blockchain obtains the decentralized transaction validation that results in cost-saving and mitigates the bottleneck that occurred at the trusted centralized third party. Furthermore, the transaction in the blockchain is immutable hence all the participating nodes in the network maintain the updated committed transactions. Meanwhile, the various cryptographic mechanisms such as the encryption algorithms, hash functions, and digital signature ensure the integrity and consistency of data blocks available in the blockchain network. Thus, it ensures traceability and nonrepudiation of each transaction since every participant can track the process with the fixed historic timestamp. Blockchain is an appropriate complement in IoT systems with the enhanced reliability, scalability, interoperability, security, and privacy.

The cryptographic mechanisms in blockchain ensure the integrity of each block in the committed transactions of blockchain. Hence, every transaction in the blockchain is traceable with the fixed timestamp and also guarantees nonrepudiation of the committed transactions. Some of the benefits of blockchain and IoT are traceability, interoperability, reliability, and automatic interactions with smart contract featured blockchain. The architecture of blockchain and IoT comprises of perception layer, communication layer, blockchain composite layer, and industrial application layer. The blockchain composite layer acts as a middleware between IoT devices and industrial applications in order to provide services to the users. The composite layer comprises several sublayers namely, data sublayer, network sublayer, consensus sublayer, incentive sublayer, and service sublayer. Nowadays, health care becomes a social-economic concern since new challenges are arising every day which is not handled properly in the traditional health care systems because of limited resources.

Recent advancement in wearable systems promotes the health care services in handling medical data at the clinic or at home as well. The wearable devices measure and gather data including blood sugar, blood pressure, heartbeat rate, etc. The health care data can be accessed by physicians and doctors anywhere, anytime in the health care network. IoT stores patients' data generated by wearable IoT devices that support medical caregivers in making efficient decisions by maintaining health care records. However, the medical data generated by IoT sensors are gathered automatically and transmitted through smart contract greatly supports real-time monitoring of patients. Privacy and security are considered as the primary concern while assessing clinical data. Heterogeneity networks and vulnerability in health care IoT devices pose severe challenges in ensuring security and preserving the privacy of medical data. Incorporating blockchain overcome these challenges in security assurance and privacy preservation in the health care network. Using a signature scheme, the authenticity of clinical data and authentication of the user can be verified. According to a Gartner report, 86% of industrial projects incorporate both IoT and blockchain, which plays a significant role in the industrial digital transformation. Moreover, the adoption rate of blockchain would be much faster than expected in the health care and pharmaceutical industries.

This chapter describes the architecture of blockchain and IoT convergence with the device deployment. The major challenges that emerge in converging blockchain and IoT such as resource constraints, security vulnerability, privacy leakage, scalability, etc., are also described. The potential of blockchain in offering clinical services using data-sharing facilities in the health care ecosystem is also presented. The chapter provides a conceptual framework of IoT and blockchain; insightful discussions are carried out to identify the potentials of these technologies by presenting in-depth analysis. A brief explanation of the taxonomy of blockchain systems with their features are outlined. The significance of using the smart contract in health monitoring and security related issues are also discussed in detail.

1.1 Overview of Internet of Things and its challenges

Nowadays, smart industries experience a paradigm shift by integrating the recent technologies, namely, IoT and big data analytics. In particular, IoT plays a vital role in extracting hidden values from a large amount of data and assists in making intelligent decisions. IoT consists of connected smart objects which offers several industrial services [2]. An IoT system comprises the layered subsystems as depicted in Fig. 9.1 and are described in detail in this section.

Perception Layer: The heterogeneous IoT devices namely sensors, actuators, RFID tags, controllers, smart meters, Quick Response Code (QR Code), bar code, and other wired or wireless devices are handled in this layer. The devices sense the physical environment and gathers data to proceed with further actions which can be stimulated by an actuator or a controller.

FIGURE 9.1 Layered architecture of IoT system.

Communication Layer: Several wired and wireless devices including sensors, actuators, RFIDs, controllers are connected with base stations (BS), access points (APs), and IoT gateways to construct a network. The connection could be established using several communication protocols such as near field communication (NFC), Bluetooth, Low-Power Wireless Personal Area Network (6LoWPAN), Low-Power Wide Area Network (LPWAN) such as LoRa and Sigfox, Narrowband IoT (NB-IoT), Wireless Highway Addressable Remote Transducer (WirelessHART), and Ethernet.

Application Layer: IoT is widely used in the huge number of applications due to its intrinsic features. The typical applications include supply chain, smart grid, food industry, manufacturing, connected vehicles, and health care.

IoT system connects diversified smart objects ("things") which consist of electronic sensors, and actuators that can sense and gather information in the surroundings and makes responsive actions accordingly. Some of the distinct features make IoT to face the research challenges in the subsequent perspectives.

- Resource Constraints: IoT devices include RFID tags, sensors, actuators, and smart meters are facing problems due to the limited resources of battery power, storage, and computing resources. For instance, the passive RFID tags are not having any battery power which can collect the energy from the ambient environment or using RFID readers.
- Network Complexity: The availability of heterogeneous communication protocols cooperates and coordinates with each other in IoT systems. The communication protocols such as Bluetooth, NFC, WirelessHART, 6LoWPAN, LoRa, Sigfox, and NB-IoT provide disparate network services. For instance, WirelessHART and 6LoWPAN are typically utilized for short-range communications (<100 mts.) but LPWAN can be used to offer the coverage up to 10 km [3].
- Heterogeneity: In IoT systems, the data will be available in the form of structured, semistructured, and nonstructured. And, the communication protocols and IoT devices are heterogeneous in nature. It also imposes several challenges namely, security, privacy, and interoperability.
- Interoperability: The information is exchanged and collaborated among hardware and software IoT systems. Due to the heterogeneity and decentralized nature of IoT systems, makes extremely challenging to transfer the data across strategic centers, various application sectors, and IoT systems. Hence, it is highly difficult to achieve interoperability among IoT systems.
- Security: The heterogeneity and decentralization of IoT systems do not guarantee security, although it is significantly important in an enterprise field. Encryption, authentication, and authorization may not be suitable for IoT systems owing to the fact of resource-constrained factor. However, IoT systems are more vulnerable to several types of malicious attacks because of lack of a security firmware updation [4].

- Privacy: It ensures the appropriate utilization of data without any disclosure of private information by an unauthorized user. It is merely challenging to preserve it because of decentralization, heterogeneity, and complexity of IoT systems. However, the IoT should be integrated with cloud computing for storage capabilities. The confidential data uploaded in the cloud server which is maintained by the third-party eventually compromises the privacy of confidential data [5].

1.2 Technical analysis of blockchain and its key characteristics

A blockchain maintains a distributed ledger, and it consists of a set of consecutive blocks in the blockchain system. Each block holds a hash value that refers to the previous block (parent block), block i holds the hash value of block $i-1$. The initial block is known as the genesis block that does not have any parent block. The structure of a block includes (1) block version, (2) the hash value of parent block, (3) timestamp, (4) nonce value, (5) the total number of transactions, and (6) Merkle Root, the root node of the Merkle tree combines all the transactions in every block that promotes easy verification. A blockchain size grows depending on the transactions which are being executed. Once a new block is created, the nodes in the blockchain network will automatically involve in the process of block validation. After validation, the block will be added at the end of the blockchain. An unauthorized modification can be easily identified by verifying the hash value of the block, which is tampered with [6].

The mechanisms that ensure data integrity in the blockchain network are (1) the blocks that are maintained in an ordered linked list structure, in such a way the newly added block will have the hash value of the previous block; (2) Merkle tree structure, is a binary tree that contains a root hash that combines all transactions. If any transaction is modified, and after recomputation, it would result in an incorrect root hash which can be easily identified. The block is validated in a trusted way in a decentralized environment without any third-party intervention. In particular, it is highly difficult to obtain a consensus for the newly created block in the distributed environment as it could be compromised or biased indeed of malicious nodes. Hence, using consensus algorithms such as proof of stake (PoS), proof of work (PoW), and Practical Byzantine Fault Tolerance (PBFT), a trustful validation can be obtained in this decentralized environment.

Some of the key characteristics of blockchain technologies are described in this section.

Immutability: The blocks that exist in the blockchain network cannot be tampered with or modified easily because each link refers to the hash value of the previous block. If any modification or tampering occurs, all the subsequent newly created blocks will be invalidated instantaneously. Moreover, the root node of the Merkle tree stores the hash value of all the transactions.

Traceability: Every transaction is fixed with a timestamp which assist users to verify and track the historical data by checking the blockchain data along with the timestamp.

Transparency: In the public blockchain systems, all users are provided with equal privilege to interact among themselves in the blockchain network. The newly created transactions are validated and stored in the blockchain; hence it is readily available for the users in the network. Thus, the blockchain data is completely transparent to the users who are accessing and verifying the transactions in the network.

Decentralization: In the traditional systems, the transactions are validated by a trusted centralized agency (e.g., government or bank). The centralized approach results in a performance bottleneck, single-point failure, and expensive. However, the transactions in the blockchain are validated among peer nodes without jurisdiction, or intervention by any centralized third party. Thus, easily mitigates the performance bottleneck, reduces the single-point failure risk and service cost.

Nonrepudiation: The private key is used to sign the transaction and it could be verified using appropriate public keys. The transaction cannot be refused by the originator since it is signed using cryptographic techniques.

Pseudonymity: In the blockchain, privacy can be preserved at a certain level because the blockchain addresses are traceable. The blockchain data could be analyzed in order to identify illegal and fraudulent transactions.

2. Role of IoT and blockchain in health care

The recent advancement in wearable devices promotes health care services, which reduce the utilization of hospital resources potentially. IoT devices are anticipated to enhance the performance of conventional health care systems using sensors in the wearable devices that can be worn in the human body which does health care monitoring, real-time analysis, emergency alert, etc. IoT has transformed the challenges faced in health care systems in measuring the body temperature, blood pressure, and monitoring breathing activities, aged people. IoT converged with blockchain affords efficient data accessing, storing, and sharing medical data in a secured way. Blockchain provides potential solutions for several health care issues that are faced in traditional health care services. Moreover, the integrated paradigm facilitates adaptable, intelligent, and reliable services in the health care sector.

2.1 Integration of IoT and blockchain

IoT systems are generally facing several challenges such as resource constraints, heterogeneity, poor interoperability, security, and privacy issues [7]. Therefore, blockchain can improve scalability, reliability, interoperability, security, and privacy. The blockchain data can be verified and tracked anywhere at any time with the stored historical transactions maintained in the blockchain. With traceability, suppliers, and retailers in the blockchain system can verify the originality and quality of the product [8]. Moreover, the immutability feature ensures the reliability of data in IoT systems since it is highly difficult to modify any stored transactions in the blockchain network. Distributed Autonomous Corporations (DACs) work is proposed to automate the transaction process without any intervention of third-party companies or government in the payment process [9].

A smart contract is implemented, thus the proposed system functions automatically without involving any third parties that subsequently saves the cost. Blockchain significantly improves interoperability by storing IoT data in the blockchain network. The heterogeneous data generated by IoT sensors are transformed, processed, fetched, compressed, and maintained in the blockchain. The data is transferred through several fragmented networks, interoperability feature supports universal access. IoT data is highly secured because it is saved as blockchain transactions that are encrypted and signed using the elliptic curve digital signature algorithm (ECDSA) [10]. Improves security using smart contracts by updating firmware of IoT devices automatically, thus prevents vulnerable breaches [11]. The potential advantages of integrating IoT and blockchain are described in the following:

- Reliability concerns the quality of data generated by IoT sensors, which are considered for further processing needs to be trustworthy. The data integrity is guaranteed by enforcing cryptographic mechanisms such as digital signature and hash functions which are the inherent features in the blockchain network.
- Traceability of data exploited in IoT systems means it is capable to track and verify the spatial and temporal data available in each transaction block. The data block in the blockchain network is commonly attached with a timestamp that subsequently assures the traceability of data.
- Interoperability among IoT devices, systems, and industrial environment is achieved by facilitating the interaction between the IoT systems and physical systems in order to exchange the information. This intrinsic feature is accomplished by including a composite layer that enables uniform accessibility across heterogeneous IoT systems.
- Interaction refers to the communication among IoT systems without any third-party intervention. It can be achieved using a smart contract, in particular, the clauses in the smart contract would be executed only the certain conditions are satisfied. It does not allow any kind of breaching activities; the particular contract will be assigned with a fine if such kind of activities is suspected.
- The adaptability of IoT devices is complicated because of heterogeneous nature in turn limits interoperability. However, the blockchain acts as a control mechanism that increases the level of adaptability of IoT devices.

Moreover, blockchain technology is proven to execute in a heterogeneous environment and adapt to several use-cases in order to meet the growing demand of IoT devices.

- Resilience feature is essential to maintain data integrity; the framework needs to be resilient to breakages and leakages of data. Blockchain maintains replicated copies among peer nodes which promotes data integrity by providing resilience to the IoT environment.
- Fault tolerance is somewhat compromised where the centralized servers are involved. Generally, smart devices gather data and perform the functionalities automatically. Blockchain mechanisms are commonly Byzantine fault-tolerant which maintains records that support to identify failures using distributed consensus mechanisms.
- Trust is a significant factor provided using blockchain technology among transacting peer nodes. Thus, it prevents malicious users to gather user's private information and also, it permits the rapid completion of automated contracts without any trusted intermediaries.
- Maintenance costs are significantly reduced that are incurred to maintain dedicated servers. Instead, storage platforms could be rented to exploit storage and computational capabilities. Moreover, it avoids single-point of failure occurrence by not depending on a centralized dedicated server.
- Privacy and security are the major concern with smart devices, blockchain implements distributed consensus protocols to obtain immutability. Data modification attacks are not possible in the public blockchain network since this attack emerges from several locations.

2.2 Reference architecture of IoT and blockchain

A composite layer of blockchain acts as an interface between IoT and applications. This interface has two benefits:

(1) provides an abstract view of lower-level functionalities and
(2) users are offered with more secured services.

Moreover, it supports more services through application programming interfaces (APIs) for several applications. The complexity in obtaining services for different applications is minimized because of abstraction obtained in the composite layer. The reference architecture of IoT and blockchain is shown in Fig. 9.2. The composite layer comprises five sublayers which are described in this section.

Data sublayer: The IoT data, which is retrieved from the perception layer will be encrypted using digital signature and hash functions. The interconnected blocks are being constructed as a blockchain after successful validation. The different hash functions and cryptographic algorithms may be appropriate for varied blockchain platforms. In particular, the bitcoin platform proceeds with the signature algorithm called ECDSA and hash function known as SHA-256.

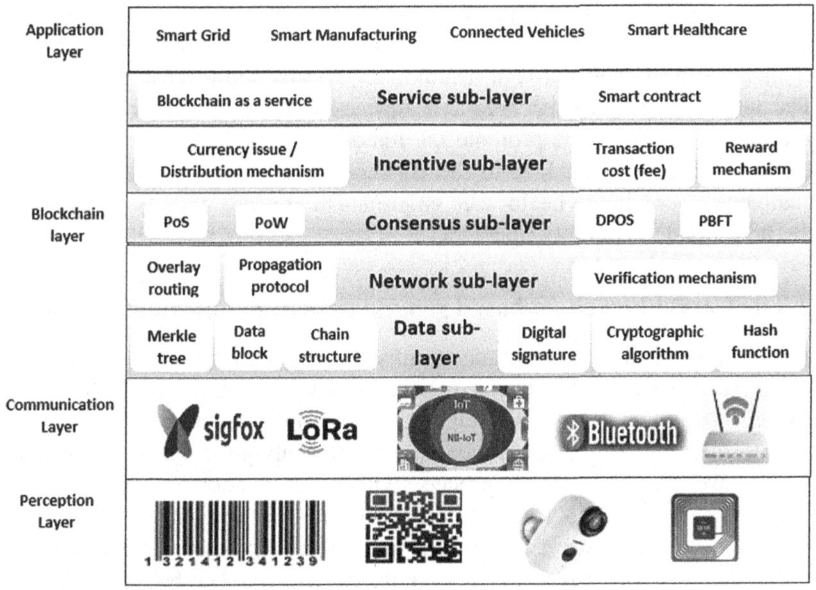

FIGURE 9.2   Reference architecture of IoT and blockchain.

Network sublayer: It functions above the communication layer which consists of physical or virtual connections among nodes in the wired or wireless network. A single node broadcasts the transaction to the connected peer nodes. Then the verification of transaction will be performed by all the participating peer nodes with the local copies that are maintained locally. Once the block is validated, it can be further proliferated to all remaining nodes in the overlay network.

Consensus sublayer: It focuses on the distributed consensus to achieve the trustfulness of every block and it is obtained using consensus algorithms such as PoS, PoW, DPOS, and PBFT. The advertisement and relay-based network propagation are considered as the prerequisite for consensus protocols during block propagation.

Incentive sublayer: The responsibility of this layer is (1) issuing digital currencies, (2) distribution of digital currencies, (3) preparing a reward mechanism for miners, (4) managing transaction cost. Moreover, it is significant to define monetary policy in terms of providing digital currencies and distributing rewards for the participants involved in the mining processes.

Service sublayer: The desired services for different sectors such as supply chain, logistics, manufacturing, food industry, health care, etc. are provided in a more trusted manner. The Blockchain as a Service (BaaS) is mainly achieved because of incorporating smart contracts that automatically triggers the responsive actions for any event is occurred. For instance, the payment contract is triggered once a product is delivered to a customer.

Recently, health care is perceived as the common social-economic issue owing to the exponential growth of the aging population that arises various challenges in the conventional health care services due to limited resources. The latest advancement in wearable devices and the vital role of big data analytics in clinical data promote the remote monitoring and services at the clinic or at home [12]. Due to the convergence of IoT and blockchain, the challenges faced in the health care environment are substantially reduced and the performance is highly upgraded.

2.3 IoT and blockchain-based remote patient monitoring system

With IoT medical devices, the quality of remote patient monitoring is enhanced and the monitoring of patients remotely increases the patient care and decreases the expense as well as time. The main functionality of remote monitoring is occurred using wearable devices; health readings are transmitted for disease diagnosis and to proceed with the treatment process [13]. The health care medical devices are categorized into stationary, embedded, and wearable. In stationary, the devices are located physically (e.g., remote chemotherapy). Embedded devices are the devices that can be implanted in a human body (e.g., deep brain stimulation). Wearable devices are body-worn devices (e.g., insulin pumps). Remote patient monitoring is rapidly growing across the world, and the transmission of health care data also occurs frequently. The health data which are maintained in electronic health records (EHR) can be available to unauthorized persons also. However, the medical data should be secured to prevent unauthorized access and smart contracts can be exploited to secure it.

The personal data of doctors and patients are protected in the aspect of privacy. The health data is analyzed and the respective doctors, health centers, and patients are alerted. Smart contracts can be initialized when the devices are manufactured and the device log facilitates the continuous tracking process. The licenses and log records are maintained with IoT devices in a decentralized way thus eliminates third-party intervention. The patients are closely integrated with wearable devices that gathers the health data such as walking distance, sleeping conditions, heartbeat, and so forth. However, the patient is solely in authority to grant, refuse, or revoke the data accessibility to health care providers. In case, a patient requires any treatment and the relevant data can be shared with the concerned health care providers. Once the treatment process is completed and the data accessibility can be removed from the network.

Wearable devices that are embedded with the microcontroller can be attached to the human body or could be fixed to the clothing also. These devices could easily interact with mobile devices or wireless transmission medium in order to provide real-time feedback. The wearable devices require the support of smart contracts for transmitting data in the network. In remote patient monitoring, data collected from the IoT device is managed using smart contract. It makes a specific contract to obtain the data from every IoT device. Due to the existence of a smart contract, only authorized doctors are permitted to access the medical data of patients, and they are allowed to alter thresholds. For example, the blood pressure data obtained from the body sensor will be transmitted to the smart contract and the corresponding functions are invoked to accomplish the monitoring task. Later, the maximum and minimum values of blood pressure are considered for analysis and evaluation. If any difference in the blood pressure value is detected, then the patient is alerted for treatment. The smart contract can be used to monitor heart rate, temperature, brain inflammation, blood pressure, glucose, and blood oxygen level. The smart contracts are very simple to maintain and error-free which allows a customized approach where the subcontract can be modified without disturbing the functionality of other devices [14].

The smart contracts are used for device authorization, IoT devices are the wearable sensors fixed on the human body. The patient who owns the device should get the device registered and the credentials of the device should also be legalized. Using a smart contract, the credential management for smart devices is performed without any third-party intervention. A contract is registered in the medical device and it can be initiated directly by the user. The users can assign the access condition and the device possession can be transferred to other participants in a decentralized way. If there are any updates in the credentials, that will be informed to the health center instantly. The blockchain-based remote patient monitoring IoT system is depicted in Fig. 9.3.

3. Monitoring health care data using smart contracts

The purpose of blockchain in the initial stage is the source of managing the cryptocurrency in the decentralized environment. But the range of applications is widespread in ensuring the maintenance of sensitive data in a decentralized manner. The various stakeholders involved in the health care industry is doctors, patient, health care providers, lab technicians, insurance, and so on [15]. The main criteria of all the involved are smooth handling of data in a secured manner. Smart contract backbone of the blockchain that is enabling the green flag for any transaction by writing a set of code based on the scenario. Creating permission to the eligible stake at the perfect time to view or edit the data is a challenging task. Monitoring and tracking are the key terms that provide the success factor of the health care industry. The relationship between the doctor and patient relies on continuous follow-up, gauge the efficiency of the system. Data are stored in the versatile region in the form of EHR and grant to access is the term that makes the channel of decision making. In the case of pharma, providing the medicine to the needy and procuring from remote is optimized by the way of making the smart contracts is used. The problem of delay or counterfeit is addressed by the way of the

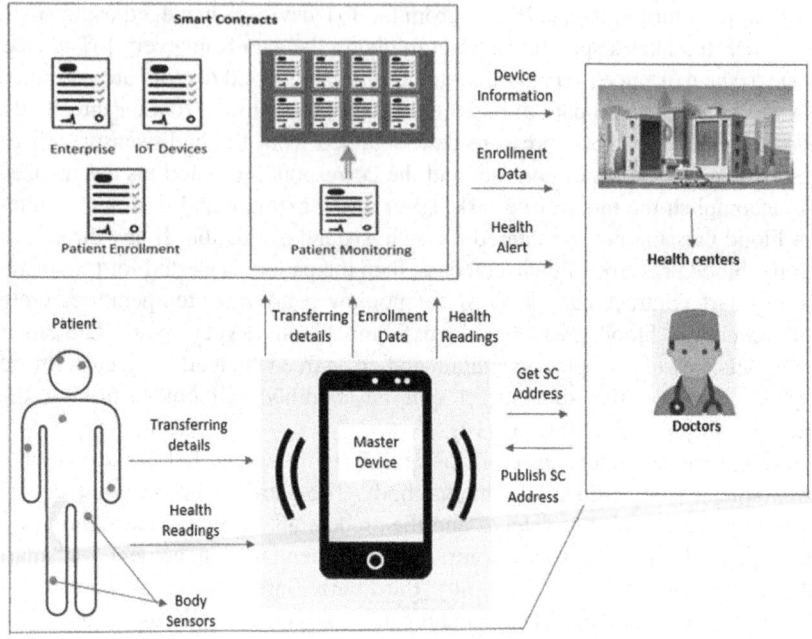

FIGURE 9.3 Blockchain-based remote patient monitoring IoT system.

exploitation of smart contracts in place [16,17]. Another main zone is claimed by the individuals for the treatment and it is eased by the way of smart contracts. The perplexing situation at any stage of the entire health care system is tackled by cleverly designing the system of digital granting without human requirement at that point of time. The codes executing behind the screen makes this system so knowledgeable. The purpose of health care data is optimized by intervening smart contracts [18]. Fig. 9.4 illustrates the trends of smart contracts in the past five years.

Utilization of the smart contract is higher in countries like China, Nigeria, South Africa, Singapore, and so on. Fig. 9.5 exemplifies the regions that are having higher working potential using smart contracts across the globe by intensified colors.

3.1 Health care IoT devices in remote medical care

In the medical field, the remote level of working is increasing day by day, and during the COVID-19 pandemic situation, the disease is increased exponentially, which alerts us to avoid direct contacts. Both patients and doctors are feared by the coronavirus and interested to go on with remote consultations [19]. The trend is quite high and it is not new. To address this remote medical care for the needy people, the IoT platform supports different perspectives. IoT

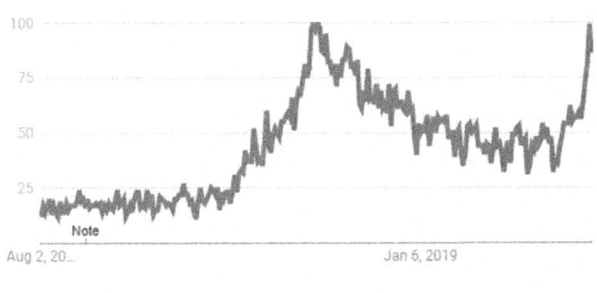

Worldwide. Past 5 years. Web Search.

FIGURE 9.4   Trend of smart contract.

FIGURE 9.5   Usage of smart contract across the globe.

is the network of integrating the sensors, patient, and device that holds the data transmitted from the sensors with the help of radio frequency identification technique [14].

According to the Pakistan-based organization, the digital information world estimated that IoT will surpass 30 billion units in 2020, and by 2025, it will be 75 billion. IoT based health care will reach $534 billion by 2025 is the forecast given by them and its 19.9% of growth in an annual manner. IoT in the medical field fine-tuned as the Internet of Medical Things (IoMT) [20].

Smart health care in the entire world that makes to build the system to remotely monitor and track the health issues with the help of ICT components like big data, cloud computing, sensors, and other related devices with the smart networking at the back end to meet all the perspectives [21]. The application of IoT in health care is large and to list a few.

Chronic patients are monitored continuously to provide great support in physical, psychological, and moral perspectives.

➢ Available heterogeneous data helps in
  - Comparison of treatments
  - Assessing novel therapeutics
  - Competence of medical trials
  - Assess the recovery rate of sick
  - Satisfaction level of patient
  - Medication utilization
  - Estimate diagnostic test
  - Settlement of claims status
  - The severity of the disease
  - Statistics helps in preplanning during the time of pandemic
  - The ratio of surplus or deficiency with available resources

IoTNOW and business insider illustrated the interesting article pertaining to IoT in health care and listed the scenario of usage in various health care related sectors. The revolution of IoT in health care is tremendous [22,23].

➢ Smart beds: Based on patient health condition and comfort automatic adjustment of the beds without human intervention.
➢ Smart medication dispenser: Remind the doctor when the patient is not taking medicine in correct time and with incorrect quantity. Very helpful in the case of the chronic patients.
➢ Apple watch: Advances its technique to track the fitness of the wearer by tracking Electrocardiogram (EKG).
➢ Electronic vaccination: In India, an intelligence network established with the help of mobile applications and vaccination provided to 27 million children.
➢ Alipay health code: In China, Skynet, Alibaba, and some operators help in health care industry by measuring the EKG and temperature. Code served the purpose of getting the information about COVID-19 and tackle the epidemic.
➢ Semtech LoRa: In Europe, with this LoRa, LPWANs, the real-time alert is given when the older people fall down or any criticality occurs, by sending the message via the cloud network.
➢ Eversense continuous glucose monitoring system: In the United States, to help and monitor the diabetics patient, the sensor is placed under the skin and it monitors 180 days continuously.

➢ Kinsa smart thermometer: In the United States, to measure the temperature of an individual, detects the temperature in one second, and pass it to provide the related medication.
➢ Vonage telemedicine API: In the United States, with API the hospital can serve a patient with remote voices and video capture.

It can be stated that without any ambiguity, IoT is making a greater impact in health care in providing quality health benefits, decreased hospital stays with the help of remote monitoring, satisfaction for all the involved stakeholders, and data, which acts as insight for making decisions. In the organization, resource allocation at the various places of health center based on the number of patients in real-time manner is facilitated. To encourage the use of IoT, some insurance agencies used it to provide the incentives for their claims and obviously flawless, and a faster process is attained. Ina nutshell, this helps in reduction of cost, optimized treatment, faster and early disease diagnosis, pharma management, and minimizes errors.

3.2 Monitoring patient's using smart contract

In general, blockchain usage answers many challenges faced by the health care sector. For the challenges like fragmented data, timely access, system interoperability, security, cost-effectiveness is catered by the decentralization principle of distributed data in a secure manner and smart contract. The early 1990s, Nick Szabo, introduced the term smart contract, which ensures the grants accepted when promises are met between, the parties involved in digital format. Ethereum is the famous platform on which smart contracts are built in many organizations [24,25]. Fig. 9.6 illustrates the smart contract with potential stakeholders get benefitted in all the way.

With the emerging industrial growth, cloud computing is incorporated to store big data remotely and access them for querying based on the requirement. Smart contract helps in making the process of smooth handling of all the activities across the stakes in an efficient manner. Monitoring of the patient is proposed by constructing the model holding patient, doctor, and nurse as a participant, sensor, health record, vital sign recording as the assets with the transaction [26,27].

4. Privacy-preserving of health care data

In health care, preserving data is vital that is required for the system to show its accuracy and people involved in it, utmost have the tendency that the stored data are not utilized or seen by other unauthorized people. Fig. 9.7 provide the cooccurrence map constructed from research articles published in the Scopus with the search word "Privacy-preserving + health care" using VOSviewer.

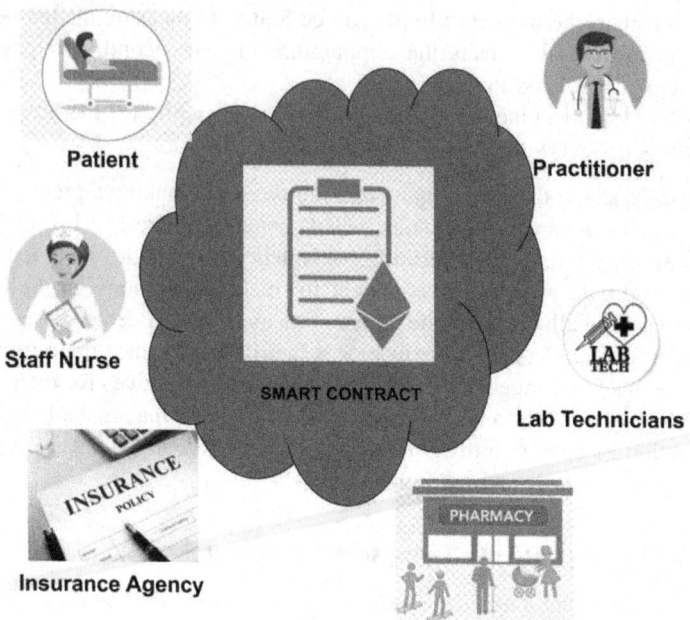

FIGURE 9.6 Smart contract: health care sector.

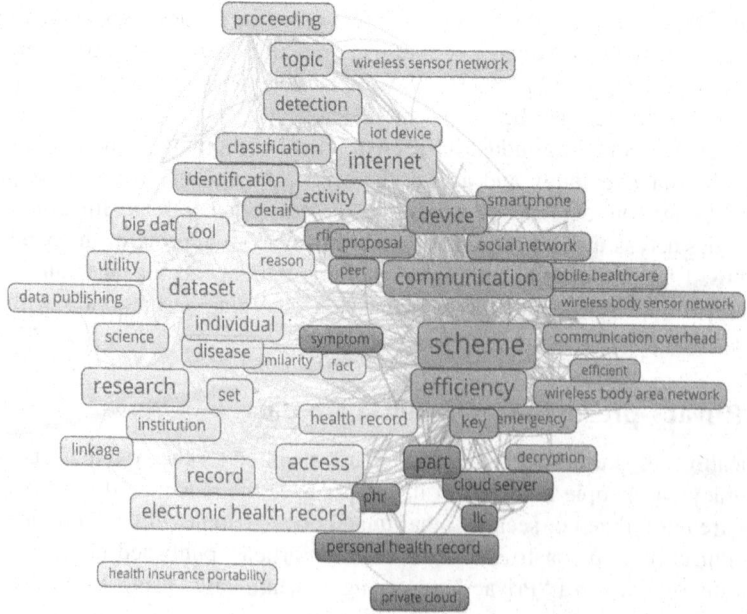

FIGURE 9.7 Coocurrence map.

4.1 System and threat model of smart health care

The primary concern of the doctor and patient is to ensure data privacy. Normally, users would not like to reveal their sickness or disorder to other people in society. In the smart health care system, there is a possibility for various attacks, namely,

- Phishing
- Pretexting
- Baiting
- Tailgating
- Denial of services
- Viruses, trojans, and worms
- Tampering
- Jamming
- Physical damage
- Flooding attack in cloud
- Authentication and Authorization
- Eavesdropping
- Spoofing attack
- Sybil attack
- Sniffing attack
- Web browser attack
- Sinkhole attack

There are a huge number of threats, and attacks are emerging every day, and thus constructing a highly reliable system is mandatory. Strategies to be considered to build a safe environment in the health care system [28,29].

- Install and update the antivirus software
- Bound network access
- Use an effective firewall
- Restrict access; provide access to protected health information based on the requirement
- Strong passwords with regular updating are mandatory
- Learn and create ransomware policy
- The threat entry point to be analyzed
- Emergency responsible plan is to be created
- Endpoint protection
- A backup copy of important data at regular interval
- Secured Wi-Fi network

4.2 Design objectives to achieve privacy-preserving

Serving patient from remote is a tough task in the smart health care system where all the related stakeholders are in different locations. Based on this perspective, a number of sensors and wearable devices were introduced in the market in the past decade. Data processing is the next task that requires at most attention. Delays in the transmission sometimes makes life unstable for the patient. This combination of stuff is grouped as the IoMT. Depending on the nature of the location, sensing, identification, and connectivity of the device, the devices are classified as stationary medical devices, wearable health monitoring devices, medical embedded devices and medical embedded devices. Discussed the various security issues in the previous session and it is mandatory to provide a protected health information system [30].

In 2014, Hoepman provided an eight-design strategy that provides data guide to make privacy and protection of data [31]

- minimize the data collection and make with only intended people;
- hide by means of encrypting, pseudonymizing, and similar methods to ensure confidentiality;
- separate the content based on the type of data;
- aggregate the data;
- inform the respondent about the purpose of data collection;
- control should be given for modification on the restricted manner;
- enforce the required privacy policy; and
- demonstrate the setup to the people involved.

Normally hefty data is generated and it is cumulated for deriving the knowledge. Based on the application, the privacy of data is required to be maintained. A number of stakeholders involved in the process of data cleaning, data segregating, and gathering directly or indirectly have an impact on the privacy maintenance of the data. Particularly in the health care field, a patient wants to keep their private and revealed only to the intended persons like doctors and practitioners. Passing the data from one place to the other based on the analysis and reports helps in taking the decision to treat the affected person. So precisely can be mentioned as hiding the sensitive data. Various techniques used in the process of privacy-preserving are data perturbation, cryptographic approach, blocking based technique, hybrid method and condensation approach. As illustrated in Fig. 9.8 the medical data applied with data mining and imbibed with mentioned privacy preserving technique produces the output for analysis.

5. Comparative analysis of existing models to secure health care data using IoT and blockchain

Multiple data are generated by health care industry that includes the data recorded by patient monitoring sensors, patients' medical records, radiography descriptions, clinical trial data, etc. E-health care is the process of digitizing these manual documents that are stored in a secured manner and can be accessed anytime anywhere. All this information is sensitive information and

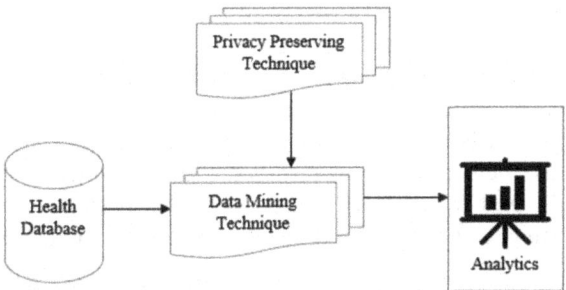

FIGURE 9.8   Privacy-preserving data mining model.

any modifications or leakage of these data may lead to greater risks. Hence blockchain technology can be employed to secure these health care—related data. Certain features of blockchain technology like distributed nature, immutability, cryptographical storage, etc. make it suitable for securing sensitive health care data. Various architectural models are proposed to secure health care data using blockchain in IoT. Here, we are going to discuss a few of those architectures.

MediBChain protocol [32] provides a patient-centric approach to health care data management using blockchain. The properties assured in this protocol are pseudonymity, integrity, accountability, privacy, and security. Here, patients send data to the system which is stored in an encrypted way. The registration unit is used to authenticate the users and the system. Private accessible unit (PAU) plays as an intermediary unit to interact between the different user and the system. The data is stored in blockchain and elliptical curve cryptography (ECC) is used to encrypt data.

In [19], IoT smart monitoring devices are used to monitor patients' health and these devices generate protected health information (PHI). Blockchain-based smart contracts are used to provide security to PHI based on Ethereum protocol. This system supports patients, hospitals, and health care professionals to communicate with each other. These sensors automatically send data through applications. Also, a notable point is that this system generates health related notifications that comply with HIPAA.

The architecture of medical blockchain [33] encompasses hospitals, patients, and third parties as to the transaction entities. Medical data are encrypted and are stored in the cloud platform. Access to these data is controlled by access control permissions. The patients' data is completely controlled by the patients themselves and they can control who can access their data. The blocks in the blockchain are validated and then they are added to the chain for preserving patient data. A hash function is utilized to the data stored in the blockchain to ensure security. This medical blockchain architecture follows the Merkle tree structure. Medical data release, storage, and sharing are the main functions of medical blockchain. Health care professional generate patient's examination reports and generates a hash for the data that which is stored in a block. Also, they digitally sign the report. Patients, while accessing the report, verify the sign. The patient decides who can access their medical report and can even give access to third parties.

MedBlock [34] is another solution to handle patients' information in a secure and distributed manner. A distributed ledger, which is a property of blockchain, is used to store and retrieve the electronic medical records of patients. In this architecture, consensus (proof of work) is given due importance and this system tries to achieve consensus with minimum energy and network congestion. MedBlock architecture uses a Certificate Authority to generate certificates, distribute certificates, and generate public and private key pair for the patients. Compared with MedShare and MedRec systems, MedBlock provides more privacy and security. Also, MedBlock is efficient in retrieving information and performs well even if the number of data increases.

BHEEM which is a blockchain-based framework for securing EHR [35], is yet another solution for securing health care information. BHEEM considers multiple stakeholders of EHR such as insurance companies, health care professional, care providers, patients, etc. in order to interact with multiple stakeholders, BHEEM differentiates three categories of nodes such as full nodes, light nodes, and archive nodes. Full nodes store the complete blockchain and hence requires high computational power and large storage. Light nodes contain only the block headers and any required data can be accessed through them. Archive nodes are used for information retrieval.

Database managers store the health records of patients by creating hashes. The cipher manager handles cryptographical issues like enciphering and deciphering. BHEEM uses five contracts: Classification contract classifies and relates the different types of nodes. A consensus contract is used to validate new nodes when they are added to the block. It has the registered list of users and ethereal addresses of the nodes. The service contract specifies the relationship of the nodes in the block. Patients' permission is needed to access his/her medical information. Owner contract lists which records belong to which patient and it also tracks who is using the data now. Permission contract gives the link address between the blocks and allows access to the data stored in the blocks based on access levels like Read, Write, Transfer, and Owner. This architecture preserves privacy using Ethereum addresses and ensures the security and integrity of data using cryptography.

More medical images related to radiography are stored in a distributed blockchain in Ref. [36]. These images can be shared with doctors as well as patients. Patients define the access permissions and can deny access to third parties. However, images are not directly stored in the blockchain. Instead, the blocks will hold the access permissions to different stakeholders. Images are actually stored in the URL and upon granting access permission, the image can be viewed by others. A large amount of data is handled throughout the process.

Health care monitoring architecture [37] focus on remote monitoring of patients out of hospitals using smart IoT devices. Patients are monitored using wearable sensor devices that can monitor patients' health continuously including, blood pressure, pulse rate, temperature, etc., and can generate reports. These reports are periodically uploaded to a remote database server for permanent storage. These sensitive data can be used for analyzing the patients' health. Hence, there should not be any security breaches and must be confidential. This architecture uses two blockchains, one for consultation and the other for medical devices.

Consultation blockchain is used for storing patients' history and based on these, treatments to the patients can be refigured. Medical devices blockchain is used for storing the reports that are generated by wearable devices. Medical device blockchain uses hyperledger fabric to provide maximum privacy, scalability, and flexibility. Application SDK is used for interacting with peer groups including physicians, lab technicians, health care workers, and so forth. The transactions carried out in this architecture are carried out in a private subnet.

Optimized blockchain model for IoT based health care applications enables patients' health data to create alerts in a secured manner [38]. These alerts are directed to health care professionals. There are many benefits in integrating IoT with blockchain for health care applications like privacy, confidentiality, smart contracts, etc. Similarly, the constraints in bandwidth, memory, and connectivity in IoT devices make it difficult to integrate with blockchain. Smart contracts are terms that are set between hospitals and patients and they are triggered upon a condition is met. Triggering happens based on the value generated by the wearable devices. This makes the data to be stored in the cloud storage by adding a digital signature to them. An overlay network is used for a distributed architecture and it ensures proof of authority. This architecture uses both types of cryptographical techniques—symmetric and asymmetric, for its transactions.

A hybrid framework for multimedia health care data processing in IoT using blockchain considers both medical images and records of patients [39]. A hash function is used for each data that is being stored in the blockchain. This framework considers many possible attacks on blockchain including wormhole attack, falsification attack, etc. The communicating parties use web-based applications to glue up the users and the blockchain. Each block in the blockchain has three entities as hash, current data, and hash of previous data block. This architecture proposes to produce 86% success rate compared to existing models.

Healthchain [40] is a blockchain-based scheme for preserving large scale data that aims to provide high efficiency. Here large scale refers to the multiple IoT devices that can be supported for health care applications. This architecture proposes to provide accountability. The users are able to revoke the access rights provided earlier. The architecture of Healthchain includes five layers as: application layer, incentive layer, consensus layer, network layer, and data layer. The application layer provides IoT data security, key management, and disease diagnosis. This layer is handled by both users and physicians. The incentive layer promotes multiple users to use the Healthchain architecture for the purpose of mining. The consensus layer determines when a node can add a new block. Healthchain is based on the Internet and hence the network layer provides a means of communication between IoT devices, doctor nodes, user nodes, accounting nodes and storage nodes. The data layer holds two data structures known as Dblock and Ublock. It also includes a few enciphering algorithms and digital signatures. The proposed scheme proposes to produce less computational cost and communication cost over traditional methods.

Remote patient monitoring is considered based on blockchain smart contracts. Both patients and doctors participate in the remote monitoring system. Alerts are generated by IoT devices when there is an emergency. Smart contracts are used for authorizing the smart devices. Blockchain reduces forgery attacks and ensures the privacy of data. Smart contracts are executed between smart monitoring devices that are embodied in the body of the patients and the health centers. SCs are used for patient monitoring, enrollment, enterprise, and device authorization. The cost incurred for implementing smart contracts is less compared with the loss faced in traditional systems.

MedChain is another initiative in combining blockchain with IoT and it works in a point-to-point network for sharing data. MedChain follows a distributed architecture that connects all health care points including hospitals, clinics, etc. The peer nodes are classified into super peers and edge peers. Super peers are the servers that belong to huge health care providers that are capable to process and store data. Edge peers are the servers that belong to small health care centers that are capable to store patients' data. MedChain uses a Certificate Authority (CA) to verify and validate the public keys [41].

MedChain architecture uses blockchain service subnetwork and directory service subnetwork. Super peers have three modules to manage health care data securely. The primary advantage of MedChain is it uses the existing framework of medical practitioners in storing the data. That is, health care professionals do no migrate their data to the new model. Only the reference link for the existing data is included in the new model. The evaluation results show that MedChain is resilient toward masquerade attack and replay attack, and assures data integrity, privacy, and security.

5.1 Comparative study

As noted, there are multiple architectures and frameworks proposed for implementing blockchain in IoT. These architectures assure to provide valuable solutions to security and privacy issues. Though multiple works are considered, there are a few notable features and setbacks in each of them. Table 9.1 shows the comparative study of these architectures focusing on the key features provided by these architectures and limitations of them.

TABLE 9.1 Architectural frameworks for blockchain in IoT.

Architecture	Key features	Drawbacks
MediBChain protocol [32]	Ensures pseudonymity, integrity, accountability, privacy and security	A clear explanation about implementation is missing
Health care blockchain system using smart contracts [19]	Provides fault tolerance, availability and traceability. Ensures security by giving anonymous addresses to patients' identity	Incurs delay when verifying a block
Medical blockchain [33]	Gives complete ownership of data to the patients. Provides security, interoperability, and privacy	The distribution of keys is not discussed
MedBlock [34]	The consensus is considered and proves to be more efficient than MedShare and MedRec architectures	Cas may be compromised by intruders
BHEEM [35]	Relies on two authorities for certificate management and use contracts for ensuring security	The searching time of data increases with the number of data
Medical Image Sharing [36]	Focuses on medical imaging sharing and provides interoperability	Complex privacy and security model
Health care monitoring architecture [37]	Uses hyperledger fabric to store medical device data that provides privacy, flexibility, and scalability	Less focus on security concerns
Optimized blockchain model [38]	Assures proof of authority, security, and privacy	Does not address scalability issues

Continued

TABLE 9.1 Architectural frameworks for blockchain in IoT.—cont'd

Architecture	Key features	Drawbacks
Hybrid health care framework [39]	Handles both medical imaging and medical reports	Practical implementation is unclear. Less focus is given to security. No methods are defined to know how large data sets are stored
Healthchain [40]	Computation and communication cost are considered	No focus is given on medical imaging as the large set data
Trusted remote patient monitoring [40]	Major focus on smart contracts. Incurs low cost	Capable to handle few sets of data
MedChain [41]	No migration of data from the old system is required Provides a clear picture of how data storage is performed	Manual uploading of data and overload to patients

6. Open research challenges and future directions

The studies show that there are various valid reasons to integrate blockchain with IoT for health care data. Blockchain is estimated as the next-generation technology that can be relied on for health care. Many multinational companies are going into contracts with popular health care providers to streamline the way the health care records are maintained. This initiative can reduce many administrative errors.

Based on the comparative study, few research challenges are encountered in deploying IoT with blockchain for health care systems. Integrating IoT with blockchain for health care gives many benefits; it cannot be denied that there are certain limitations too. The open research challenges are

- Constraints in IoT devices: IoT devices have multiple constraints including memory, power, resource, and network connection. A huge amount of data is dealt with by blockchain. Integrating blockchain with IoT might result in infeasible solutions and a practical difficulty may arise in handling a large amount of data.
- Security breach: Though blockchain assures security using hash functions and cryptographical techniques, there are few attacks made on blockchain due to defects in the programming structure of smart contracts and Ethereum. Similarly, IoT devices are also vulnerable to attacks namely

eavesdropping, denial of service attack, jamming attack, replay attack, etc. This shows that integrating IoT with blockchain may attract more attention in handling sensitive data.

- Problems in interoperability: Multiple IoT devices are used in different platforms. Also, there are many aspects of blockchain like distributed ledger, smart contracts, etc. Hence, IoT devices need to find a feasible method to interoperate. Hospital driven interoperability is taking a shift to patient-driven interoperability when blockchains are used. So, few alterations are needed in terms of policies and rules for handling medical data.
- Scalability issue: When the number of IoT devices increases, the blockchain must also be equipped to support the scaling devices. But when more workload is given to the blockchain, the performance of the blockchain in terms of data retrieval and access diminishes. This issue needs to be addressed or else it may lead to poor support for IoT devices.

In the coming years, systems can be devised to address the aforesaid issues. Also, artificial intelligence techniques can be equipped with these proposed health care solutions. When the blockchain is integrated with IoT to its perfection, the probability of errors will greatly reduce. This will be a notable aspect in the health care industry, as a small mistake in health care data may lead to the most serious consequences that may endanger a person's life.

7. Conclusion

Many diseases are spreading across the world, and there is a necessity in the health care industry to adopt new technologies. The health care industry includes remote patient monitoring, insurance claims, pharmaceutical supply management, hospital machinery management, clinical trails' data, health care data mining, etc. Multiple researches are carried out in these areas by integrating developing technologies like IoT, blockchain, machine learning, etc. This chapter discussed how blockchain technology can be integrated with smart IoT devices for carrying out multiple works in the health care industry. Blockchain exhibits potential features as a decentralized network. immutability, smart contract, distributed ledger, etc. that makes it convenient for deploying in health care. Security and privacy are major concerns as the medical industry handles numerous amount of sensitive data. Blockchain technology ensures the security of data by applying hashing and cryptographical techniques. The work investigated the current trends of blockchain and IoT in health care. The reference architecture describes how blockchain and IoT can be integrated for remote patient monitoring. Smart contracts play a significant role in securing data and a patient-centric data management can be implemented when smart contracts are used in health care data. The chapter also discusses the various threats that can be launched on smart health care and possible design objectives to preserve privacy. Finally, a comparative study is made on all the existing models of health care that use IoT and blockchain. Based on the study, it can be concluded that more focus is given to remote patient monitoring. There are many other scenarios in the health care industry that need research attention in the integration of new technologies.

Appendix A. Supplementary data

Supplementary data to this article can be found online at https://doi.org/10. 1016/B978-0-12-824038-0.00001-8.

References

[1] Chen M, Miao Y, Hao Y, Hwang K. Narrow band Internet of Things. IEEE Access 2017;5:20557–77.

[2] Khutsoane O, Isong B, Abu-Mahfouz AM. IoT devices and applications based on LoRa/LoRaWAN. In: Iecon 2017 - 43rd Annual conference of the IEEE industrial electronics society; 2017. p. 6107–12.

[3] Dai H-N, Wang H, Xu G, Wan J, Imran M. Big data analytics for manufacturing Internet of Things: opportunities, challenges and enabling technologies. Enterprise Information Systems; 2019.

[4] Roman R, Zhou J, Lopez J. On the features and challenges of security and privacy in distributed Internet of Things. Comput Netw 2013;57(10):2266–79.

[5] Zhou J, Cao Z, Dong X, Vasilakos AV. Security and privacy for cloud-based IoT: challenges. IEEE Commun Mag 2017;55(1):26–33.

[6] Poongodi T, Sujatha R, Sumathi D, Suresh P. Balamurugan balusamy, cryptocurrencies and blockchain technologies and applications: decentralized and smart contracts, blockchain in social networking. John Wiley & sons, Inc; 2020. p. 55–75.

[7] Poongodi T, Lucia agnesbeena T, Janarthanan S, Balamurugan B. An industrial IoT approach for pharmaceutical industry growth, accelerating data acquisition process in the pharmaceutical industry using Internet of Things. Elsevier, Academic Press; 2020. p. 117–52.

[8] Lu Q, Xu X. Adaptable blockchain-based systems: a case study for product traceability. IEEE Software 2017;34(6):21–7.

[9] Zhang Y, Wen J. An IoT electric business model based on the protocol of bitcoin. In: Proceedings of 18th international conference on intelligence in next generation networks (ICIN); 2015. p. 184–91.

[10] Johnson D, Menezes A, Vanstone S. The elliptic curve digital signature algorithm (ECDSA). Int J Inf Secur 2001;1(1):36–63.

[11] Christidis K, Devetsikiotis M. Blockchains and smart contracts for the Internet of Things. IEEE Access 2016;4:2292–303.

[12] Poongodi T, Krishnamurthi R, Indrakumari R, Suresh P, Balusamy B, Wearable devices and IoT. A handbook of Internet of Things in biomedical and cyber physical systems. Springer ISRL series; 2020. p. 245–73.

[13] Indrakumari R, Poongodi T, Suresh P, Balusamy B. The growing role of Internet of Things in healthcare wearables, emergence of pharmaceutical industry growth with industrial IoT approach. Academic Press, Elsevier; 2019. p. 163.

[14] Poongodi T, Rathee A, Indrakumari R, Suresh P. IoT sensing capabilities: sensor deployment and node discovery, wearable sensors, wireless body area network (WBAN), data acquisition, principles of Internet of Things (IoT) ecosystem: insight paradigm. Cham: Springer; 2020. p. 127−51.

[15] Meinert E, Alturkistani A, Foley KA, Osama T, Car J, Majeed A, Van Velthoven M, Wells G, Brindley D. Blockchain implementation in health care: protocol for a systematic review. JMIR Res Protoc 2019;8(2):e10994.

[16] Abou Jaoude J, Saade RG. Blockchain applications−usage in different domains. IEEE Access 2019;7:45360−81.

[17] Kuo TT, Kim HE, Ohno-Machado L. Blockchain distributed ledger technologies for biomedical and health care applications. J Am Med Inf Assoc 2017;24(6):1211−20.

[18] O'Donoghue O, Vazirani AA, Brindley D, Meinert E. Design choices and trade-offs in health care blockchain implementations: systematic review. J Med Internet Res 2019;21(5):e12426.

[19] Griggs KN, Ossipova O, Kohlios CP, Baccarini AN, Howson EA, Hayajneh T. Healthcare blockchain system using smart contracts for secure automated remote patient monitoring. J Med Syst 2018;42(7):130.

[20] n.d. https://www.digitalinformationworld.com/2020/02/iot-in-healthcare-expectations-for-2020.html#:~:text=It%20is%20estimated%20that%20the,annual%20growth%20rate%20of%2019.9%25.

[21] Lionel MN, Zhang Q, Tan H, Luo W, Tang X. Smart healthcare: from IoT to cloud computing. Scientia Sinica Inf 2013;43(4):515−28.

[22] n.d. https://www.businessinsider.in/science/news/iot-healthcare-in-2020-companies-devices-use-cases-and-market-stats/articleshow/74126142.cms.

[23] n.d. https://www.iot-now.com/2020/04/23/102387-iot-in-healthcare-8-examples-from-around-the-world/.

[24] Khatoon A. A blockchain-based smart contract system for healthcare management. Electronics 2020;9(1):94.

[25] Rabah K. Challenges & opportunities for blockchain powered healthcare systems: a review. Mara Res J Med & Health Sci 2017;1(1):45−52.

[26] Gordon WJ, Catalini C. Blockchain technology for healthcare: facilitating the transition to patient-driven interoperability. Comput Struct Biotechnol J 2018;16:224−30.

[27] Jamil F, Ahmad S, Iqbal N, Kim DH. Towards a remote monitoring of patient vital signs based on IoT-based blockchain integrity management platforms in smart hospitals. Sensors 2020;20(8):2195.

[28] Tripathi G, Ahad MA, Paiva S. S2HS-A blockchain based approach for smart healthcare system. Healthcare 2020;8(1):100391.

[29] Rizvi S, Pipetti R, McIntyre N, Todd J. Threat model for securing Internet of Things (IoT) network at device-level. Internet of Things 2020:100240.

[30] Dwivedi AD, Srivastava G, Dhar S, Singh R. A decentralized privacy-preserving healthcare blockchain for IoT. Sensors 2019;19(2):326.

[31] Hoepman JH. Privacy design strategies. In: IFIP international information security conference; 2014. p. 446−59.

[32] Al Omar A, Rahman MS, Basu A, Kiyomoto S. Medibchain: a blockchain based privacy preserving platform for healthcare data. In: International conference on security, privacy and anonymity in computation, communication and storage; 2017. p. 534−43.

[33] Chen Y, Ding S, Xu Z, Zheng H, Yang S. Blockchain-based medical records secure storage and medical service framework. J Med Syst 2019;43(1):5.

[34]   Fan K, Wang S, Ren Y, Li H, Yang Y. Medblock: efficient and secure medical data sharing via blockchain. J Med Syst 2018;42(8):136.

[35]   Vora J, Nayyar A, Tanwar S, Tyagi S, Kumar N, Obaidat MS, Rodrigues JJ. BHEEM: a blockchain-based framework for securing electronic health records. In: 2018 IEEE globecom workshops (GC wkshps); 2018. p. 1–6.

[36]   Patel V. A framework for secure and decentralized sharing of medical imaging data via blockchain consensus. Health Inf J 2019;25(4):1398–411.

[37]   Attia O, Khoufi I, Laouiti A, Adjih C. An IoT-blockchain architecture based on hyperledger framework for healthcare monitoring application. In: 2019 10th IFIP international conference on new technologies, mobility and security (NTMS); 2019. p. 1–5.

[38]   Dwivedi AD, Malina L, Dzurenda P, Srivastava G. Optimized blockchain model for Internet of Things based healthcare applications. In: 2019 42nd international conference on telecommunications and signal processing (TSP); 2019. p. 135–9.

[39]   Rathee G, Sharma A, Saini H, Kumar R, Iqbal R. A hybrid framework for multimedia data processing in IoT-healthcare using blockchain technology. Multimed Tool Appl 2019:1–23.

[40]   Kazmi HSZ, Nazeer F, Mubarak S, Hameed S, Basharat A, Javaid N. Trusted remote patient monitoring using blockchain-based smart contracts. In: International conference on broadband and wireless computing, communication and applications; 2019. p. 765–76.

[41]   Shen B, Guo J, Yang Y. MedChain: efficient healthcare data sharing via blockchain. Appl Sci 2019;9(6):1207.

Chapter 10

Environmental pain with human beauty: emerging environmental hazards attributed to cosmetic ingredients and packaging

Kartick Chandra Pal
Karimpur Pannadevi College, Karimpur, West Bengal, India

1. Introduction

Physical and facial beauty produces an overall pleasant appearance, which influences the self-esteem of a person. Universally, beauty has an important positive effect on social proficiency perception. Besides enhancing the facial reliability and charm, most of the time cosmetics are used as visual super-stimulous for eliciting response to opposite sex [1]. Therefore, use of cosmetics has an important role in enhancing the attractiveness. The beauty concept and use of cosmetics started from the dawn of mankind and civilization. Around 10,000 BCE perfumed oils, ointments, and masks were used as cosmetics by the Egyptians for cleaning and softening their skin and for body aroma [2]. The information about the use of cosmetics from around 3000 BCE is documented in ancient texts and artifacts of Mesopotamia and Egypt [3]. Some historians defended the concept that the use of cosmetic body art started from the earliest African Middle Stone Age [4]. They conferred this theory on the evidences of utilized crayons and red ochre (mineral pigments) in ancient paintings which are still used by the Himba women of northwestern Namibia in Africa for hair and body painting [5–7]. Some women in Roman antiquity invented make up for whitening their skin and kohl (*stibium*) to use as eyeliner [8]. The word "cosmetic" was derived from the Greek word *cosmos*, giving the sense of order or adornment and from the adjective *cosmeticos*, having the meaning embellishment or the art of beautifying [2]. Cosmetics cleanse, beautify, sustain charm, and modify body appearance [9]. According

Cognitive Data Models for Sustainable Environment. https://doi.org/10.1016/B978-0-12-824038-0.00010-9

to EU Regulation 1223/2009 (article 2, 1.a) [10], cosmetic is defined as: "Cosmetic product means any substance or mixture intended to be placed in contact with the external parts of the human body (epidermis, hair system, nails, lips and external genital organs) or with the teeth and the mucous membranes of the oral cavity with a view exclusively or mainly to cleaning them, perfuming them, changing their appearance, protecting them, keeping them in good condition or correcting body odors". The term personal care product (PCP) is inclusive of the term cosmetics, which are actually the collection of some organic compounds used in daily human life in considerable quantities for personal hygiene and beautification [11].

At present the pervading use of cosmetics is a common practice in societal daily life. The projected compound annual growth rate (CAGR) of global cosmetics market is going to record 4.3% during the forecast period 2016 to 2022 and is projected to reach $429.8 billion by 2022 [12]. The United States is one of the most beauty-obsessed countries among the world and the average daily usage of cosmetics per men and women in this country are 6 and 12, respectively [13]. These cosmetics or PCPs consist of about 12000 artificial chemical ingredients for the formulation and preservation of cosmetics, but only 20% of these ingredients are considered to be safe to utilize [14].

After the use, an outsized percentage of those cosmetics products are often rinsed off from our body and then it enters into the wastewater treatment plants (WWTP) via drainage systems. Many of the cosmetic ingredients or chemicals like synthetic musks [15,16], perfluoroalkyls compounds [17], some organic UV filters [18] and microplastics [19] is not removed effectively by WWTP [20]. Every day many cosmetics like body wash, shampoo, and hand cleanser are used for cleaning purposes contributing a massive amount of surfactants is discharged into aquatic and terrestrial environments. These surfactants are widely recognized for their toxic effects on flora and fauna with adverse effects on human health imposing environmental hazards [21]. Therefore, ultimately they end up in the aquatic environment. Besides, these compounds can be accumulated in the environment when the sewage sludge is applied as manure in crop fertilization [22] and enter into food chain via bioaccumulation in plant [23].

Since these chemical ingredients are intended to be used in external surfaces of human body, not to enter in metabolic transformation, cosmetic products pose the foremost tenacious ecological concern for our beauty obsession [24]. Therefore the environment is exposed to large quantities of chemical substances everyday emerging from the cosmetics which can potentially harmful to aquifer particularly along with soil and air to some extent. Worldwide several studies has been carried out in recent years on personal care and cosmetic products (PCCPs) as a group of organic pollutants ubiquitously distributed in the aquatic environment [15,25,26], even in the raw water sources of drinking water treatment plants [27,28].

In this chapter an overview is illustrated to provide some suspected adverse environmental effects of some universally used cosmetic components and packaging that are not traditionally measured to be of much serious ecological concern but are now being speculated. This chapter focuses particularly on microplastics, which have received a greater importance in recent days and became a budding area of research [29–31] along with other ingredients like UV filters, parabens, triclosan, and so forth as emerging environmental contaminants.

2. Global scenario of PCCP production

Intensity of ecological concern arose from cosmetics utilization is depicted by the growing production of cosmetics which again is reflected by its market and the increasing tendency of market is evaluated by CAGR. According to the report published by FIOR [link: https://www.fiormarkets.com/report-detail/407144/request-sample] markets in 2018, the global beauty and personal care market which includes cosmetic market had reached 493.34 billion USD and growing at a CAGR of 5.81% in the year 2018. On the basis of Fung Business Intelligence Center (FBIC) report 2015, the Asia pacific region appeared to be the largest market for beauty and PCPs and held a market value of USD 138.5 billion (Fig. 10.1).

Over the last 10–15 years, the steady growth in global cosmetic market is observed and is predicted to raise the CAGR by 5%–6% in the next two decades due to enhanced beauty and personality consciousness and rising GDP.

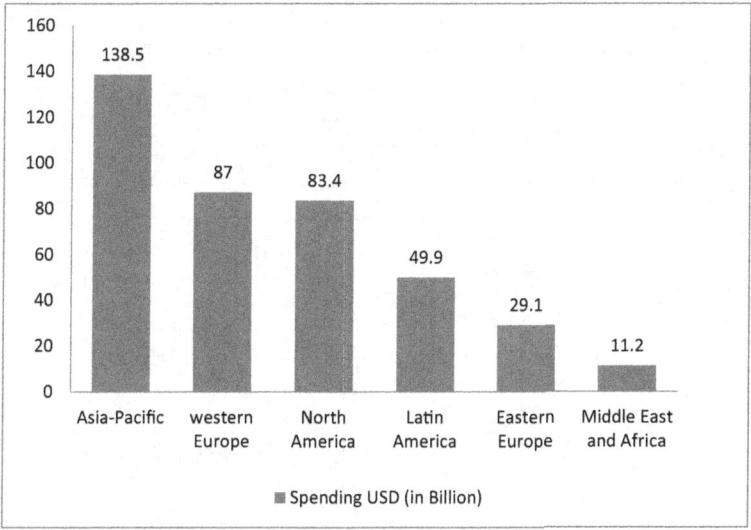

FIGURE 10.1   Global expenditure on cosmetic (in billion USD).

The global cosmetic market can be classified into various product categories like skin care, eye care, hair care, color products, and fragrances, among which skin care group is the most highly developed, diversified category and has the leading market share. Skin care products can be categorized by different subcategories like facial care, sun care, body care, hand care, and others, among which only facial care category contributes 65% of market share according to the Procter & Gamble Market Assessment Survey, 2013. Most of the times, the consumers are trapped by the unrealistic and misleading claim of the cosmetic manufacturers to have fair, younger and beautiful face and therefore the face care products are now being leading concern not only for the dermal health but also for the environment.

3. Fate of cosmetics and related hazards

Once released into the environment, cosmetic becomes a horse of long run due their low volatility, high polarity and hydrophilicity nature and go to a possible long range transport via aquatic system, even to the food chain, spreading a viable amount of cosmetic ingredients depending on their physicochemical properties and characteristics of the receiving environment [32]. After utilization, the rinsed off cosmetics find their routes of dispersion into the environment (Fig. 10.2). Washed-out cosmetics comes to the household wastewater stream or to the sewages, which is used as influent in sewage treatment plants (STPs) or WWTPs and these STPs are the prime sources of PCPs to the environment [33], including WWTPs, and landfill leaching. After

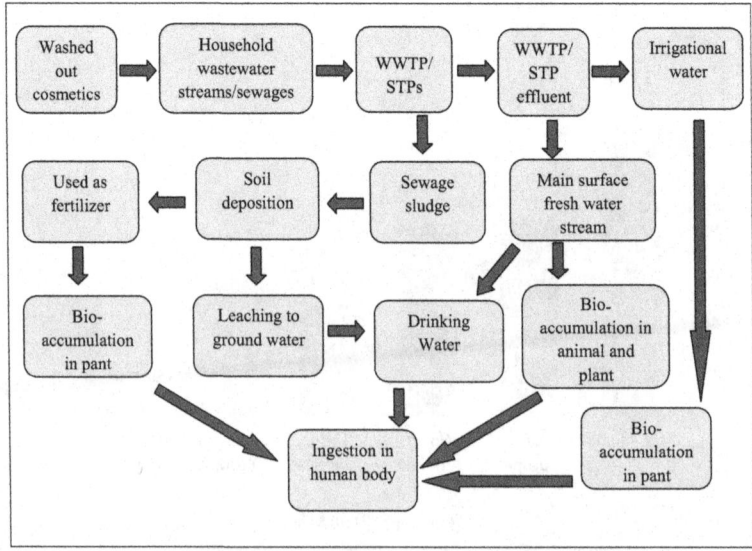

FIGURE 10.2 Illustration of cosmetic flow into the environment.

conventional wastewater treatment processes WWTP or STP effluent comes to the main surface freshwater stream, but often there is an inconsistency in the removal of the PCPs, and often they are not removed completely [34–36]. Sometimes the effluent obtained after the treatment of sewage or wastewater is used for irrigational purposes, and treated sludge is generally used as manure in agriculture [37]. This ecological exposure of cosmetics ultimately leads to ingestion in human body via bioaccumulation in vegetations. The WWTP or STP-emerged effluents contaminated main surface freshwater stream also ends up to the cosmetic ingestion in human body by entering in food chain through bioaccumulation in aquatic animals like fish and lobsters. Another important source of cosmetics or its ingredients to the environment is the effluents of manufacturing units that directly goes to the STPs [38]. Again this fresh stream may be used as drinking water after treatment which also cannot remove the cosmetic ingredients with cent percent efficiency. After treatment, the sewage sludge is deposited on the soil with a probable leaching into the groundwater, which is mostly used as drinking water. Therefore cosmetic contamination into the environment is posing impending threat to the human being as well as to the aquatic ecosystem. Despite of low level ecological exposure of cosmetics in the range of ng/L to µg/L concentration, it is of significant concern for steady and perpetual supply of cosmetics ingredients in aquatic systems, which is described as "pseudopersistent" in nature [39].

During last few decades due to enormous use of cosmetics, contamination in the environment came forth as a new potential risk to the aquatic environment for routinely released high amount of low biodegradable substances [40]. Monitoring of these cosmetic ingredients is rarely included in environmental legislations of different countries around the world [41]. Biogeochemical cycles of food chain are closely related to the production and usage of organic matters which may be hampered to some extent due to daily use of PCPs and cause perturbations on aquatic ecosystems. Hydrophilic and lipophilic compounds of cosmetics can be responsible for a metabolic breakdown and rapid elimination [42,43]. An associated process to bioaccumulation is biomagnification which has been explained in some previous studies of interaction of hydrophobic cosmetic-chemicals on fish [44,45] which may arise. Again, the availability of pollutants can be increased due to altered degradation processes and different environmental phase distribution with variation in temperature [46]. During WWT processes PCPs may be subjected to microbial biodegradation reactions [47] and transformed to somewhat less toxic product for the environment. Therefore natural microbial flora or microorganisms should have a therapeutic use. But according to Onesois et al. intended remediation may not be achieved, because biodegradation entail substrate–specific enzymatic reactions which are not predictable for the variety of cosmetic ingredients [48]. Again, the biodegradation may be inhibited by the increased PCP concentrations for its toxic effect on the microorganisms [49].

Change in behavioral pattern of an organism or abolition from an ecosystem acts as bio-indicator of environmental change [50] which is a difficult assess. Therefore to conjecture the fate and related hazards of cosmetic contamination needs a rigorous study.

4. Cosmetic ingredients and their environmental impact

4.1 Ingredients

The major classes of vital ingredients used in PCPs according to their purpose of use are: antimicrobials (biocides), UV filters (sunscreens), preservatives, fragrances, surfactants, siloxanes and microplastics.

During 1970, PCPs Council in the United States established the International Nomenclature of Cosmetic Ingredients (INCI) as a dictionary for cosmetic ingredients which has been adopted by many countries in the world and they register all new ingredients used in cosmetics. A view of some types of cosmetics on the basis of function and maximum allowed limit in the European Union (EU) can be obtained from Table 10.1.

4.2 Environmental impact

4.2.1 Biocide compounds

Soap, Antiseptic and disinfectant are of this category and the most commonly used biocide compounds in these products are benzotriazole, triclosan and triclocarban. Benzotriazole is a very polar substance and is hardly removed by conventional wastewater treatment technologies [51] with an ultimate fate of drinking water contamination. Popular antimicrobial agents, Triclosan and triclocarban are widely used in various types of products like, toothpastes, soaps, deodorants, many other cosmetics, fabrics and plastics. Due to their hydrophobic properties [52,53] triclosan and triclocarban can be removed more easily from aqueous phase than the other hydrophilic substances and they show lower intensity to be accumulated in sludge and sediments [54,55].

4.2.2 UV filters (sunscreens)

Intensive use of UV filters or sunscreen agents in cosmetics have become a common phenomena to reduce photoaging, and to protect the skin from photocarcinogenesis and photo immune control promoted by UV sun radiation [56–58]. Beside PCPs, these are also used in variety of industrial commodities like plastics, textiles and paints [59]. Many UV filters are precipitated in sewage sludge during wastewater treatment [60] for their poor biodegradability and high lipophilicity and they are accumulated in sediments [61,62] or biota [63,64].

TABLE 10.1 Maximum allowed limit of some commonly used PCPs (according to regulation 1223/2009/EC).

Type of cosmetic (according to the purpose of use)	INCI name	Maximum limit allowed (according to regulation 1223/2009/EC)
Talc	Boric acid	5% (not to be used in products for the children under 3 years of age)
Bath products	Boric acid	18% (not to be used in products for the children under 3 years of age)
	1-Hydroxyethylidene-diphosphonic acid and its salts	0.2%
Oral products	Boric acid	0.1% (Not to be used in products for the children under 3 years of age)
	Hydrogen peroxide	0.1%
	Ammonium monofluorophosphate	0.15% (as F^-)
	Calcium fluoride	0.15% (as F^-)
	Sodium fluoride	0.15% (as F^-)
	Strontium acetate hemihydrate	3.5%
Hair products	Boric acid	8%
	Thioglycolic acid	8% (for general use) 11% (for professional use)
	Oxalic acid	5% (for professional use)
	Hydrogen peroxide	12%
Biocides	Triclosan	0%
	Triclocarban	0.20%
	Butyl methoxydibenzoylmethane	5%
	Ethylhexyl triazone	5%
	Phenylbenzimidazole sulphonic acid	8%
	Terephthalylidene dicamphor sulphonic acid	10%
	Methylene bis-benzotriazolyl tetramethylbutylphenol	10%
	Drometrizole trisiloxane	15%

Continued

TABLE 10.1 Maximum allowed limit of some commonly used PCPs (according to regulation 1223/2009/EC).—cont'd

Type of cosmetic (according to the purpose of use)	INCI name	Maximum limit allowed (according to regulation 1223/2009/EC)
UV filter	Benzyl salicylate	0.001% in leave-on products, 0.01% in rinse-off products
	4-Methylbenzylidene camphor	4%
	Ethylhexyl salicylate	5%
	Benzophenone 4	5%
	Ethylhexyl dimethyl PABA	8%
	Octocrylene	10%
	Benzophenone 3	10%
	Ethylhexyl methoxycinnamate	10%
Preservative	Methyl paraben	0.4% (as acid) for single ester, 0.8% (as acid) for mixtures of esters
	Ethyl paraben	0.4% (as acid) for single ester, 0.8% (as acid) for mixtures of esters
	N-propyl paraben	0.4% (as acid) for single ester, 0.8% (as acid) for mixtures of esters
	I-propyl paraben	0.4% (as acid) for single ester, 0.8% (as acid) for mixtures of esters
	N-butyl paraben	0.4% (as acid) for single ester, 0.8% (as acid) for mixtures of esters
	Phenoxyethanol	1%
Fragrance	Musk xylene	(a) 1.0% in fine fragrance (b) 0.4% in eau de toilette (c) 0.03% in other products
	Musk ketone	(a) 1.4% in fine fragrance (b) 0.56% in eau de toilette (c) 0.042% in other products
	Acetyl hexamethyl indan	2%
Surfactant	Nonylphenol	0%
	Dibutyl phthalate	0%

4.2.3 Preservatives

The most commonly used preservatives in cosmetics like pharmaceuticals, soaps, gels, creams etc. are parabens. Parabens are the family of odorless and colorless compounds derived from parahydroxybenzoic acid and do not cause any kind of discoloration or hardening in the cosmetics. They are used widely for their exclusive feature of antibacterial and fungicidal function, low production cost and low toxicity. Although in several studies, low concentrations of parabens in WWTPs and surface water were found [65–67], but there are some growing evidences of their role in endocrine disruption [68].

4.2.4 Fragrances

From the ancient days aesthetic sense was developed in human society and used fragrances, mostly prepared from floral and animal extracts to enhance prettiness of people. As artificial and some natural fragrances contain organic lipophilic compounds, they are mostly absorbed in sludge, sediments and biota [69]. Due to close exposure of fragrances to human body with perfumed products, more consciousness is necessary about their nature and utilization. For this reason a group of synthetic nitromusk fragrances have been withdrawn from the European market for its transformation into aniline products, which are harmful both for environment and biologic metabolism [70]. In spite of their detection in wastewaters [71] and sludge [72], it is difficult to realize the providence of fragrances in the environment for the unavailablity of analytical methods.

4.2.5 Surfactants

Surfactants are the main source of environmental pollution arose from the cosmetics as it is disposed in huge amounts directly into the aquatic environment without any proper treatment or through the WWTPs. Because surfactants have a wide range of applicability such as detergent manufacturing, industrial use in textile, herbicides formulation, and fragrances stabilizing in cosmetics [40]. They are amphoteric in character and therefore accumulated easily in sediments, sludge and biota, creating a potential environmental concern [73].

4.2.6 Siloxanes

Siloxanes consisting of a polymeric organic silicone are used in a large variety of cosmetic products like skin care creams, hair conditioners, color cosmetics, and antiperspirants for their physiologic inertness, high thermal stability, low surface tension, and even surface [16]. These are also used in industrial segments as automotive polish, fuel additives and antifoaming agents, and so forth. Some of the siloxanes have global annual production up to several thousand tons (45–227) and these are discharged into sewage systems and being adsorbed retain in to sludge in WWTPs for their high Octanol-water partition coefficient and ultimately released to the aquatic environment [74–76]. A high concern stems from recent research reports which established the potential toxic effects of cyclic siloxanes [77].

4.2.7 Microplastics

Microplastics, familiar with several names like microbead, microplastics, microspheres, nanospheres, and plastic particulates are spherical or amorphic small plastic fragment or particulates having length less than 5 mm [78]. Beside their several implements in clothing and some other industrial purpose, microplastics are used in PCCPs as ingredients for film formation, exfoliation or sorbent phase, regulation of viscosity etc. Although, small fraction (<10%) of total plastic garbage floating on the sea is contributed by the MPs in the environment [79], but they are increasingly incorporated in food chain when they are eaten up by the fish and other aquatic animals creating a bigger concern. This eventually enters in our endocrine system when the fish or other aquatic animals come to our food plate.

5. Plastic in cosmetics

Light weight, durability and cost-effectiveness are the main features of plastics which made them so important in our daily life [80]. The global plastic production increased 189 times between the year 1950–2015 from 1.7×10^6 ton to 3.2×10^8 ton [81,82]. In cosmetics plastic may be used as ingredient as well as for packaging. Both the ingredients in microform and packaging in macro form enters ultimately to the marine aquatic system. According to Jambeck et al., 2015, $4.8–12.7 \times 10^6$ ton of plastic waste came into the oceans from 192 coastal countries in 2010 [83]. The macrosized plastic packagings can be weathered, resulting in micro- to nanosized fragments after a long-time exposure [84,85].

5.1 Microplastic

From 1960s plastics have been used as cosmetic element for quite a few decades. With the innovation of new PCPs plastic has become an important ingredient and used mostly as exfoliating plastic microbeads known as microplastic in face and body washes for scrubbing (Fig. 10.3). The spherical or amorphic plastic particulates having size less than 5 mm known as microplastics, include mainly three variants-, microfibers, microbeads, and micropellets or nurdles. On the basis of their origin microplastics are further categorized as primary or secondary microplastics [31]. Primary microplastics are the small-sized particles prepared deliberately to serve the industrial purpose; mainly for the cosmetics, textile and drug industries. Secondary microplastics

Tan removal scrub   Plastic micro beads

FIGURE 10.3 Magnified plastic microbeads isolated from a tan removal scrub.

include solid plastic fragments, microfibers, peeled off coatings etc. derived from any organic synthetic polymer or macroplastic products by virtue of degradation through natural weathering process [80]. Both types of microplastics are entering persistently at huge levels in aquatic and marine ecosystems. Most of the MPs have lesser density; these low-density microplastics remain buoyant on the sea surface and can take a worldwide trip driven by wind and oceanic currents [86].

Plastic is made of certain types of synthetic polymers and sometimes additives are added to achieve the desired properties of the cosmetic material. Mainly two types of plastic materials are employed to produce plastic beads for the formulation of cosmetics, namely, (1) thermosets and (2) thermoplastics. Thermoset plastics include primarily polyurethanes and certain polyesters and thermoplastic materials are of the kind: polyethylene, polytetrafluoroethylene (Teflon), polypropylene, polystyrene, polyamide poly (methyl methylacrylate) etc. These microplastic are water soluble or water dispersable, but the microbead formed from some amorphous silicon polymers are sparingly soluble in water [87]. Plastic microbeads (μBs) are utilized as ingredients in various types of leave-on formulations like deodorant, insect repellent, moisturizers, antiwrinkle creams, sunscreen, lipstick, nail polish, eye shadow, facial masks, mascara hair color, and rinse-off formulations like shower gel, conditioner, shampoo, hair spray, shaving cream, and so forth. Depending on the function of cosmetics the applied size and amount of the microbeads are determined. For example, higher amount of microplastics is observed in a typical exfoliating shower gel [88].

Most of the plastic ingredients in PCCPs are made of nonbiodegradable polymers and for this reason MPs in particular, is considered to be the most relevant topic of study [89].

According to the study report of Gouin et al., 2015, more than 4000 tons of microbeads are used per year in cosmetics in the countries of European Union [90]. Cheung and Fok in 2017 reported that, the aquatic environment in mainland China is contaminated by average 209.7 trillion (306.9 tons) microbeads [91]. From the study of Boucher and Friot (2017), a region-wise contribution of the environmental input to the global picture of the micro-plastic pollution, for utilization of μBs in PCCPs are illustrated below [92] in the Fig. 10.4.

South Asia appears to contribute the most to the global environmental pollution arising from cosmetic microbeads (Fig. 10.4).

Microplastics entered in the aquatic environment can be ingested to the small aquatic or marine organisms like invertebrates [93], fish [94], sea turtles [95], and marine mammals [96], causing severe adverse biological effects [30]. Ingestion of primary MPs in different species of zooplankton, worms, mussels, and vertebrates is well accepted in various literatures [97–99] for their similarity in size to the natural food of these consumer species. This phenomenon has important implications to human health when these aquatic organisms, mainly the fishes, come to our food menu. The tiny sized MPs become indistinguishable from the foods and therefore ingested easily by various aquatic organisms. Dermal uptake mostly, through the gills and, direct ingestion may cause accumulation of MPs in fishes [94].

Feeding behaviors determines the level of MPs in different fish species [100]; unselective filter feeding behaviors of fish such as mackerels, leads to higher ingestion of MPs in their body. According to the study of Mazurais et al. in 2015, the mortality rate of microplastic ingested in young fish is considerably higher in comparison to the control under laboratory conditions [101].

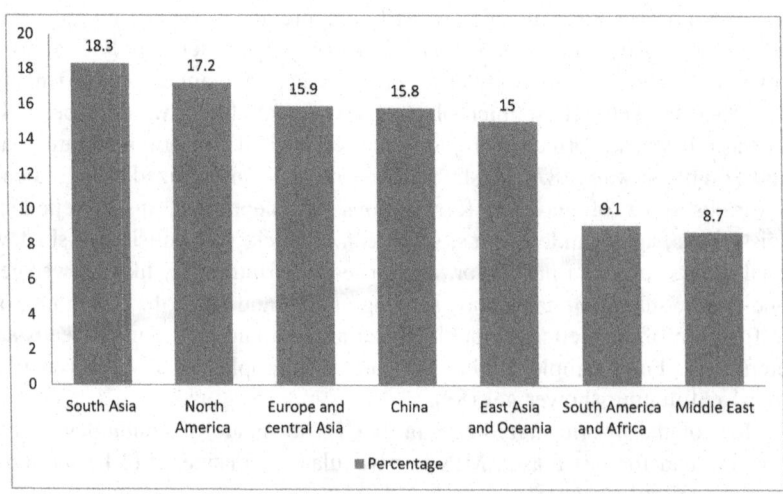

FIGURE 10.4 Global region-wise utilization of microbeads in PCPs.

Sometimes color of microplastic may play a role in varied occurrence of microplastics ingestion by fish. Researches on epipelagic and mesopelagic fishes showed their tendency to ingest microplastics which had identical color as their prey items [102]. Again, several studies proved a peculiar tendency of fish to intake dark colored plastics. For example, the shark *Galeus melastomus* [103], demersal fish from the Spanish Atlantic and Mediterranean coasts [104] and from the English Channel [102] ingested blue or black MPs most frequently than the light-colored MPs.

The danger of MPs in marine environment is augmented by their hydrophobic character favoring the superficial deposition of organic toxicants from the surrounding water [105,106]. The sorption capacity of microplastics are higher than macroplastics for their larger surface area. Therefore MPs can play a role of carrier in transport of chemicals from aqua-system to the body of biota assisting the entrance of harmful chemicals including trace metals and organic chemicals into the food chain [107].

Corals can also intake MPs as their food mistakenly and mesenterial tissues within the coral gut cavities are affected severely damaging the corals' health [108].

Diverse species of zooplanktons in the Baltic Sea, like shrimps, copepods, worms, cladocerans, ciliates, and polychaete larvae were found to be ingested with microplastics [109], which accumulate in their digestive tract leading to a disturbance in feeding and digestion.

The benthic worm, *Arenicola marina* is very important member in marine food chains for their high lipid content but due to indirect ingestion of polystyrene MPs [110] reduction in feeding aptitude and in weight of *A. marina* was observed [111].

The accumulation of microplastic mostly (30%–35%) in the form of pellets [112,113] are found in sea birds like shearwaters, albatross, northern fulmar, and petrels, which are fed from the sea surface.

The human being, the offender of this microplastic pollution, is exposed to primary microplastics in toothpastes, scrubs, cosmetics, and hand washes. Unconscious brushing of teeth may lead to swallowing toothpaste containing microplastics or microbeads and are absorbed via the gastrointestinal tract [114]. The indirect way of microplastics ingestion in the human body is via sea foods, such as fish mussel, oyster, crab, and sea cucumber may create a hazardous health effect on human beings.

Due to consistent sources of MPs it has now been a pervasive and resilient pollutant and of high concern beyond control. Therefore it is essential to ban the use of μBs in cosmetics globally. An all-out ban has been imposed on the microbead-containing products in many European countries. In the United States, President Barack Obama signed the Microbead-Free Waters Act of 2015 [115].

5.2 Plastic packaging

Plastics are made of lightweight, nontoxic, and long-lasting polymer compounds, which are cheap in production and purification, can be tailored easily for making materials according to demand for their flexibility, can be shaped or tined to suit structural design and can form nonporous coating. Therefore plastic got the prime role in packagings for last few decades.

According to E. Pongr'acz, 1998, packaging systems can be defined as "a set of operations that fulfill the function of creating sales units of the product" [116]. Cosmetic products are made of costly and easily perishable, therefore proper packaging is needed for preservation, storage, transport, and distribution [117].

Plastics have the most desirable properties of packaging and storage such as transparency, non-porosity, softness, inertness, and good strength-to-weight ratio. Therefore most of the cosmetics are packaged using rigid and flexible plastics. For the difficulties in cleaning of cosmetic packagings which are strongly held by the residual dirt of greasy and creamy cosmetic products, they are rarely reused. Around 25.8 million tons of plastic waste is generated in Europe every year of which only 30% is collected for recycling. The rigid packaging like bottles, or pots and caps made of high density poly(ethylene) (HDPE); poly(propylene) (PP) for non-transparent packages; and poly(ethylene terephthalate) (PET) for transparent packages are generally recovered from waste-plastics. Flexible packagings made of LDPE, PP or PET consisting single material are affected by residual liquid or cream products and is usually managed by incineration.

Most of plastic packagings are thrown away as garbages after utilization, sometimes to the nearby ponds with other household wastes creating a sedimental plastic layer (Fig. 10.5).

Floating cosmetic packagings   Plastic sedimental layer

FIGURE 10.5 Creation of plastic sedimental layer in ponds from floating cosmetic packagings due to long term disposal of household plastic wastes.

The plastic packaging is one of the emerging most concerned pollutant as cosmetic production is an unstoppable business. The packaging used in cosmetics is hardly recycled and therefore creates a potential threat to the environment.

Since the 1960s, global production of plastics has reached to 322 million tons in 2015, and which is expected to be doubled in the coming next 20 years [118]. Only in Europe, turnover of plastic production was EUR 340 billion in 2015 [119].

Normally, high-density plastics made from PVC, polyester, and so forth settle down in the sea bed near the source whereas low-density plastics (e.g., polystyrene, polyethylene) floats on the surface with a probability of vertical mixing by freshwater inputs, storms, and biofilm formation [120,121].

In most of the cases the cosmetic packagings are thrown to and fro in the garbage. The used bottles of cosmetics are seen to be spread over the sea beaches which ultimately received by the ocean with tides. These unplanned disposal of cosmetic packagings block sewage system, leaching of ground water and ultimately enter in food chain after formation of secondary microplastic by fragmentation. Some suitable innovative bio-degradable materials for cosmetic packaging are already developed and some are in advanced stage of implimentation having very promising properties and perspective. These bioplastics indeed, should be the future packaging material to mitigate this huge ecological concern. Polylactic acid (PLA), Polyhydroxyalkanoates (PHAs) and polysaccharides are some of the promising biodegradable packaging polymers having high biocompatibility. Increase in biocompostable cosmetics packaging is urgently required to save our earth from the detrimental effects of plastic.

6. Conclusion

As the cosmetic production is growing in an exponential rate and the beauty mania mainly in the women society is growing day by day. With increasing GDP, the beauty products are becoming a symbol of standard living self-confidence. Along with all ingredients used in cosmetic formulation and packaging, microplastic and plastic packaging is one of the most ecological concern as these plastic products are nonbiodegradable and persists for a long time in the environment. Ingestion of microplastics are of serious concern for all type of aquatic animals as well as for the human being. These are entering in food chain, which is alarming situation for our ecology as the number MP-sources are increasing day by day. Fragmentation by weathering after long exposure of large plastic substances made from synthetic polymers causes secondary microplastic, which is very much relevant for aquatic environment, the ultimate destination of discharged plastic debris. Therefore the use of developed innovative biodegradable materials for cosmetic packaging is strongly recommended banning the use of plastic packagings. Microplastic and harmful ingredients in cosmetics should be strictly debarred by the government; because beauty is becoming the beast to encroach our mother nature.

Appendix A. Supplementary data

Supplementary data to this article can be found online at https://doi.org/10. 1016/B978-0-12-824038-0.00010-9.

List of abbreviations

CAGR Compound annual growth rate
INCI International nomenclature of cosmetic ingredients
MPs Microplastics
PCCPs Personal care and cosmetic products
PCPs Personal care products
STPs Sewage treatment plants
WWT Waste water treatment
WWTP Waste water treatment plants
μBs Microbeads

References

[1] Etcoff NL, Stock S, Haley LE, et al. Cosmetics as a feature of the extended human phenotype: modulation of the perception of biologically important facial signals. PLoS One 2011;6(10):e25656.

[2] Barabasz W, Pikulicka A, Wzorek Z, Nowak AK. Ecotoxicological aspects of the use of parabens in the production of cosmetics. Tech Trans 2019;12:99−124.

[3] Massoume Price. Cosmetics, styles & beauty concepts in Iran. Available 22 January 2016 at: http://www.iranchamber.com/culture/articles/cosmetics_beauty.php.

[4] Power C. Cosmetics, identity and consciousness. J Conscious Stud 2010;17(7−8):73−94.

[5] Power C. Women in prehistoric art. In: Berghaus G, editor. New perspectives in prehistoric art. Westport, CT & London: Praeger; 2004. p. 75−104.

[6] Watts I. Red ochre, body painting and language: interpreting the Blombos ochre. In: Botha R, Knight C, editors. The cradle of language. Oxford: Oxford University Press; 2009. p. 62−92.

[7] Watts I. The pigments from Pinnacle Point Cave 13B, Western Cape, South Africa. J Hum Evol 2010;59:392−411.

[8] Olson K. Cosmetics in Roman Antiquity: substance, remedy, poison. Classical World 2009;102(3):294−8.

[9] Malik V. The Drug and Cosmetics Act. 18th ed. New Delhi: Eastern Book Company; 1940. p. 5−6.

[10] Regulation (EC) No 1223/2009 of the European Parliament and of the Council of 30 November 2009 on cosmetic products. Available online: http://eur-lex.europa.eu/LexUriServ/LexUriServ.do?uri=OJ:L:2009:342:0059:0209:en:PDF. [Accessed on 10 March 2017].

[11] Tolls J, Berger H, Klenk A, Meyberg M, Beiersdorf AG, Müller R, et al. Environmental safety aspects of personal care products − a European perspective. Environ Toxicol Chem 2009;28(12):2485−9.

[12] Market V. Global opportunity analysis and industry forecast, 2017−2023. Pune, India: Allied Market Research; August 2016.

[13] Chen X, Sullivan DA, Sullivan AG, Kam WR, Liu Y. Toxicity of cosmetic preservatives on human ocular surface and adnexal cells. Exp Eye Res 2018;170:188−97.

[14] Karr S, Houtman A, Interlandi J. Toxic bottles? On the trail of chemicals in our everyday lives. In: Kate Ahr P, editor. American environmental science for a changing world. New York, NY: W. H. Freeman; 2013. p. 54.

[15] Carballa M, Omil F, Lema JM, Llompart M, García-Jares C, Rodríguez I, et al. Behavior of pharmaceuticals, cosmetics and hormones in a sewage treatment plant. Water Res 2004;38:2918—26.

[16] Liu N, Shi Y, Li W, Xu L, Cai Y. Concentrations and distribution of synthetic musks and siloxanes in sewage sludge of wastewater treatment plants in China. Sci Total Environ 2014;476—477:65—72. https://doi.org/10.1016/j.scitotenv.2013.12.124.

[17] Campo J, Masiá A, Picó Y, Farré M, Barceló D. Distribution and fate of perfluoroalkyl substances in Mediterranean Spanish sewage treatment plants. Sci Total Environ 2014;472:912—22.

[18] Ramos S, Homem V, Alves A, Santos L. A review of organic UV-filters in wastewater treatment plants. Environ Int 2016;86:24—44.

[19] Browne MA, Galloway T, Thompson R. Microplastics — an emerging contaminant of potential concern? Integrated Environ Assess Manag 2009;3:559—61.

[20] Conley K, Clum A, Deepe J, Lane H, Beckingham B. Wastewater treatment plants as a source of microplastics to an urban estuary: removal efficiencies and loading per capita over one year. Water Res X 2019;3:100030.

[21] Jardak K, Drogui P, Daghrir R. Surfactants in aquatic and terrestrial environment: occurrence, behavior, and treatment processes. Environ Sci Pollut Res 2016;23:3195—216.

[22] Díaz-Cruz MS, García-Galán MJ, Guerra P, Jelic A, Postigo C, Eljarrat E, et al. Analysis of selected emerging contaminants in sewage sludge. TrAC Trends Anal Chem 2009;28:1263—75.

[23] Gao P, Lei T, Jia L, Yury B, Zhang Z, Du Y, et al. Bioaccessible trace metals in lip cosmetics and their health risks to female consumers. Environ Pollut 2018;238:554—61.

[24] Ternes TA, Joss A, Siegrist H. Peer reviewed: scrutinizing pharmaceuticals and personal care products in wastewater treatment. Washington, DC, USA: ACS Publications; 2004.

[25] Kasprzyk-Hordern B, Dinsdale RM, Guwy AJ. The removal of pharmaceuticals, personal care products, endocrine disruptors and illicit drugs during wastewater treatment and its impact on the quality of receiving waters. Water Res 2009;43:363—80.

[26] Lishman L, Smyth SA, Sarafin K, Kleywegt S, Toito J, Peart T, et al. Occurrence and reductions of pharmaceuticals and personal care products and estrogens by municipal wastewater treatment plants in Ontario, Canada. Sci Total Environ 2006;367:544—58.

[27] Radjenovic J, Petrovic M, Ventura F, Barcelo D. Rejection of pharmaceuticals in nanofiltration and reverse osmosis membrane drinking water treatment. Water Res 2008;42:3601—10.

[28] Vieno NM, Härkki H, Tuhkanen T, Kronberg L. Occurrence of pharmaceuticals in river water and their elimination in a pilot-scale drinking water treatment plant. Environ Sci Technol 2007;41:5077—84.

[29] Sutherland WJ, Clout M, Côté IM, Daszak P, Depledge MH, Fellman L, et al. A horizon scan of global conservation issues for 2010. Trends Ecol Evol 2010;25:1—7.

[30] Wright SL, Thompson RC, Galloway TS. The physical impacts of microplastics on marine organisms: a review. Environ Pollut 2013;178:483—92.

[31] GESAMP Sources, fate and effects of microplastics in the marine environment: a global assessment. In: Kershaw PJ, editor. Reports and studies 90. London: IMO/FAO/UNESCO-IOC/UNIDO/WMO/IAEA/UN/UNEP/UNDP joint group of experts on the scientific aspects of marine environmental protection; 2015.

[32] Caliman FA, Gavrilescu M. Pharmaceuticals, personal care products and endocrine disrupting agents in the environment — a review. Clean 2009;37:277—303.

[33] Daughton CG, Ternes TA. Pharmaceuticals and personal care products in the environment: agents of subtle change? Environ Health Perspect 1999;107:907—38.

[34] Chen W, Xu J, Lu S, Jiao W, Wu L, Chang AC. Fates and transport of PPCPs in soil receiving reclaimed water irrigation. Chemosphere 2013;93:2621—30.

[35] Castiglioni S, Bagnati R, Fanelli R, Pomati F, Calamari D, Zuccato E. Removal of pharmaceuticals in sewage treatment plants in Italy. Environ Sci Technol 2006;40:357—63.

[36] Santos JL, Aparicio I, Alonso E. Occurrence and risk assessment of pharmaceutically active compounds in wastewater treatment plants. A case study: Seville city (Spain). Environ Int 2007;33:596—601.

[37] Yang Y, Toor GS, Reisinger AJ. Contaminants in the urban environment. Pharmaceuticals and personal care products (PPCPs) Part 2. University of Florida Extention; 2015.

[38] Fick J, Söderström H, Lindberg RH, Phan C, Tysklind M, Larsson DGJ. Contamination of surface, ground, and drinking water from pharmaceutical production. Environ Toxicol Chem 2009;28:2522—7.

[39] Daughton CG. Cradle-to-cradle stewardship of drugs for minimizing their environmental disposition while promoting human health: I Rationale for and avenues toward a green pharmacy. Environ Health Perspect 2003;111:757—74.

[40] Molins-Delgado D, Díaz-Cruz MS, Barceló D. Introduction: personal care products in the aquatic environment. In: Díaz-Cruz MS, Barceló D, editors. Personal care products in the aquatic environment series title: the handbook of environmental chemistry. Series ISSN: 1867-979X. ISSN: 1616-864X. Springer International Publishing; 2015.

[41] Zenker A, Cicero MR, Prestinaci F, Bottoni P, Carere M. Bioaccumulation and biomagnification potential of pharmaceuticals with a focus to the aquatic environment. J Environ Manag 2014;133:378—87.

[42] Meador J. Rationale and procedures for using the tissue-residue approach for toxicity assessment and determination of tissue, water, and sediment quality guidelines for aquatic organisms. Hum Ecol Risk Assess 2006;12(6):1018—73.

[43] Wennmalm Å, Gunnarsson B. Pharmaceutical management through environmental product labeling in Sweden. Environ Int 2009;35(5):775—7.

[44] Bruggeman WA, Opperhuizen A, Wijbenga A, Hutzinger O. Bioaccumulation of super-lipophilic chemicals in fish. Toxicol Environ Chem 1984;7(3):173—89.

[45] Thomann RV. Bioaccumulation model of organic chemical distribution in aquatic food chains. Environ Sci Technol 1989;23(6):699—707.

[46] Sweetman AJ, Valle MD, Prevedouros K, Jones KC. The role of soil organic carbon in the global cycling of persistent organic pollutants (Pops): interpreting and modelling field data. Chemosphere 2005;60(7):959—72.

[47] Helbling DE, Hollender J, Kohler HPE, Singer H, Fenner K. High-throughput identification of microbial transformation products of organic micropollutants. Environ Sci Technol 2010;44:6621—7.

[48] Onesios KM, Yu JT, Bouwer EJ. Biodegradation 2009;20:441.

[49] Ebele AJ, Abdallah MAE, Harrad S. Pharmaceuticals and personal care products (PPCPs) in the freshwater aquatic environment. Emerg Contam 2017;3:1—16.

[50] Van Der Oost R, Beyer J, Vermeulen NPE. Fish bioaccumulation and biomarkers in environmental risk assessment: a review. Environ Toxicol Pharmacol 2003;13(2):57—149.

[51] Reemtsma T, Miehe U, Duennbier U, Jekel M. Polar pollutants in municipal wastewater and the water cycle: occurrence and removal of benzotriazoles. Water Res 2010;44(2):596—604.

[52] Stasinakis AS, Petalas AV, Mamais D, Thomaidis NS, Gatidou G, Lekkas TD. Investigation of triclosan fate and toxicity in continuous-flow activated sludge systems. Chemosphere 2007;68(2):375−81.

[53] Ying G-G, Yu X-Y, Kookana RS. Biological degradation of triclocarban and triclosan in a soil under aerobic and anaerobic conditions and comparison with environmental fate modelling. Environ Pollut 2007;150(3):300−5.

[54] Heidler J, Halden RU. Mass balance assessment of triclosan removal during conventional sewage treatment. Chemosphere 2007;66(2):362−9.

[55] Heidler J, Sapkota A, Halden RU. Partitioning, persistence, and accumulation in digested sludge of the topical antiseptic triclocarban during wastewater treatment. Environ Sci Technol 2006;40(11):3634−9.

[56] Whitmore SE, Morison WL. Prevention of UVB-induced immunosuppression in humans by a high sun protection factor sunscreen. Arch Dermatol 1995;131(10):1128.

[57] Seite S, Colige A, Piquemal-Vivenot P, Montastier C, Fourtanier A, Lapiere C, et al. A full-UV spectrum absorbing daily use cream protects human skin against biological changes occurring in photoaging. Photodermatol Photoimmunol Photomed 2000;16(4):147−55.

[58] Liardet S, Scaletta C, Panizzon R, Hohlfeld P, Laurent-Applegate L. Protection against pyrimidine dimers, P53, and 8-hydroxy-2-deoxyguanosine expression in ultraviolet-irradiated human skin by sunscreens: difference between UVB & Plus; UVA and UVB alone sunscreens. J Invest Dermatol 2001;117(6):1437−41.

[59] Lowe NJ. Sunscreens: development: evaluation, and regulatory aspects. Boca Raton: CRC Press; 1996.

[60] Gago-Ferrero P, Diaz-Cruz MS, Barceló D. Occurrence of multiclass UV filters in treated sewage sludge from wastewater treatment plants. Chemosphere 2011;84(8):1158−65.

[61] Barón E, Gago-Ferrero P, Gorga M, Rudolph I, Mendoza G, Zapata AM, Diaz-Cruz M, Barra R, Ocampo-Duque W, Páez M. Occurrence of hydrophobic organic pollutants (BFRs and UV-filters) in sediments from South America. Chemosphere 2013;92(3):309−16.

[62] Gago-Ferrero P, Diaz-Cruz MS, Barceló D. Fast pressurized liquid extraction with in-cell purification and analysis by liquid chromatography tandem mass spectrometry for the determination of uv filters and their degradation products in sediments. Anal Bioanal Chem 2011;400(7):2195−204.

[63] Buser H-R, Balmer ME, Schmid P, Kohler M. Occurrence of UV filters 4-methylbenzylidene camphor and octocrylene in fish from various Swiss rivers with inputs from wastewater treatment plants. Environ Sci Technol 2006;40(5):1427−31.

[64] Fent K, Zenker A, Rapp M. Widespread occurrence of estrogenic UV-filters in aquatic ecosystems in Switzerland. Environ Pollut 2010;158(5):1817−24.

[65] Jonkers N, Sousa A, Galante-Oliveira S, Barroso CM, Kohler H-PE, Giger W. Occurrence and sources of selected phenolic endocrine disruptors in Ria De Aveiro, Portugal. Environ Sci Pollut Res 2010;17(4):834−43.

[66] Lee H-B, Peart TE, Svoboda ML. Determination of endocrine-disrupting phenols, acidic pharmaceuticals, and personal-care products in sewage by solid-phase extraction and gas chromatography-mass spectrometry. J Chromatogr A 2005;1094(1):122−9.

[67] Loraine GA, Pettigrove ME. Seasonal variations in concentrations of pharmaceuticals and personal care products in drinking water and reclaimed wastewater in Southern California. Environ Sci Technol 2006;40(3):687−95.

[68] Regueiro J, Llompart M, Psillakis E, Garcia-Monteagudo JC, Garcia-Jares C. Ultrasound-assisted emulsification microextraction of phenolic preservatives in water. Talanta 2009;79(5):1387−97.

[69] Kannan K, Reiner JL, Yun SH, Perrotta EE, Tao L, Johnson-Restrepo B, et al. Polycyclic musk compounds in higher trophic level aquatic organisms and humans from the United States. Chemosphere 2005;61(5):693−700.

[70] Gatermann R, Biselli S, Hühnerfuss H, Rimkus GG, Hecker M, Karbe L. Synthetic musks in the environment. Part 1: species-dependent bioaccumulation of polycyclic and nitro musk fragrances in freshwater fish and mussels. Arch Environ Contam Toxicol 2002;42(4):437−46.

[71] Vallecillos L, Pocurull E, Borrull F. Fully automated determination of macrocyclic musk fragrances in wastewater by microextraction by packed sorbents and large volume injection gas chromatography-mass spectrometry. J Chromatogr A 2012;1264:87−94.

[72] Vallecillos L, Pocurull E, Borrull F. A simple and automated method to determine macrocyclic musk fragrances in sewage sludge samples by headspace solid-phase micro-extraction and gas chromatography-mass spectrometry. J Chromatogr A 2013;1314:38−43.

[73] Olmez-Hanci T, Arslan-Alaton I, Basar G. Multivariate analysis of anionic, cationic and nonionic textile surfactant degradation with the H_2O_2 UV-C process by using the capabilities of response surface methodology. J Hazard Mater 2011;185(1):193−203.

[74] Sparham C, Van Egmond R, O'Connor S, Hastie C, Whelan M, Kanda R, et al. Determination of decamethylcyclopentasiloxane in river water and final effluent by headspace gas chromatography/mass spectrometry. J Chromatogr A 2008;1212(1):124−9.

[75] Richardson SD. Environmental mass spectrometry: emerging contaminants and current issues. Anal Chem 2010;82(12):4742−74.

[76] Sanchís J, Martínez E, Ginebreda A, Farré M, Barceló D. Occurrence of linear and cyclic volatile methylsiloxanes in wastewater, surface water and sediments from Catalonia. Sci Total Environ 2013;443:530−8. https://doi.org/10.1016/j.scitotenv.2012.10.047.

[77] Horii Y, Kannan K. Survey of organosilicone compounds, including cyclic and linear si-loxanes, in personal-care and household products. Arch Environ Contam Toxicol 2008;55(4):701−10.

[78] Arthur C, Baker J, Bamford H. Proceedings of the international research workshop on the occurrence, effects and fate of microplastic marine debris. NOAA Technical Memorandum; 2009.

[79] Eriksen M, Lebreton LC, Carson HS, Thiel M, Moore CJ, Borerro JC, et al. Plastic pollution in the world's oceans: more than 5 trillion plastic pieces weighing over 250,000 tons afloat at sea. PLoS One 2014;9:111913.

[80] Shim WJ, Hong SH, Eo S. Marine microplastics: abundance, distribution, and composition. In: Zeng EY, editor. Microplastic contamination in aquatic environments: an emerging matter of environmental urgency. Amsterdam, Netherlands: Elsevier; 2018. p. 1−26.

[81] Plastics Europe. Plastics − the facts 2013: an analysis of European latest plastics production, demand and waste data. Belgium: Plastics Europe; 2013.

[82] Plastics Europe. Plastics − the facts 2016: an analysis of European plastics production, demand and waste data. Belgium: Plastics Europe; 2016.

[83] Jambeck JR, Geyer R, Wilcox C, Siegler TR, Perryman M, Andrady A, et al. Plastic waste inputs from land into the ocean. Science 2015;347:768−71.

[84] Koelman AA, Besseling E, Shim WJ. Nanoplastics in the aquatic environment: critical review. In: Bergmann M, Gutow L, Klages M, editors. Marine anthropogenic litter. New York: Springer; 2015. p. 245−307.

[85] Song YK, Hong SH, Jang M, Han GM, Jung SW, Shim WJ. Combined effects of UV exposure duration and mechanical abrasion on microplastic fragmentation by polymer type. Environ Sci Technol 2017;51:4368−76.

[86]   Maximenko N, Hafner J, Niiler P. Pathways of marine debris derived from trajectories of Lagrangian drifters. Mar Pollut Bull 2012;65:51−62.

[87]   Cosmetic Ingredient Review. Silylates and surface modified siloxysilicates as used in cosmetics. Washington D.C: Final safety assessment; 2011. p. 25.

[88]   Leslie HA, UNEP. Plastic in cosmetics. 2015.

[89]   Anderson AG, Grose J, Pahl S, Thompson RC, Wyles KJ. Microplastics in personal care products: exploring perceptions of environmentalists, beauticians and students. Mar Pollut Bull 2016;113:454−60.

[90]   Gouin T, Avalos J, Brunning I, Brzuska K, de Graaf J, Kaumanns T, et al. Use of microplastic beads in cosmetic products in Europe and their estimated emissions to the North Sea environment. Int J Appl Sci SOFW-J 2015;141:40−6.

[91]   Cheung PK, Fok L. Characterisation of plastic microbeads in facial scrubs and their estimated emissions in Mainland China. Water Res 2017;122:53−61.

[92]   Boucher J, Friot D. Primary microplastics in the oceans: a global evaluation of sources. 2017. https://doi.org/10.2305/IUCN.CH.

[93]   Davidson K, Dudas SE. Microplastic ingestion by wild and cultured manila clams (*Venerupis philippinarum*) from baynes sound, British Columbia. Arch Environ Contam Toxicol 2016;71(2):147−56.

[94]   Lusher AL, O'Donnell C, Officer R, O'Connor I. Microplastic interactions with North Atlantic mesopelagic fish. ICES J Mar Sci 2016;73:1214−25.

[95]   Tourinho PS, Ivar do Sul JA, Fillmann G. Is marine debris ingestion still a problem for the coastal marine biota of southern Brazil? Mar Pollut Bull 2010;60(3):396−401.

[96]   Besseling E, Foekema EM, Van Franeker JA, Leopold MF, Kℰuhn S, Bravo Rebolledo EL, et al. Microplastic in a macro filter feeder: humpback whale *Megaptera novaeangliae*. Mar Pollut Bull 2015;95:248−52. https://doi.org/10.1016/j.marpolbul.2015.04.007.

[97]   Welden NAC, Cowie PR. Environment and gut morphology influence microplastic retention in langoustine, *Nephrops norvegicus*. Environ Pollut 2016;214:859−65.

[98]   Renzi M, Guerranti C, Blašković A. Microplastic contents from maricultured and natural mussels. Mar Pollut Bull 2018;131:248−51.

[99]   Scopetani C, Cincinelli A, Martellini T, Lombardini E, Ciofini A, Fortunati A, et al. Ingested microplastic as a two-way transporter for PBDEs in *Talitrus saltator*. Environ Res 2018;167:411−7.

[100] Rummel CD, Loder MG, Fricke NF, Lang T, Griebeler EM, Janke M, et al. Plastic ingestion by pelagic and demersal fish from the north sea and Baltic Sea. Mar Pollut Bull 2016;102:134−41.

[101] Mazurais D, Ernande B, Quazuguel P, Severe A, Huelvan C, Madec L, et al. Evaluation of the impact of polyethylene microbeads ingestion in European sea bass (*Dicentrarchus labrax*) larvae. Mar Environ Res 2015;112:78−85.

[102] Lusher AL, McHugh M, Thompson RC. Occurrence of microplastics in the gastrointestinal tract of pelagic and demersal fish from the English Channel. Mar Pollut Bull 2013;67:94−9. https://doi.org/10.1016/j.marpolbul.2012.11.028.

[103] Alomar C, Deudero S. Evidence of microplastic ingestion in the shark *Galeus melastomus* Rafinesque, 1810 in the continental shelf off the western Mediterranean Sea. Environ Pollut 2017;223:223−9.

[104] Bellas J, Martnez-Armental J, Martnez-Camara A, Besada V, Martnez-Gomez C. Ingestion of microplastics by demersal fish from the Spanish Atlantic and Mediterranean coasts. Mar Pollut Bull 2016;109:55−60.

[105]   Lee H, Shim WJ, Kwon JH. Sorption capacity of plastic debris for hydrophobic organic chemicals. Sci Total Environ 2014;470:1545—52.

[106]   Ziccardi LM, Edgington A, Hentz K, Kulacki KJ, Driscoll SK. Microplastics as vectors for bioaccumulation of hydrophobic organic chemicals in the marine environment: a state-of-the-science review. Environ Toxicol Chem 2016;35:1667—776.

[107]   Wang F, Fei Wang F, Zeng EY. Sorption of toxic chemicals on microplastics. In: Zeng EY, editor. Microplastic contamination in aquatic environments: an emerging matter of environmental urgency. Amsterdam, Netherlands: Elsevier; 2018. p. 1—26.

[108]   Hall NM, Berry KLE, Rintoul L, Hoogenboom MO. Microplastic ingestion by scleractinian corals. Mar Biol 2015. https://doi.org/10.1007/s00227-015-2619-7.

[109]   Setala O, Fleming-Lehtinen V, Lehtiniemi M. Ingestion and transfer of microplastics in the planktonic food web. Environ Pollut 2014;185:77—83.

[110]   Besseling E, Wegner A, Foekema EM, van den Heuvel-Greve MJ, Koelmans AA. Effects of microplastic on fitness and PCB bioaccumulation by the lugworm *Arenicola marina* (L.). Environ Sci Technol 2013;47:593—600.

[111]   Wright SL, Rowe D, Thompson RC, Galloway TS. Microplastic ingestion decreases energy reserves in marine worms. Curr Biol 2013;23:1031—3.

[112]   Robards MD, Piatt JF, Wohl KD. Increasing frequency of plastic particles ingested by seabirds in the subarctic North Pacific. Mar Pollut Bull 1995;30:151—7.

[113]   Blight LK, Burger AE. Occurrence of plastic particles in seabirds from the eastern North Pacific. Mar Pollut Bull 1997;34:323—5.

[114]   Lassen C, Hansen SF, Magnusson K, Noren F, Hartmann NIB, Jensen PR, Nielsen TG, Brinch A. Microplastics: occurrence, effects and sources of releases to the environment in Denmark. The Danish Environmental Protection Agency; 2015. http://www.eng.mst.dk/.

[115]   Pallone FHR. Microbead-Free Waters Act of 2015, Public Law No. 114-114. https://www.congress.gov/bill/114th-congress/housebill/. [Accessed on 14 August 2017]. 1321.

[116]   Pongrácz E. The environmental effects of packaging. Licentiate thesis. Tampere University of Technology, Department of Environmental Technology, Institute of Environmental Engineering and Biotechnology, Tampere, Finland; 1998.

[117]   Said P, Pradhan R, Sharma N, Naik B. Protective coatings for shelf life extension of fruits and vegetables. J Bioresour Eng Technol 2014;1:1—6.

[118]   Cinelli P, Coltelli MB, Signori F, Morganti P, Lazzeri A. Cosmetic packaging to save the environment: future perspectives. Cosmetics 2019;6:26. https://doi.org/10.3390/cosmetics6020026.

[119]   A sustainable bioeconomy for Europe: strengthening the connection between economy, society and the environment. Available online: https://ec.europa.eu/research/bioeconomy/pdf/ec_bioeconomy_strategy_2018.pdf. [Accessed on 2 February 2019].

[120]   Lattin GL, Moore CJ, Zellers AF, Moore SL, Weisberg SB. A comparison of neustonic plastic and zooplankton at different depths near the southern California shore. Mar Pollut Bull 2004;49(4):291—4.

[121]   Lobelle D, Cunliffe M. Early microbial biofilm formation on marine plastic debris. Mar Pollut Bull 2011;62(1):197—200.

Chapter 11

Indian rural housing: an approach toward sustainability

Ajay Kumar

Department of Architecture, National Institute of Technology, Patna, Bihar, India

1. Introduction

Currently around 44% of the total populace lives in rural territories, while in India approximately 66% of the total populace lives in rural territories [1]. India is also known as country of villages. The deficiency/shortages of housing in the rural areas in India was assessed 4.00 Crores in 2011 [2]. To address these gaps in rural area the government of India is working to provide pucca house in rural zones under plan of Pradhan Mantri Awaas Yojna-Gramin (PMAY-G). PMAY-G aims to provide 2.95 Crores pucca houses to low-income rural houseless unit with fundamental necessities living in kutcha and dilapidated/ill-structured house by 2022. It has been claimed that 1.00 Crore houses has been built so far [2]. It empowers development of value houses by the recipients utilizing neighborhood materials, plan, and prepared artisans. There is a requirement to provide housing in rural areas. The traditional and local construction materials could be interesting alternatives for low-income household in rural areas because it is cheaper than industrial materials. Neighborhood development materials and procedures can possibly alter the house development on ways of life, geology, atmosphere, and protection from normal catastrophes [3]. Indigenous structure materials are less expensive, promptly accessible and required less preparing before use, it required neighborhood work, regularly restricted to family or more distant family part or nearby network individuals for development measure, which is saving the extensive charges of labor and work costs [4,5].

Mud or mud brick (adobe) has been used as construction materials for houses in Russian Turkestan from 8000 to 6000 BCE [6]. Even today it is estimated that approximately 33% of the world population live in houses having at least some part made of earth [7]. In developing nations the level of mud house is assessed a lot higher. Mud houses burn through less energy for material creation and less transportation costs due nearby asset used [8,9].

Mud houses are natural well-disposed and moderate when contrasted with mechanical material, it likewise improves indoor air quality and warm solace [10]. The increasing price of industrial materials revives the interest in mud construction globally [11].

India has many regions based on languages, cultures, and geographical basis. Although the rural India of different regions have different culture and local practices but one common thing is sustainable vernacular building construction techniques. The use of local building material is being adopted for house construction right from the early Vedic ages. The most common materials are clay, mud, stone, grasses leaves, bamboo and wood. These materials are used both for wall and roof covering of the house. Vernacular architecture is an informal architectural style which reflects local traditions and based on the local needs of people and functions. The design skills are locally developed and transferred to next generation. The vernacular buildings show simplistic construction system and minimum use of local material. The vernacular architecture is an approach toward the harmonious balance between man and nature. The application makes it sustainable. The simple form, types of materials use in vernacular architecture provide exemplars and model for upcoming architects. The climate, geography, social and cultural factors dictate the spatial vocabulary and material selection in vernacular architecture [12]. This chapter concludes that use traditional locally available building materials in rural Indian housing are an approach toward the sustainability and sustainable development.

2. Concept of sustainability

In 1992, the United Nations Rio Earth Summit made the world aware of various ways of achieving sustainable development. This highest point likewise brought the worry up in a dangerous atmospheric deviation and environmental change. Prior to that in 1987, at the 42nd UN congress, the then Norwegian Prime Minister, Gro Harlem Brundtland, delivered a report named "Our Common Future" presented the thought of feasible turn of events. Brundtland report characterized "sustainable development is development that meets the needs of the present without compromising the ability of future generations to meet their own needs" [13].

The Johannesburg Declaration on "Sustainable Development, 2002" further presented the three "Sustainable Development turn of events— Economical events, social turn of events and ecological insurance at the neighborhood, public, local and worldwide levels". This can achieve by sustainable approach toward the use of natural resources and environmental protection. The building construction uses the significant number of natural resources and energy. So the application of concept of sustainability in building construction leads to path of sustainable development. The most important thing is the consideration of the whole life cycle of material while selecting the material. The use of natural raw material and renewable energy source is second most important thing. And last but not the least is the reduction in the use of material and energy used in raw materials extraction [13].

Despite various endeavors, supportable acceptable houses in the rural sector have remained hard to describe, yet it should be lucid to most of the aspects covering sustainable development [14]:

- to give support to the underprivileged rural population they are left with no decision other than to crush their atmosphere;
- the credibility of unforeseen development, inside trademark resource objectives;
- financially smart headway, which implies in this manner that improvement should not ruin normal quality, nor should it decrease productivity as time goes on;
- the issues of irresistible anticipation, reasonable developments, food security, clean water, and cover for all; and
- he belief that people-centered participatory exercises are required; individuals as such are the resources.

This perspective the affordable and sustainable housing procedure ought to unite three objectives [14]:

The first of these is that future plans should give the reason to nuclear family improvement. The second objective of the systems which could achieve reasonable lodging improvement is stressed over the reinforcing of penniless people. The third objective of such methodologies should be to intellectually give the lower segment of the metropolitan culture an impression of confidence. As needs be, to be viable, lodging exercises should be monetarily sensible, socially sufficient and moderate, indeed pragmatic and biologically very much arranged. Likely commitment of housing to sustainability, it is appropriate to take note of that housing area can altogether add to manageability in light of its nearby relationship with natural angles:

- Houses takes-through a ton of standard and man-made resource being developed, upkeep and continued with the utilization for the common civilization.
- Houses are an immovable asset through an extensive operative and functional future.
- Structures are among the vital requirement for a nice individual fulfillment, and consequently have recommendations past lodging impacting transport, prosperity, business and organization.
- Houses in development are pleasing to different habits by which reused resources could be recycled reused for improvement.
- Huge quantity of progressions is available aimed at viably working the structures tallying usage of reused material for improvement, wastewater treatment and use, energy capability, sun arranged warming, and inactive sun controlled warming, making metropolitan open as well as Green spaces in region toward minor rural housing needed.

"National Urban Housing and Habitat Policy, 2007" states comment on various aspects of environment sustainability, a para has been given specifically [15]:

> *Development of sustainable habitat is closely related to the adoption of 'the Regional Planning approach' while preparing Master Plans of towns/ cities, District Plans and Regional/Sub-Regional Plans. It involves maintenance of the ecological balance in terms of a symbiotic perspective on rural and urban development while developing urban extensions of existing towns as well as new integrated townships. Promotion of sustainable habitat is closely linked with reserving a significant proportion of the total Master Plan area as 'green lungs of the city' (e.g., Master Plan for Delhi 2021 provides 20% of green areas), protecting water bodies with special emphasis on the flood plains of our rivers and developing green belts around our cities. It will be desirable to pursue a goal of 20%–25% recreational land use area (excluding water bodies) which has been prescribed for Metro-cities by the Urban Development Plan Formulation and Implementation Guidelines (UDPFI) in order to enhance the sustainability of human settlements. Recreational land use refers to parks, playfields and other open space such as specified park, amusement park, maidan, a multipurpose open space, botanical garden, zoological parks, traffic parks, etc. It is also necessary to estimate the Gross Geographic Product (GGP) of a given sub-region and endeavour to enhance it while developing new urban settlements. The new Habitat Policy recognizes the sustainability limits of existing urban settlements. It also seeks to emphasize the mutual interdependence between towns and villages.*

3. Indian rural housing

Houses and housing are a fundamental necessity alongside food and garments for social means. Satisfactory safe house for every single family is an essential for sound living in each general public. A house gives huge monetary security also, status in the public arena. The home gives the house dwellers both mental and physical power and psychological base on which they can substitute while getting to other crucial necessities, for example, food, garments, and so on for poor family units a house establishes a resource. The households can make this resource as a guarantee safe keeping for acclamation during troublesome period [16]. The rural building technology is vernacular technology which has evolved in a particular region over more than hundreds of years. It transferred from one generation to another generation. In this chapter a case study of North and Northeast India has been made. North India Includes the state of Uttar Pradesh, Bihar, Bengal and Orissa while northeast India covers the states of Arunachal Pradesh, Tripura, Assam, Manipur, Nagaland, Meghalaya, and Mizoram. The tradition, socioculture, climate, landscape, and settlement form are the basic factor of architectural types and building materials of vernacular

Housing. The climate of North India region can be classified as composite climate. Hot summer months run from April to June after which the monsoon rains strike from July to September in North India. The rains last until September after which autumn begins to fade into a mild winter. Winter is very cold during month of December and January. The spring season run from February to March which is sunny and pleasant. The climate of northeast India is mild to cold with very heavy rainfall. Rural houses in India are mostly made of local materials. India has diversified natural resourced and geographical conditions. A large number of building materials are obtained from vegetation sources and geological sources. The clay, mud, sundry clay brick, stone, grasses leaves, reeds, bamboo and wood are used both for the wall and roofs of the house. Some building materials of industrial origin like galvanized iron sheet, metal sheet, polyethylene, cement concrete, asbestos sheet, kiln burn clay brick, and factory-made tiles are also used for walling material and roofing material. But these materials are preferred by rich people or provided by government.

3.1 Rural housing schemes and programs in India

This part plots the significant of the housing plans which are actualized in India subsequent after Independence of 1947. It evaluates the plan of these plans to decide if housing destitution of urban areas has been fittingly managed by the system characterizing such plan. Introducing to developed communities as "Cities are the Growth Engine" infers desires that they are well-performing. What are the qualifications of well-performing metropolitan focuses, in the first place? The broadly acknowledged model of urbanization characterizes it as a development of a crude agricultural economies in progressive contemporary; the advancement is powered by an adjustment in the example of business age—a move of excess work from agribusiness to industry and administrations. The capacity of a metropolitan place as a motor of development starts at business age. The development of the business and administrations areas launches the cycle wherein work creation supports total interest, adding to increment in GDP which, thusly, makes more positions. The resulting metropolitan development turns into the wellspring of agglomeration economies that quicken industrialization and the extension of administrations area. Urbanization additionally includes a change from the casual to the conventional area [17].

The movement of urbanization in India has been continuous. The mechanical system set up following freedom smothered the market influences administering the economy. Being unnecessarily directed, Indian industry endured stagnation because of glaring shortcomings, nonattendance of rivalry, deterrents to part of new firms and exit of old ones, administrative mass required for making sure about authorizations, and charge and nondemand blocks. A thoughtful view of the imperative association among financial advancement and mechanization impelled the public power to profit to their procedure for mechanization. Thus, progression changes were started in the mid-1980s and accepted force in the mid-1990s [18].

Schemes and programs	Objectives
"Water Conservation Stories"	Refining the efficiency of land and expanding water accessibility.
"Gram Swaraj Abhiyan"	Engaging Panchayati Raj Institutions (PRIs) for achieving Sustainable Development Goals
"Antyodaya Mission"	It is an assembly and responsibility structure meaning to bring greatest use and the board of assets distributed by 27 Ministries/Department of the Government of India under different projects for the improvement of rural areas.
"PMAY-G"	To progress the suitability and affordability of housing for rural India
"PMGSY"	It is a cross country intend to give great all-climate street availability to detached towns villages
"NSAP"	It gives monetary help to the old matured, widows and people with incapacities as social annuities.
"RURBAN (NRuM)"	To convey multicultural undertaking based foundation in provincial territories, which will likewise incorporate the advancement of monetary exercises and expertise improvement.
"Diksha (Training Portal)"	It is a stage that offers instructors, understudies and guardians drawing in learning material identified with the endorsed school educational plan.
"SAGY"	The advancement of model towns and making models of nearby improvement which can be recreated in different towns.
"DDUGKY"	It is a part of the National Rural Livelihood Mission (NRLM), tasked with the dual objectives of adding diversity to the incomes of rural poor families and cater to the career aspirations of rural youth.
"DISHA"	For effective development coordination of almost all the Central Government's programs, whether it is for infrastructure development or Social and human resource development.
"Sabki Yojna Sabka Vikas"	It aims to draw up **Gram panchayat development Plans** (GPDPs) in the country and **place them on a website** where anyone can see the status of the various government's flagship schemes.
"Swachh Gram"	This plan has been started for improving the degrees of neatness in rural zones through strong and fluid waste administration exercises.
"DAY-NRLM"	It aims at offering effective and efficient institutional platforms to empower the rural poor to increase their household income by means of sustainable living improvements and better access to financial services.
"MGNREGA"	It is Indian labor law and social security measures that aim to guarantee the "right to work".

Underneath is the list of housing schemes that have been launched by the govt. till date:

Schemes of housing	Position
"MGNREGA, (Active Workers)"	127,202,000
"PMAY-G (Houses Sanctioned)"	14,159,830
"NRLM (Households Mobilized)"	68,445,155
"PMGSY (Completed Road Lengths Kms)"	621,429
"NSAP (Total Data Digitized)"	32,823,008
"SAGY GPs (Identified)"	1832
"DDUGKY (Candidates Trained)"	982,907
"MISSION ANTYODAYA (GPs Completed)"	262,303
"RURBAN (Clusters Allocated)"	300

Rural Housing Shortage: Working Group Method—2012-17.

Equation no.	Factors taken into account for accessing housing shortage	Computation	Shortage (in millions)
a	No. of dwelling units not having houses in 2012	no. of dwelling units existing stock of houses (in numbers)	4.10
b	No. of dwelling units temporary (kutcha) houses in 2012	no. of dwelling units existing stock– no. of permanent (pucca) and semipermanent (semi pucca) houses	20.20
c	No. of dwelling units with the shortage due to congestion 2012	No. of households in 2012 was of 6.5%	11.30
d	No. of dwelling units with the shortage due to obsolescence 2012	No. of households in 2012 was of 4.3%	7.50
t^1	The total rural housing shortage - 2012	A + B + C + D	43.10
e	The additional housing shortage arising between 2012 and 2017	Increase in no. of households between 2012 and 17, increase in stock of houses between 2012 and 2017	0.50
t^2	Total rural housing shortage - 2017	T1+E	43.60

All numbers for 2012 were projections based on increased growth rates between the Censuses of 1991–2001.
Reproduced from Planning Commission, Working group on rural housing for the twelfth five year plan, MRD. New Delhi: GOI; 2011:7.

3.2 Problems of housing for the rural deprived

The places of the provincial poor in India are discovered to be lacking differently. The significant housing issues of the poor identify with the accompanying.

- The houses in the provincial zones need insurance to the occupants against wind, downpour and cold.
- They need appropriate course of action in well-lit and natural air.
- Countryside households don't have distinct strategy for possession of individuals.
- No legitimate plan aimed at essential sterilization and drinking water.
- The environmental factors of provincial houses need prerequisites for cleanliness.
- Rural houses are plagued with bugs, rodents, and so forth which mess wellbeing up.
- Rural houses include high repeating costs (support) which the helpless inhabitants can't bear.
- Rural houses are unequipped for giving insurance against common catastrophes like floods, typhoons, and so forth.

4. Need for a separate rural housing policy

The aforementioned conversation shows the emergency of the rural housing and scant supporting of it in the plans and arrangements of India. It likewise epitomizes the disregard what's more, low need concurred to rural housing. Within excess of 43 million housing deficiencies and wide holes in the states of housing and housing comforts, rural area actually doesn't have any critical approach to address these difficulties. It has been contended by a couple of specialists that there can't be a different rural housing strategy [20]. Nonetheless, this examination firmly suggests a different rural housing strategy for India to oblige the breaking down rural housing circumstance. There are numerous reasons.

At first, the rural housing deficiencies in the country territories are very nearly over multiple times more than that of the metropolitan territories. Also, the deficiencies of rural housing have expanded ceaselessly throughout the long term. However, the worry of the Government is plainly slanted toward the metropolitan zones. The last two housing approaches—"*National Urban Housing and Habitat Policy (2007)*"and Draft "*National Rental Housing Policy (2015)*"—were only aimed at the metropolitan territories. The last housing strategy which had a few arrangements for the rural area was the National Housing Policy (1998), however for the most recent 18 years; provincial housing doesn't have any approach bearing. Regardless of the quickly expanding housing deficiencies in the rural and country regions, a selective provincial housing strategy is still not a need before the government.

Also, neediness and hardship of country India is moderately higher as looked at to metropolitan India. This has direct bearing on the housing situation. A helpless living space has a long-lasting unfriendly effect on the advancement of its occupants. Destitution and hardship powers a higher extent of the country poor to wind up living in kutcha houses. In addition, neediness additionally antagonistically influences the buying intensity of the country families which keeps them from developing or buying their own home.

The financing structure and organizations for rural housing are not just lacking yet additionally skeptical in giving advances. Also, the security and plenty of documentation needed to look for advances from the proper financing organizations are hard to outfit. Under these conditions, the rural populace wind up taking credits from casual sources, for example, cash moneylenders at extravagantly high places of interest which further powers them into neediness and abuse. While defining plans for improving these houses, notwithstanding, it is important to oppose the allurement of forcing metropolitan qualities (fundamentally working-class esteems) on provincial zones [21]. Hence, the issues of rural housing must be tended to through a select strategy on the grounds that the nature and degree of the housing emergency in provincial zones is very not the same as the metropolitan zones. Handling the circumstance with a typical rural housing strategy (rather without strategy) for both the areas is anything but a practical methodology. A free provincial housing strategy subsequently turns into a need.

5. Components of rural housing policies

An independent housing strategy should help in the formation of housing stocks that would not exclusively be moderate yet in addition sufficient. Property land for the landless cultivators and farmers should be given through the public authority activity and the housing strategy ought to encourage the satisfactory progression of assets to help poor people and the minimized segments in the country zones. India has resolved to satisfy manageable improvement objectives and housing is one of its segments. For an arranged provincial territory improvement with community conveniences and vocation framework, a different housing strategy, subsequently, is attractive. Development expenses and accessibility of building materials, which are climate agreeable and fiasco safe, should be affirmed through an appropriate component. The strategy should help in recognizing the function of different partners, for example, the Panchayati Raj foundations, private area, public area and the cooperatives. The public authority needs to outline the part of different foundations to give cover in the rural regions in a comprehensive and practical way.

5.1 Land availability

It is significant that the public authority needs to dispense property lands for the country family units or manage the current housing locales. Land cost, which is one of the significant worries in metropolitan zones, isn't a lot of an issue in the towns. With the participation of town panchayats, the public authority can without much of a stretch secure property lands for the rethinking of towns and development of housing stocks [22]. Group approach should be embraced in creating estate lands for vagrants, which would be financially savvy and it will be simpler to give the essential framework to the bunched houses.

Guidelines for the residence grounds can give Rural Housing in India, housing locales to the assessed four million landless family units the nation over. With sufficient accessibility of assets, these terrains can be purchased by the public authority with no political or managerial challenges. Assignment of residence terrains can be fused into the current rural housing plans or the coordinated rural advancement programs.

The minimum size of the plots should be remembered in light of the fact that restricted territory prompts clog and unreasonable residence, which will just prompt further deficiencies. Absence of protection is one of the central points in rural family units where three to four people need to share a solitary room, more among the lower pay gatherings [23]. The current size of the norm staying unit in country zones has been expanded to 25 square meters under the PMAY-G, which is as yet not adequate for sufficient housing. We propose at any rate 30 square meters or more for a sound environment in country zones. Truth be told, the creature shed/poultry ranch, which is a basic piece of rural housing requests generally bigger size of abiding units than in metropolitan territories.

5.2 Housing finance

GDP and housing are interlinked and add to every other's development. Interest in rural housing area won't just assist with handling the issue of rural housing deficiencies yet it will likewise give occupation to countless populace and encourage the financial advancement of the nation. Cost of development materials have expanded over the long haul. The public authority has not had the option to give less expensive structure materials and innovation. Thus, the cost of a house is getting excessively expensive for the country individuals. Absence of sufficient monetary help has stayed essential explanation behind the housing deficiencies in India. Big share of the housing accounts in rural regions come from the casual sources such as cash loan specialists, companions, and family members, which is essentially shifty in nature. The government funds have so far stayed blocked off to the rural housing masses. Further, the inclination to save in rural territories is less as larger part of the profit goes toward immoderate things like food and wellbeing offices. Furthermore, dominant parts of rural poor stay jobless for an enormous span in a year. With no close to home resources or a kept acquiring to show as home loans, admittance to credit offices from formal sources turns out to be troublesome. Despite the fact that the rise of network based monetary sources which gives credit at an insignificant rate is promising, however they also have restricted assets available to them which is probably not going to take care of the issue at mass scale.

Interest sponsorship/subsidy plans for the EWS and the LIG must be presented for the provincial masses also. The Rural Housing Fund needs to give monetary help to essential loaning establishments to loan for rural housing at a financed rate. The danger weight on the Rural Housing area must be limited by the public authority with protection of both the resources and the securitization of credits. Another activity which the public authority can do is that the miniature money establishments in provincial territories must be supported through associations with monetary foundations so the casual credit assets are decreased. Reimbursement of advances which is a significant issue for the monetary organizations can be handled through the Productive Housing plans which are a decent idea and need to be advanced as it can help in the reimbursement of advances. Distinguishing proof of recipients is additionally a difficult issue as they are frequently barred because of nepotism and managerial inconsistencies. Straight forwardness is of most extreme significance for the correct execution of the housing plans.

Significant level checking and intermittent examination of plans is additionally important for smooth cycle of usage plans. While the money sponsorship plans should be restricted to the EWS and the LIG individuals, the interest for country housing for the APL and the MIG which needs monetary instruments for redesigning or fixing of houses should likewise be dealt with. In spite of the fact that, there are different state-run plans running in different areas however they have not had the option to show the ideal yield, basically in light of low assignment of assets.

The state-run plans should be given further catalyst by the association government through bigger asset portion and its coordination with the focal plans as both targets giving housing to all.

6. Identification of rural areas for different aspects in housing

Prior to arranging specialists start the way toward drafting the composed assertion of the advancement plan, as illustrated in this segment, it is crucially significant that a cycle of assessment and investigation be done into populace and advancement patterns in rural territories. This analysis ought to incorporate the distinguishing proof of the area and degree of the rural territory types set out in segment:

1. **Rural zones under solid metropolitan impact:** These regions will show attributes, for example, vicinity to the quick environs or close driving catchment of huge urban communities and towns, quickly rising populace, proof of impressive weight for improvement of housing due to vicinity to such metropolitan territories, or to significant vehicle passageways with prepared admittance to the metropolitan zone, and weights on foundation, for example, the nearby street organization.
2. **Stronger Rural territories:** In these zones' populace levels are by and large steady inside a very much created town and town structure and in the more extensive provincial zones around them. This steadiness is upheld by a generally solid agrarian monetary base and the degree of individual lodging improvement action in these regions will in general be moderately low and kept to specific zones.
3. **More vulnerable country regions structurally:** These regions will show qualities for example, relentless and huge populace decay just as a more fragile monetary dependent structures on records of pay, business and monetary development.
4. **Areas with bunched settlement designs:** Where there are nearly less town or more modest town type settlements contrasted and other provincial territories; rather there tends in those zones to be a commonness of lodging bunches, gatherings of bunches and sometimes straight turn of events. It is normal that all arranging authority territories which are overwhelmingly rural will contain, to fluctuating degrees, in any event three of the rural territory types characterized.

The three concerned are (1) territories under solid metropolitan impact; (2) regions with a generally solid horticultural base; and (3) fundamentally frail regions. In regions where the region town and different towns are similarly more modest in populace terms, zones displaying the qualities of being under the impact of metropolitan zones will be very restricted with the issue here being one basically of evading lace advancement stretching out along outspread streets from the town.

7. Locally available building materials

In parts of northern India, a clayey top soil makes the bricks blocks and mud grout or mortar plays a significant role for a usually used subject building for wall. Other main building materials in the plains are earth, stone, brick, thatch and timber. In the hills it is restricted to stone and timber depends on local availability. The northeast has an abundance of grasses, bamboo and timber for its own distinctive type of architecture. Bamboo-framed walls with woven mats and thatched roofs are common. Fig. 11.1 shows the use of bamboo as a predominant material for wall and roof in Indian rural housing. All the building materials are available without any manufacturer. Earth has always been the most widely used material for building in India.

FIGURE 11.1 Bamboo as a predominant material used in rural housing. *Source: https://en. wikipedia.org/wiki/Indian_vernacular_architecture#/media/File:House_construction_bamboo_ wall_IMG_8946.jpg.*

Practical rural housing innovation is a strategy for housing development that includes the utilization of modest, climate amicable and privately sourced materials, for example, bamboo, bagasse loads up, fly debris-based blocks, mud and lime for building cost-effective, agreeable, and catastrophe safe houses, which give satisfactory norms of living.

It coordinates the utilization of neighborhood development procedures appropriate for the climatic furthermore, geological states of a locale with the components of current design furthermore, innovation, as, simple to-utilize development hardware, premanufactured parts, utilization of rural, and mechanical side-effect-based development materials, and progressed compositional highlights that upgrade the quality and supportability of the houses. The work serious nature of development exercises including privately sourced materials and indigenous structure methods can assume an instrumental part in creating work openings. Associations, for example, BMTPC and NBCCL are liable for advancing feasible housing advancements in India through innovation scattering and circulation of modest and maintainable development materials [24].

As of now rural housing in India is seriously missing on the specialized front. The housing development designs have changed significantly throughout the long term and that's only the tip of the iceberg what's more, more individuals are currently choosing block and cement based pucca houses over the mud and covered kutcha houses. Nearby workers and craftsman undertaking the development exercises have irrelevant specialized ability about savvy and climate amicable structure materials and development strategies.

These pucca structures are built with no specialized help as there is a serious lack of prepared experts and architects in the country zones. Such pucca houses made without specialized direction can represent a natural danger and may imperil the life of their occupants because of their weakness to common cataclysms like earthquakes, flood, and cyclones [24].

7.1 Predominant walling materials

The alluvial plains provide mud, clay, sundry clay brick, kiln burn clay brick, grasses, leaves, and bamboo for wall materials locally. The hilly terrains provide stones and wood for wall materials locally. The northeastern region provides wood and bamboo for wall materials locally. The reed matting with mud plaster is also used for wall material in rural houses. According to the 2011 census, 42% of total rural houses were with walls made of grass, thatch, bamboo, wood, mud, and unburned brick. The 43% of total rural houses had walls made of burned brick, concrete, GI metal, and asbestos sheet. The 10% of total rural houses were with walls made of stone packed with mortar and 4% with stone packed without mortar and rest 1% were with walls made of other categories of materials. Approximately 50% of total rural houses were with wall made of traditional local unprocessed materials which has negligible embodied energy [25]. Grass, thatch, bamboo, wood, mud and unburnt brick/ burnt brick are major building material of wall in Indian rural housing as shown in Fig. 11.2.

7.2 Predominant roofing materials

The alluvial plains provide mud, clay, sundry clay brick, burned brick, grasses leaves, and bamboo for roof materials locally. The hilly terrains provide stones

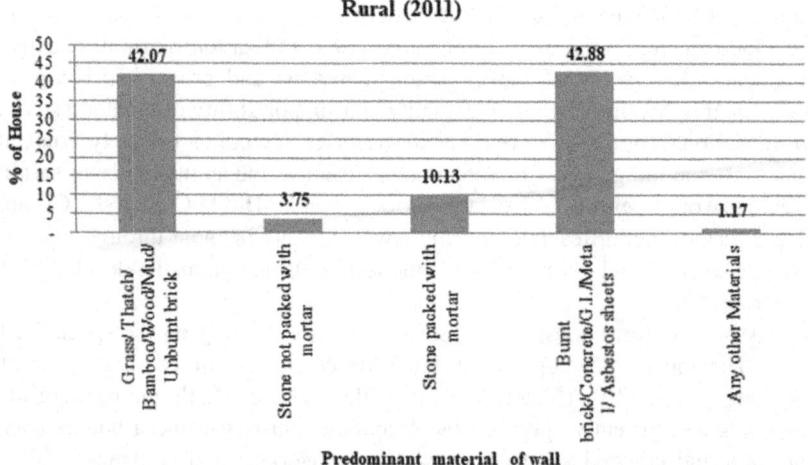

FIGURE 11.2   Predominant material of wall for the rural housing in India.

Rural (2011)

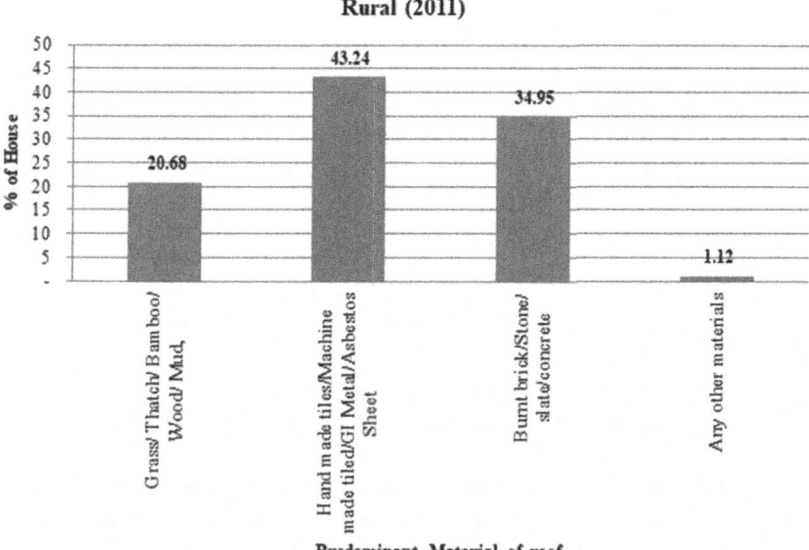

FIGURE 11.3   Housing based on predominant material of roof in rural India.

and wood for roof materials locally. The northeastern region provides wood and bamboo for roof materials locally. The reed matting with mud plaster is also used for roofing material in rural houses. As per 2011 census, approximately 21% of total rural houses were with roof made of grass, thatch, bamboo, wood and mud. The 43% of rural houses had roof made of asbestos sheet, GI metal, made by hand tiles and tiles made by machinery. The 35% of rural houses were with roof made of burned brick, stone, slate and concrete. Approximately 1% were with roof made of other categories of materials [25] Fig. 11.3 shows that grass, thatch, bamboo, wood, mud, asbestos sheet, G.I. metal and handmade/machine-made tiles are predominant material of roof in Indian rural housing.

7.3 Predominant flooring materials

As per 2011 census, approximately 63% of total rural houses floor was made of mud, wood and bamboo. The 33% of rural houses floor was made of cement, burned brick and stone. 4% of rural house floor was made of mosaic, floor tiles and rest of houses had flooring made of other materials [25]. Mud, wood, bamboo burnt brick and stone are major building materials of floor in Indian rural housing as shown in Fig. 11.4.

8. Construction techniques

The Indian rural building construction techniques are vernacular in nature (passed down orally generation to generation), which has evolved in a particular region over a long period of time. The techniques are indigenous in

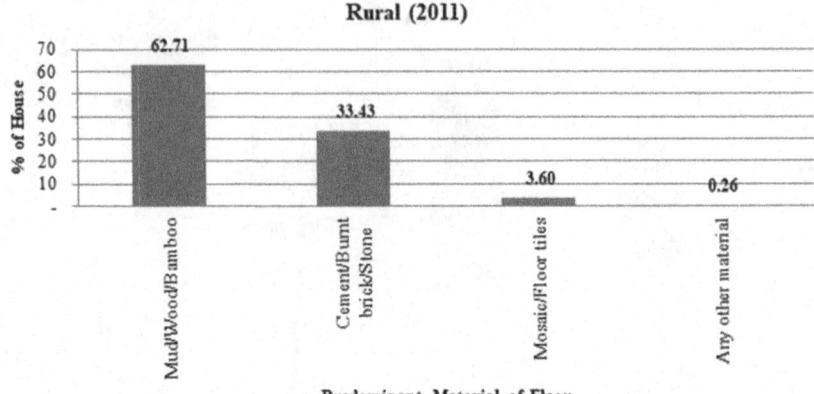

FIGURE 11.4   Predominant material of floor in Rural India Housing.

nature. The people learned these techniques from past generations, and they also transfer these techniques to next generation. But by the experiences and knowledge it evolves and improves day by day. They are very much concerned about the environment, local climate, soil characteristics, geological condition, and their culture while selecting the building materials. They also try to identify the various building needs of a given area and try to give the optimum solution for that. The building materials are identified such that it should be locally available. The materials are mostly ecofriendly which can be repair, reuse, recycle or lastly decomposed naturally. Fig. 11.5 shows that how rural

FIGURE 11.5   Rural houses made of locally available materials in Natural Setting (Dharamsala, H.P.).

houses are constructed in natural setting. The fittings and other hardware are also selected such that it should be locally available or some time designed locally and produced locally by skilled persons like lohaar (blacksmith) and badhee (carpenter).

The construction techniques also vary from region to region based on the availability of building materials, local climate, soil characteristics, geological condition and their culture. It is varying from "Bunga" (round shape mud house) of Kutchh, Gujarat to stone houses in Rajasthan to mud and thatch houses in Jharkhand, Orissa and Bihar to bamboo houses in northeastern states of India. These are very famous for their constriction techniques.

8.1 Wall construction techniques

Mud wall construction techniques are classified in four different types, namely cob, adobe, rammed earth, and different varieties of wattle and daub. Cob is one of the simplest and oldest wall construction techniques of rural India. The mixture of the soil, clay, cow dung, chopped straw, cow urine, and some lime are used for wall construction. The proportion of mixture depends on soil characteristic and also on the strength you want. Small balls are prepared from the properly kneaded mixture and then putting together construct the wall. It is very easy to give any shape one wants by the method. It is constructed in layers of 1 m high in one go. Adobe wall are also constructed by same mixture as in cob, but construction method is different. The mixture is putted into blocks and after that it dried in the sun to form sundried bricks. These mud bricks are placed one over others to make walls, and are plastered with the same mud mix. Rammed earth walls are also constructed by same mud mixture by putting two wooden planks on the both side of the wall and filled with mud mix and pushed from the top by walking or by a wooden beam. The region having abandon bamboo, wattle, and daub method is used. In this method frame structure is first made with wood or bamboo and then flattened bamboo strips are interweaved in the gap. Finally, mud is daubed over it from both sides. Mud comes straight from nature; hence mud wall construction is sustainable and ecofriendly methods. Fig. 11.6 shows that how locally available mud can be used in wall for multi storey house.

In North India, walls, which are generally composed of clay, chopped straw, bits of broken terracotta, and cow dung are built up in layers 1 m high. All the components have a functional role—the straw distributes the cracks evenly through the clay while drying, the earthenware gives the strength, and the dairy animals compost has glue and fungicidal properties. At some point this blend is framed into blocks which are then positioned like brick work. It is put with an earth and dairy animals dunk blend regularly finished with designs in rice paste. In British period the walls are built of brick and finished in white or cream plaster with cornices.

FIGURE 11.6 Double story mud house (Dharamsala, H.P.).

8.2 Roof construction techniques

Most of the rural houses have pitched roofs. The single piched (Ekchala), two-plane (Dochala), and four plane (Charchala) roof with a curve bamboo ridge is very common throughout the India. Some circular pitched roof is also used in Kutchh region. The pitched roof is to deals with the monsoon season. The various materials are used for roof covering. The local grass, rice straw, wheat straw and wild grasses are very common in thatched roof. Fig. 11.7 shows that how locally available slate can used as roof covering. Terracotta roof tiles of different form are also used over bamboo frame and flattened bamboo strips as roof covering. In typical house of Bankura West Bengal the roof of the house has a twofold bended formed like an improved boat, since a rooftop is classified "Bangla." The top of cover which is built in situ is made firmly bound bamboo covered by rice straw or indigenous grass.

9. Settlement forms

The villagers are primarily agrarian. The rural settlement in North India is surrounded by agriculture fields and orchards. The settlements are arranged around caste delineated and often containing members of the same caste based on their occupation. The courtyard clusters commonly found in North India. Houses are arranged in a linear pattern around a courtyard which is aligned along the more or less straight main street. Fig. 11.8 shows that how locally available stone can be used for outdoor sitting space. The courtyard is the focus of family activities and rituals. Settlements in the northeast are

FIGURE 11.7   Use of local material slate as roof covering (Dharamsala, H.P.).

FIGURE 11.8   Use of local stone for Outdoor Sitting (Dharamsala, H.P.).

characterized by villages built atop hills and ridges out of fear of attacks by rival tribes. Instead of the caste system, the community is centered on the hereditary chief or a group of village elders. Fig. 11.9 shows that how locally available stone can be used for beautiful pathway.

FIGURE 11.9 Use of local stone for pathway (Dharamsala, H.P.).

9.1 Architectural types

The courtyard is one of most important feathers of vernacular Indian housing in North India. Local stone are being used for step and landscape as shown in Fig. 11.10. The courtyard is generally square and more enclosed in hot plain of Uttar Pradesh and Bihar. While in the eastern hot humid area, the courtyard is long and aligned to the prevailing wind direction to increase the wind speed, and penetration. In Uttar Pradesh, the houses usually have no windows to the street, and there is no direct entry to the courtyard. Fig. 11.11 shows that how locally available stone/slate can be used for wall and roof covering. The patio speaks to rectangular open space of assorted site, size, shape, circumstance, capacity and environmental factors relying on the need, accessible space or just the impulse of the inhabitants. Its circumstance and design are likewise a sign of status of the tenants. Local bamboo and wood are used for roof frame as shown in Fig. 11.12. The most widely recognized event of the yard is in the rear, where it is encircled by an inward verandah, connected to the primary or by the mass of these rooms, and an external divider, seldom having an entryway affectionate in current kind of residences. In North India houses tend to have open plans to suit the climate. At the back of the house there is often a verandah room, closed by trellised screens, which serves as the main living space in the hot afternoons. Buildings face north and south to avoid the low angle of the eastern and western sun. The wide verandah with low sweeping roofs sometime encircles the complete buildings. The verandah is protected from heavy rain by the use of tats, rolled read and cane blinds. Fig. 11.13 shows that how soil can be used for staircase step in double storey house.

FIGURE 11.10   Use of local stone for step and landscape (Dharamsala, H.P.).

FIGURE 11.11   Use of stone for wall and slate for roof covring (Dharamsala, H.P.).

FIGURE 11.12 Use of bamboo and wood for roof frame (Dharamsala, H.P.).

FIGURE 11.13 Use of soil for staircase (Dharamsala, H.P.).

FIGURE 11.14 (A) House construction in Bir worked with consideration regarding plan and designing [owned by an organization], (B) Earthen house which are traditional 'Bhunga' in Khavda, (C) Adobe house with metal rooftop in Kriparampura, (D) Cob house with asbestos rooftop in Serjitkhel, Khunti, (E) Cob house with nation terminated bended tile in Tangerpalli, Sundargarh, (F) 'Ikara' house with metal rooftop in Namchi, (G) Under development house in Delhi [split bamboo and grass noticeable on walls], (H) Earth pack house with cover rooftop in Tiruvannamalai, (I) CSEB house in Namchi, (J) CSEB curve structure close to Pondicherry [9.5 m range and 42 m length is 10 cm thick], (K) Poured earth houses close to Pondicherry and (L) pushed earth divider with terminated block tiles and cover rooftop in Tiruvannamalai [26].

In Orissa, the homes are aligned east-west, an orientation resulting from a combination of superstitious and climatic consideration. For example, it is considered inauspicious to eat food while facing west and to sleep facing west and north. In the other hand, houses are aligned to catch the prevailing wind or to allow for cross-ventilation. The courtyard arrangement is preferred with rooms looking onto the inside rather than outside. Fig. 11.14 shows the use of various local building material and construction style used in different parts of India.

10. Technical intervention aspects in rural areas

Currently, the innovative intercession in rural areas is practically insignificant. Method for development of the rural zones is currently shifting from mud and cover to block and cement with no specialized information sources. Driving toward more prominent shortcoming from the seismic perspective and rustic people don't think about it. Generally, common houses have been worked by close without building plan by craftsman's, comprehensive technique, bylaws, managers, and so forth the organizations of arranged capable, engineers, progression authority, lodging architects, financial association, and lodging pleasant social requests weren't open in provincial towns. In this manner, the improvement in houses were for the people groups measure in country zones.

Diverse kind of advancements have been created to help rural individuals, yet it has not reached to the individuals for whom it has been created because of correspondence gaps between the facilitators like the government and client who are the intact beneficiaries of the rural areas. A portion of the significant Barriers/hindrances in innovation move are:

- Insufficient awareness among the people about newer technologies.
- Lack of demonstration by groups or competent authorities to the rural groups.
- Inadequate flow and linkages between technology developer and provider for the rural economy.
- The present scenario and newer technologies may differ and find difficulty in implementation.
- Lack of the flow for Information.
- Extent of illiteracy among the rural population.
- Restricted flow of new building materials to the rural populace.
- The extents of services are lacking.
- The small-scale developers and contactors are limited.
- Market sees an extent of monopoly of certain shops, people, contactors etc.
- The lack of institution and regulatory authorities.

11. Solution by technical sustainability in rural housing

The extensive and economical world, here is a solid need of arrangement of proper innovation, answers for country housing development by utilizing the current administration of India Strategies what's more, monetary incentives. India is exceptional with both technologies furthermore, human resources. Through the process of history, it is with the assistance in innovation that has changed assets accessible to us utilization esteems [27,28]. Innovation is supposed to be essentially an integrative marvel and as such the equivalent is not out of the ordinary in housing advancement likewise [29].

NHHP [30] has recognized the role of technology for the housing sector. By embracing proper innovation, the expense of house would be decreased through diminished utilization of the two materials and capital, in this way bringing about improvement of moderateness of the rural masses. Various climate inviting, energy-proficient, financially savvy building materials and parts have been created in Ref. [31]. The Council of Logical and *"Industrial Research Laboratories, especially the Central Building Research Foundation (CBRI), have concocted a wide scope of advances for ease housing and imaginative building materials, parts and frameworks"* [32]. A few associations, nongovernmental organizations (NGOs), pioneers and innovative rehearsing experts, for example, Late Laurie Baker have concocted numerous inventive choices which can add to cost decrease in the development of houses. *"Cook has additionally done a few helpful tests in low-cost innovation and materials for rural housing dependent on locally accessible materials, for example, Mud, by product, Bamboo, and so on His commitment uncovers a comprehensive view second to none of what he calls "cost productivity" in the creation of building"* [33,34]. Thus, a broad mechanical information has been created revolved around research also, advancement research facilities and institutions. Despite the accessibility of specialized abilities, collection of creative and practical innovation, it has not been moved adequately from the generator to the client till date because of absence of data stream [35]. There are essentially no plans with the expectation of complimentary progression of data about proper innovation from Research and development foundation to the client end. The country individuals are yet to get the advantage of innovation.

12. The various technological solutions

There is for all intents and purposes no institutional structure accessible in the nation to help provincial housing activities. There is nonattendance of hierarchical instrument to gather, report, safeguard and scatter logical and specialized data for provincial housing; consequently, there is a requirement for production of an Institutional Mechanism. Institutional system is intended to guarantee conveyance of innovation, plan and plans of provincial houses, the skill levels of neighborhood craftsman, esteem expansion of neighborhood materials, direction, augmentation, connections, and organization. Plus, that will likewise assistance to create guff of structure resources, advancements, parts of building, building framework which grants neighborhood craftsman and families to assemble practically productive houses because of neighborhood needs, moderateness, and weakness to common catastrophes [36]. This would be conceivable just if the important institutional courses of action exist.

12.1 Human as a resource for technological transfer

A foundation needs a center skilled group of experts for information getting to, information sharing and information catching. Data assortment, data taking care of, and data preparing identified with suitable innovation all need HR. Further, expertise up gradation also, preparing of bricklayers, craftsman, woodworkers, and other structure related labor force including experts like planners, architects, and directors rely on the sound human asset base of the establishment. Plan of augmentation network with prepared labor for conveying innovation as a contribution to housing development additionally relies on the abilities of experts. There is no lack of skillful expert in India. There is a collection of advances accessible in various exploration foundations for appropriation in provincial housing in India. They might be caught, pro-cured and recorded with the assistance of a center skillful group of experts and Data and Communication Technology to make them more open and usable.

Ability impartation is additionally significant in light of the fact that regularly houses remain deficient and impractical because of the absence of prepared talented workers. Absence of legitimate ability is additionally a significant explanation for the low quality of houses. It likewise needs to be noticed that giving spotlight on country housing will create huge work openings in the country nonranch area as of now deficiencies are tremendous. The Government and NGOs may prepare development laborers in short courses affirmed by the Technical Education Council. Steadily, such preparing may be made required and redesigned with new information and expertise. At first such preparing may likewise be made a piece of the school educational plan [24].

12.2 ICT for rural housing

NHHP has accentuated that the data framework the board has the requirements in rural housing improvement [30]. Information and Communication Tech-nologies (ICTs) uphold is a fundamental need to assist rural individuals with taking full points of interest of accessible innovation for housing advancement [37]. To satisfy the requirements of innovation for housing development, ICTs may assume critical job. The advantages could be moved to the end clients, if recognized tool for data framework for rural housing is advanced. It would be the significant wellspring of innovation dispersal for housing development in country territories. The information and correspondence innovation ICTs is the innovation of correspondence of data, sharing of data, the speed of data stream, volume of data measure capacities and so forth [38]. Data technology is the non-HR liable for capacity of information, handling, recover, and cor-respondence of data. It isn't the substitute for land, building materials, tech-nology, and finance needed for country housing improvement. Be that as it may, it has incredible supporting force in quickening the advancement exercises identified with housing improvement. The intensity of ICTs can possibly add to provincial to housing improvement. ICTs uphold a crucial need to assist country individuals with taking full points of interest of accessible innovation for housing development. It tends to be useful in conveying the inventive innovation, development strategy, building plan, plan, and other data related with Government rural housing plans and monetary bundles to the client closes. The Government has stepped up to the plate for utilization of ICT for conveyance of data for provincial housing improvement. In the website of the Ministry of Rural Development (MoRD) of the Government of India (www.provincial.nic.in), subtleties rules, physical and monetary execution of housing plans, rural sanitation, and accelerated rural water supply programs are available. But information on fitting Innovation, for example, weakness (floods, tornado, dry spells) conditions; materials for cost successful and debacle safe housing; climate neighborly, energy-effective items for low-cost housing; financially savvy innovations created for provincial housing by CBRI and other innovative work. institutions are not accessible on the MoRD site. *"Advanced information stockroom is required on specialized part of rural housing. The data for information distribution center might be gathered from the various sources. This data space should be distinguished. The data sources are numerous research centers of Council of Scientific and Industrial Examination (CSIR), Technical Institutions, BMTPC, HUDCO, CBRI, Building Centers, Non-Governmental Organization (NGO) and so on along these lines, all the important data related with innovation would be gathered, archived and dispersed for provincial housing advancement from the single source in staggered authoritative framework"* [32]. The information and data foundation could be an achievement for openness in financial and sustainable knowledge innovation.

12.3 Research and development for rural housing

Housing advancement and its infrastructure in the country territories is as yet moping because of absence of linkages and incorporation between various innovative work foundations. Utility connections will increase interest for the entertainers, advancements at the user end for rural housing. This linkage is likewise considered to assets between the public authority foundation and different associations and encourage a two-path stream for information and data.

12.4 Extension of the network circulation of technology

The fundamental way of thinking of augmentation is self-improvement *"(individuals fix the issues themselves), participatory approach (including a two-route channel of information and experience), influence and training of the individuals"* [39]. With regards to country housing, self-improvement infers a huge commitment of families during the time spent development of staying units without anyone else. In this self-improvement measure, families additionally need assistance from the outer offices according to the proposals of [30]. The self-improvement methodologies incorporate individual contacts between provincial occupants and prepared augmentation laborers. It should be wide based, better arranged, district explicit, participatory and base up and top down to address neighborhood issues for practical country housing improvement. Hence, the expansion program should be planned counting the government agencies, PRIs, and RBCs.

12.5 Skills improvement of the of Craftsman's found locally

Preparing of artisans, development laborer's, and building experts not just builds up their aptitudes, yet additionally, engages them for job age and gives successful approaches to disperse innovation [40]. Subsequently, expenses of staying units can diminish along these lines of construction for housing more moderate to individuals in rural areas. The weakness of rural houses to common assaults could be forestalled just if the nearby abilities of craftsman, development laborers, and taught youth are created for safe development.

12.6 Emphasis on the use of materials found locally

Construction resources count almost 60% to 65% of the expense of the housing development. The expense of accommodation and housing can be diminished by 15%—40% by utilization of imaginative structure resources and innovation. In this manner, the central purpose for any country housing innovation choice would be the appropriation of neighborhood materials, which are least energy burning-through and natural touchy, nearby work furthermore, neighborhood aptitudes [41]. The neighborhood Building materials are to be cordial to the climate, yet in addition be a more advantageous spot to housing for the dwellers and inhabitants. It will likewise add to limit the Earth-wide temperature boost. The updegree or worth expansion of neighborhood materials is just conceivable by fitting innovation data sources and information alternatives for exorbitant materials like concrete, steel, and stones. The reasonable arrangement is sensibly joining more nearby materials with less imported materials, utilization of gifted craftsman, working-class working limits, and specialized sources of information given by the nodal organization through expansion organization.

13. Approach toward the sustainability

The traditional rural house construction practices are significantly more sustainable because wall, roof and floor are made of local unprocessed raw materials which have negligible embodied energy. Most of the materials are renewable. The natural raw materials pollute less in procurement and preparation. The most of the material are recyclable. It does not lead to large exploitation of natural resources due to their decentralized mode of application. The rural houses are also sustainable in terms of three pillars of sustainable development namely, social development, economic development, and environmental protection.

13.1 Socioeconomic

Socioeconomic factors of rural households are considered as the main factors that influence the choice of overwhelming structural material in housing development. The adequacy of transcendent customary structure materials like mud/unburned brick, grass/thatch/bamboo, stone and wood depends on economic situation of household and awareness of benefits of local building material. A low-income rural household choose mud/unburned brick, grass/thatch/bamboo, stone, and wood household for its economic reasons. A middle- and high-income rural household prefer mud/unburned brick, grass/thatch/bamboo, stone, and wood-dwellers for their perception toward the environment. The availability of local skill labor and professionals are also important parameters for choosing local buildings material. House construction in rural India is a culturally sensitive and highly ritualistic. The caste and occupational structures in rural India are still strong. House construction as a social event different occupational are involved in construction process which consolidate the social interaction among them. The assistance of relatives, friends, and neighbors also strengthen the social bonding and promote social development. Traditional Indian rural house construction involves low initial investment and high maintenance. This support work makes ordinary work in nonrural period for all, which is one of the significant elements of financial supportability of the network.

13.2 Environmental protection

The mud house was widely considered environmentally friendly because local availability, recyclable, reusable, and minimum procession. It can be reused numerous times. It has been reported that mud house is good for health due to comfortable indoor condition and pollution absorbent ability of soil [26]. Use of mostly recyclable local unprocessed raw materials in traditional rural houses involves negligible embodied energy, low operational energy, and low CO_2 emission. Hence, rural Indian housing is sustainable in the terms of three pillars of sustainable development.

13.2.1 Policy guidelines

In figuring approaches for rural housing that are maintainable, arranging specialists, as per these rules should:

1. Take record of the cycles that are setting off changes in settlement designs in rural zones, especially those components that are offering ascent to interest for housing in rural zones;
2. Take record of other related measurements comparable to country settlement, for example, ecological security and the need to keep up the trustworthiness of financial assets;
3. Act as a facilitator in uniting, inside existing nearby structures, the primary interests worried about provincial settlement, for example, the chosen individuals, cultivating and network associations, organization speaking to provincial tenants, area improvement sheets, natural associations, and some other applicable associations, for example, the suppliers of provincial public vehicle;
4. Develop inside the expansive interests laid out by the current realities on the ground comparable to populace and financial patterns in provincial regions, such as ecological pointers that will advise the strategy choices for the arranging authority's advancement plan; and
5. Work with interests, for example, those at (3) to make a shared perspective on how the issue of provincial settlement should be tended to in the specific authority worried through the improvement plan.

It is essentially significant that arranging specialists work to bring the chosen individuals, authorities, the more extensive public and vested parties together in building responsibility for improvement plan and its execution.

14. Inferences from Five Year Plans

1. The Rural Housing—Working Group on for 11th Five Year Plan. It determined absolute housing & accommodation deficiencies by adding equilibrium of houses left during 11th arrangement period (24.5 million), housing shortages because of outdated systems and clog at the pace of 3% and 6.5% individually (16.6 million) and expansion in the number of family units during the 12th arrangement time frame (7.7 million). The outdated nature factor was based on 65th round of NSSO which thought about houses with over 6 years as outdated and blockage factor was determined based on number of couples not having a different space for them.
2. The straightforward strategy to compute this deficiency was to add number of family units who don't have house (4.1 million), number of impermanent houses (20.2 million), lack because of clog furthermore, outdated nature at the pace of 6.5% and 4.3% (18.8 million) and extra housing deficiencies emerging during 2012—2017 (0.5 million).
3. The safe/improved wellspring of drinking water joins: separated water, diverted water into withstanding, channeled water to plot and yard, public taps/standpipes, tube wells/bore-holes, ensuring wells, guaranteed springs, and deluge water grouping.

4. Despite the fact that a draft National Policy was brought out in 1988, it couldn't get endorsed.
5. Prior the public authority in 1962 had put forth attempts to disperse the excess grounds of the land proprietors for the landless ranchers.
6. A planner of British cause who got comfortable India in 1945 and obtained Indian citizenship in the year 1989, sought after his vocation as a planner in the province of Kerala, building savvy what's more, climate well-disposed houses for the oppressed. He used conventional building construction techniques with contemporary design to ensure cost adequacy, climate responsive and essential strength of the houses. His design also protected the residents social character.
7. For instance - Utilization of rodent trap bond strategy for laying blocks which improved the underlying strength of the dividers and decreased costs, remembering twists for the dividers which upgraded solidness of the structure and furthermore filled the need of racks. He pushed the utilization of energy-effective mud dividers with openings that permitted ventilation and legitimate lighting. He recommended the utilization of Mangalore tiles which saved 30% of the development cost. Bread cook presented other creative structural highlights and development methods, for example, the utilization of frameless entryways on turns, supplanting windows with jaalis and subbing marble with red or dark oxide flooring. Scrap wood could be reused to make switchboards and lights. Cook favored the utilization of corbels or curves and block on edges rather than pillars and lintels. He additionally spearheaded water collecting innovation and put forth true attempts to limit the harm to the common environmental factors of the building site. Through his creative and inventive mix of customary and current engineering he ably mixed his structures with the encompassing common scene.

15. Conclusion

Mud houses are widely accepted as environmentally friendly, ecologically suitable, and economical for low-income households. Other traditional locally available building materials like grass, thatch, bamboo, unburned brick, and wood are also widely used in rural housing in India. However poor performance of traditional rural households like helpless water and climate obstacle, challenges and termite contamination just as continuous support give a negative standing to customary provincial lodging. To maintain rural character and sustainability in rural housing it is recommended that design guidelines should be match with tradition and use local building material and incorporate community participation in rural house construction.

The contemporary clay material such as compressed stabilized earth block (CSEB) is picking up prevalence in India because of its completing and monetarily. Its creation is efficient and maintainable. The environmentally conscious middle and high income group families preferred CSEB for house construction in rural as well as urban areas. The economic, self-help construction potential, low energy use, and recyclability aspects of traditional construction are more valuable for sustainability.

A combination of traditional and industrial materials and new techniques has also potential for economically, socially, and environmentally sustainable. India is full of examples that amply indicate that buildings were designed in conformity with local and environmental condition which is most basic principle of sustainability. If we take experience from our past vernacular architecture and adopt some modern and scientific development we can certainly have building and housing, that meet their own energy requirements for cooling, heating and lighting by optimum utilization of freely available solar energy and get sustainability.

The PMAY-G program may offer to explore locally available building materials and skill for houses in rural area for economically poor household. The future of rural housing in terms of traditional local available building materials and construction depends on researcher, designer and entrepreneurs who can develop appropriate building materials from locally available resource, designs with minimum cost and maintain supply chain and training to local people. The research work on traditional locally available material and construction in rural housing shall be founded on the mainstays of manageability as well as sustainability, strength and socially worthy for a more extensive acknowledgment.

The Government of India is making part numerous strides for the rural advancement as talked about above. Yet at the same time there are numerous loopholes clauses in this cycle. The evacuation of these loopholes clauses or issues will quicken the cycle of country advancement in India. The function of administrative and nonlegislative associations in such manner is, surely, excellent. Yet, much remaining parts to be done. In the event that we as a whole work along with full focus toward this path we can most likely make progress.

Appendix A. Supplementary data

Supplementary data to this article can be found online at https://doi.org/10. 1016/B978-0-12-824038-0.00009-2.

References

[1] Bank, World. World bank rural population (% of total population). World Bank; 2020 [Online]. Available [Accessed, https://data.worldbank.org/indicator/SP.RUR.TOTL.ZS? end=2019&name_desc=false&start=1960&view=chart. 10-July-2020.

[2] MoRD. Pradhan Mantri Awaas Yojana —Gramin (PMAY-G). New Delhi: GOI; 2018. Online Available: http://pmayg.nic.in/netiay/Uploaded/English_Book_Final.pdf 2016.

[3] IDFC-RDN. India rural development report 2012—13. Delhi: IDFC Rural Development Network; 2013.

[4] Bredenoord J. Sustainable building materials for low-cost housing and the challenges facing their technological developments: examples and lessons regarding bamboo, earth-block technologies, building blocks of recycled materials, and improved concrete panels. J Architect Eng Technol 2017;1.

[5] Schroeder H. Sustainable building with earth. Cham: Springer International Publishing; 2016.

[6] Pumpelly R. Explorations in Turkestan. 1980. Washington, USA: s.n.

[7] UNESCO. Earthen architecture: the environmentally friendly building blocks of tangible and intangible heritage. UN; 2018. Online Available: http://www.unesco.org/new/en/unesco/resources/earthen-architecture-the-environmentallyfriendly-building-blocks-of-tangible-and-int.

[8] Houben H, Guillaud H. Earth construction: a comprehensive guide. Intermediate Technology Publications; 1994. s.l.

[9] Morel J, et al. Building houses with local materials: means to drastically reduce the environmental impact of construction. Build Environ December 2001;10:1119—26.

[10] Cascione V, Maskell D, Shea A, Walker P. A review of moisture buffering capacity. From laboratory testing to full-scale measurement. Construct Build Mater July 2019;11:333—43.

[11] Baiche B, Osmani M, Walliman N, Ogden R. Earth construction in Algeria between tradition and modernity. Proc. Inst. Civ. Eng. - Constr. Mater. February 01, 2017:16—20.

[12] Arboleda G. Architecture without architects. Berkeley: UA; 2004.

[13] Gauzin-Müller, Dominique. Sustainable architecture and urbanism: concepts, technologies, examples, Birkhauser. Basel-Berlin-Boston: Publishers for Architectur; 2002.

[14] Choguill CL. The search for policies to support sustainable housing. UK: University of Sheffield, Habitat International.; 2007. p. 31.

[15] GOI. National urban housing & habitat policy. New Delhi: MHRD; 2007.

[16] Shivanna T, Kadam N, Ravindranath D. Problems and solution of rural housing, ISBN 978-93-87793-00-2. Noida.

[17] Kant A, Mehta V. Cities as engines of growth 2018. In: Revi Aromar, Koduganti Jyothi, Anand Shriya, editors. Cities as engines of inclusive development. New Delhi: Bloomberg Quint, IIHS-RF Policy Paper Series; 2014.

[18] Colmer J. Urbanisation, growth and development: evidence from India. New Orleans: Mimeo; 2015.

[19] Commission, Planning. Working group on rural housing for the twelfth five year plan. MRD. New Delhi: GOI; 2011. p. 7.

[20] Tiwary P. Rural housing in India: India infrastructure report 2007: rural infrastructure. New Delhi: Oxford University Press.; 2007.

[21] Hirway I. Housing for the rural poor. Econ & Political Weekly 1987;vol. 22(34):22. 1st ed.

[22] Iyer SK. What is holding up rural housing. Econ & Political Weekly 1965;II:32.

[23] Tiwari P, Rao J. Housing markets and housing policies in India, report no 565. Tokyo: Asian Development Bank Institute; 2016.

[24] Kumar A, Deka A, Sinha R. Rural housing in India status and policy challenges. New Delhi: Lokashraya Foundation; 2016, ISBN 978-81-933290-2-3.

[25] Census. Houselisting and housing census data. New Delhi: GOI; 2011. https://censusindia.gov.in/2011census/Hlo-series/hlo.html.

[26] Kulshreshtha Y, et al. The potential and current status of earthen material for low-cost housing in rural India. Construct Build Mater January 01, 2020:247.

[27] Drucker PF. Technology management & society. London: Pan Books; 1970. s.n.

[28] Fisher JC, Pry RH. A simple substitution model of technological change. Technol Forecast Soc Change 1972;3.

[29] Giriappa S. Some aspects of rural housing technology. In: Shelter for the rural poor. New Delhi: Ashish Publishing House; 1992.

[30] GOI. National housing and habitat policy. New Delhi: Ministry of Urban Affairs & Employment; 1998.

[31] Report, India: National. Progress of implementation of the habitat Agenda (1996–2000), istanbul + 5th UNCHS (habitat) conference. New Delhi: Ministry of Urban Development and Poverty Alleviation, GOI; 2001.

[32] CBRI. Rural building and environment. Roorkee, India: Central Building Research Institute; 1997.

[33] Achwal MB. Voluntary Agencies and housing. s.l. UNICEF; 1979.

[34] HUDCO-CASTFORD. The next step towards getting a Laurie Baker home. New Delhi, India: HUDCO-CASTFORD Publication; 1989.

[35] Kumar N, et al. Urban-rural linkages:issues of technology transfer for rural housing. New Delhi, India. In: Proceedings of second international BASIN conference; 2004.

[36] Sihag R. Housing as a resource. Lucknow: Gupta Publishing & Co.; 2000.

[37] Kumar N, et al. Rural housing: search for suitable Strategies, vol. 7. New Delhi: Shelter; 2004. No.3.

[38] Emma M-L. Changing information technology policies. USA: Sybase Magazine; 1994.

[39] Dahama OP. Extension and rural welfare. Agra: Ram Prasad and Sons; 1986. p. 91.

[40] GOI. Ninth five year plan; n.d. New Delhi: Planning Commission, Government of India, vol. 2, [Chapter 3]: 1997–2002.

[41] Suresh V. Policy and technology options for rural housing in rural housing in India problems and prospects. New Delhi: Ministry of Rural Development, Government of India; 2000. p. 81–4.

Index

Note: 'Page numbers followed by "*f*" indicate figures and "*t*" indicate tables.'

Piezoelectric device, application of smart
material in field of, 49–50
Piezoelectric substances, 46
Plant-mediated synthesis, 4–5, 4t
Plastic
in cosmetics, 240–245
packaging, 244–245
Polyacrylonitrile (PAN), 170–172
Practical Byzantine Fault Tolerance
(PBFT), 207
Pradhan Mantri Awaas Yojna-Gramin
(PMAY-G), 253
Predicted *vs.* actual plots of indoor variables
deriving through RSM, 196
Preservatives, 239
Pressure sensor, 28
Primary microplastics, 240–241
Privacy, 203
privacy-preserving of health care data,
217–220
design objectives to achieve, 219–220
system and threat model of smart health
care, 219
Proof of stake (PoS), 207
Proof of work (PoW), 207
Proteus mirabilis, 6
Provisions, 120–121
public participation, 121
resource recovery, 121
selection of treatment technology during
inception of projects, 120
Proximity sensor, 27–28
Pseudo-first order (pFO), 80–82
Pseudomonus aeruginosa mediated AgNPs, 6
Pseudonymity, 208
Pseudosecond order (pS), 80–82

Q

Quick Response Code (QRCode), 205

R

Remote monitoring, 212
Resilience, 127–128
Resource recovery, 121
Response Surface Methodology (RSM), 66,
182, 184
optimization of desirable condition by
using, 193–196
predicted *vs.* actual plots of indoor
variables deriving through, 196
RSM-based optimization, 82–87

Ricinus communis, 8–10
Risk mitigation analysis, 149–155
Risk Reduction Action Plan (RRAP), 134
Roof construction techniques, 270
Rural kitchen, 182
material and methods, 182–185
area of study and research design,
182–183
ethical permission, 185
experimental model design, 185
kitchen and living room ventilation
pattern, 184
measurement of indoor air quality, 183
RSM and CCD, 184
study period, 183
model improvements, 196–199
results, 185–196

S

Sanitation, 104–105
Sargassum polycystum mediated silver
nanoparticles, 2, 8–10
SarvSikshaAbhiyan, 146
Scadoxus multiflorus, 2
Scanning electron microscopy (SEM), 49,
167–168
Secondary microplastics, 240–241
Selected area electron diffraction (SAED), 68
Sequential Batch Reactor (SBR), 118–119
Settlement forms, 270–275
architectural types, 272–275
Severe acute respiratory syndrome
(SARS), 127
Sewage treatment infrastructure projects
in Bihar and Patna, 114–115
status of projects pertaining to sewage
treatment infrastructure in India, 113
Sewage treatment plants (STPs), 105–106,
234–235
Shape memory alloy, 54
Shewanella oneidensis mediated silver
nanoparticles, 6
Sida retusa, 4–5
Siloxanes, 239–240
Silver nanoparticles (AgNPs). *See also*
Biofabricated silver-nanoparticles
application on mosquito, 8–12
mosquito egg and adult, 12
mosquito larvae, 8–10, 9t
mosquito pupae, 10–12, 11t
synthesis, 2–7